Heat Transfer

Heat Transfer

A problem solving approach

Tariq Muneer, Jorge Kubie, and Thomas Grassie

LONDON AND NEW YORK

First published 2003 by Taylor & Francis

2 Park Square, Milton Park, Abingdon, Oxfordshire OX14 4RN
52 Vanderbilt Avenue, New York, NY 10017

Routledge is an imprint of the Taylor & Francis Group, an informa business

First issued in paperback 2019

Typeset in Sabon by
Newgen Imaging Systems (P) Ltd, Chennai, India

British Library Cataloguing in Publication Data
A catalogue record for this book is available from the British Library

Library of Congress Cataloging in Publication Data
A catalog record for this book has been requested

ISBN 978-0-415-24169-4 (hbk)
ISBN 978-0-367-39515-5 (pbk)

To Jameela, Amanda, and Gillean

Thank you for your forbearance!

Contents

Highlights of the book

Modern books on most engineering subjects are increasingly being written for a computer literate audience. Hence, nowadays, it is a common practice to use a hard–soft publishing format. This book uses the MS-Excel medium for the development of workbooks that are available on the companion CD-ROM. A sample of the workbooks is also available from the website: http://www.napierengineering.org.uk/research/heatxfer/heatxfer.html

The highlights of this book are:

(i) Clear and concise text that presents the physical theories for selected topics in thermal conduction, convection, and radiation.
(ii) Mathematical rigour maintained throughout the book.
(iii) Latest available regression models for convection heat transfer.
(iv) Use of a standard spreadsheet package for solving involved problems.
(v) Extensive use of numerical experimentation which enables the reader to develop the understanding of the critical parameters.
(vi) Thermophysical properties of common substances made available as digital files.
(vii) A comprehensive thermodynamic software package, 'Allprops' provided, courtesy of the University of Idaho.

Some of the key features of the accompanying workbooks are:

- Rapid, step-wise computations with dynamically linked graphs that enable numerical experimentation.
- All matrix operations handled with ease.
- Solution of non-linear system of equations experienced in multi-mode heat transfer.
- Solution to linear and non-linear optimisation problems.
- Electronic look-up tables for transport and thermodynamic properties.
- Facility for user-developed T–s and p–h charts for visualisation of thermodynamic property change.
- Conjoining of thermodynamic and thermal transfer regimes for heat exchanger design problems.

- Finite element approach to obtain radiation view factors.
- Use of in-built Visual Basic for Applications (VBA) language for handling complex iterative tasks.
- Use of Graphical User Interface (GUI) for solving one- and two-dimensional (Cartesian) conduction problems, producing temperature raster plots and exploring accuracy dependence on mesh size.

Preface

The main purpose of this book is to give a lucid account of the theoretical developments in the field of Heat Transfer while attempting to present the latest available computational methods, particularly within the convection heat transfer discipline. The text would be suitable for an undergraduate as well as postgraduate course.

It is fair to say that while thermal conduction and thermal radiation heat transfer disciplines have reached a plateau in terms of their evolution, convection transfer, due to its empirical nature, still enjoys an ongoing development phase with regression models being added within the forced- and free-convection domains. In all fairness to conduction and radiation disciplines their development under physical laws has helped them acquire rapid maturity. As a matter of fact even fifty years ago thermal radiation development was considered 'dead', the space exploration programme being instrumental in its brief revival. On the other hand, following the explosion of activity within solid mechanics finite element techniques, convection in particular has gained prominence with the use of Computational Fluid Dynamics (CFD) development. This has led not only to solution of convection problems for complex geometries, but also, it has been responsible for new, more involved regression models that offer ease of application and enhanced accuracy.

The new generations of engineering students need new type of working textbooks that enable their training to be completed at a faster pace without compromising the quality of education. The emphasis has therefore shifted from a teaching to a learning mode. The authors hope that by taking away the tedium of repetitive calculation, by providing general-purpose software, the reader will be able to concentrate more on the learning and numerical experimentation part.

The origins of this book lie in the work previously undertaken at Napier University. Pedagogical experiments have been devised at this institution with control groups of engineering students' with- and without access to software. In each case their performance was monitored. It was found that a significant improvement in a student's learning and problem-solving capability results with the use of software that is not too demanding in terms of its own training.

Due to its very nature engineering heat transfer poses problems that often require iterative solution. With this in view the book has a numerical solution slant. The book covers the fundamental and constituent areas of engineering heat transfer – conduction, convection, and thermal radiation. The necessary rigour for understanding the physical principles and analysis is backed up with a number of computer workbooks, each respectively dealing with a given computational aspect and catering the solution of a particular problem type. The accompanying software also allows the reader to perform numerical experimentation – changes may be made to independent variables and their effects seen readily via numerical and graphical outputs. Furthermore, identification of strong and weak variables is possible.

A contributing factor for the present development is the emerging use of computer spreadsheet environment for engineering computation. To quote E. M. Forster, 'The only books that influence us are those for which we are ready, and have gone a little farther down our particular path than we have yet gone ourselves'. The simplicity of data entry and its manipulation afforded by spreadsheets makes it an ideal design and analysis tool. Modern spreadsheet software offers very powerful numerical and decision making facilities. Almost every engineering student and professional has a spreadsheet loaded on his or her computer. This book exploits this situation to fill the gap between the ability to perform involved calculations and the desire to do so without going to the tedium of writing one's own software.

Another advantage offered by the spreadsheet medium is that each formula is readable, and fits in its appropriate position, that is, the spreadsheet cell. This enables the user to easily follow the algorithmic chain. On the other hand commercially available software, while powerful, does not allow the user the facility to examine or alter the computer code. In many cases such commercial or shareware software, due to its very design, may take several weeks and months of training. Owing to their widespread use and user-friendly layout, spreadsheets require only a few minutes training before the user is fully in charge of the situation. This point shall be demonstrated in Chapter 2 where more sophisticated features of Microsoft Excel shall be introduced.

The production of text has also provided the authors with an opportunity to explore the frontiers of the MS-Excel medium to handle quite involved problems. In this respect the authors are indeed grateful to the publishers for providing such an opportunity. Highlights of the present text are presented in a companion section.

The authors would like their readers to feel that they are participating in a hands-on seminar rather than attending a lecture. In this context it is to be emphasised that the present book is inter-active. The hard-soft combination of material presented here is, we hope, 'just the beginning'. In due course there is bound to be much more activity for this type of presentation where the boundary between the author and the reader starts to become blurred. The reader may then be able to take the book onward to a different plane! Indeed we hope that the readers will communicate with the authors and suggest ways of further improving text, particularly the material related to numerical computation and convection regression modelling part. In this respect the authors are looking forward to learning from the readers.

The solution techniques presented herein are of such a nature that they may easily be incorporated in the solution of other thermophysical problems, that is, fluid mechanics, thermodynamics, refrigeration, and air-conditioning.

The greatest problem in the preparation of this book has been to avoid the unnecessary repetition of what has been said elsewhere, while at the same time making each individual chapter comprehensive. Topics, which demand more extensive reading for complete understanding, have therefore been treated in sufficient detail, even if referenced elsewhere in a prior chapter.

Acknowledgements are gratefully made here to colleagues and others whose work has been drawn upon when preparing the text. Additional and express acknowledgement is further made to those works that have been extensively used via references and bibliography.

The authors welcome suggestions for additions or improvements.

Acknowledgements

The authors are indebted to a number of individuals for their support. Professor Steve Penoncello of the University of Idaho for furnishing the 'Allprops' thermodynamic property software for inclusion in our book's companion CD. Dr David Kinghorn prepared the first drafts of Chapters 7 and 8. Dr Nasser Abodahab and Ms Alison Murdoch furnished some elementary material and their contribution is acknowledged. Our research students Muhammad Asif and Xiaodong Zhang undertook the task of checking all Microsoft Excel workbooks. Steven Paterson efficiently prepared all drawings. The publishers Tony Moore and Sarah Kramer have been very patient with us, we owe them our heartfelt thanks.

The completion of this book was made possible due to a generous grant from Napier University. We have also been influenced by a number of concurrent heat transfer publications and it is only fair to cite those key references: Heat Transfer by J P Holman, McGraw-Hill; Fundamentals of Heat and Mass Transfer by F P Incropera and D P DeWitt, Wiley; Engineering Heat Transfer by N V Suryanarayana, West Publishing Company; Heat Transmission by W H McAdams, McGraw-Hill; Conduction Heat Transfer by Carslaw and Jaeger, Oxford University Press; Thermal Radiation Heat Transfer by R Seigel and J R Howell, Taylor & Francis.

The following references were the basis for generation of thermophysical tables that have been used within numerous Excel workbooks in this publication, the authors remain grateful to the respective sources: ASHRAE Handbook of Fundamentals (1977), American Society of Heating, Refrigerating, and Air-Conditioning Engineers; Fundamentals of Heat and Mass Transfer (2002), F P Incropera and D P DeWitt, Wiley (for properties of saturated water, based on the work P E Liley: Steam tables in SI units, Purdue University (1978); NBS/NRC Steam tables: Thermodynamic and transport properties and computer programs for vapor and liquid states of water in SI units (1984), L Haar, J Gallagher and S Kell, Hemisphere Publishing Corporation; and CRC Handbook of tables for applied engineering science (1977), R E Boltz and G E Tuve, CRC Press.

Acknowledgements

The authors are indebted to a number of individuals for their support. Professor Steve Penoncello of the University of Idaho for permitting the 'Allprops' thermodynamic property software for inclusion in our book's companion CD. Dr David Kingsmore prepared the first drafts of Chapters 7 and 8. Dr Naser Abdullah [illegible] Ms Alison Murdoch furnished some elementary material and their contributions are acknowledged. Our recent students Mohammad [illegible] and Nanda [illegible] undertook the task of converting all NIST [illegible] Excel [illegible] Steven Paterson efficiently prepared all sketches. The publishers Tony Moore and Sarah Kramer have been very patient with us, and we offer them our heartfelt thanks.

The compilation of this book was made possible through a generous grant from Napier University. We have also been influenced by a number of excellent heat transfer publications, and it is only fair to mention a few key references: Heat Transfer by J P Holman, McGraw-Hill; Fundamentals of Heat and Mass Transfer by F P Incropera and D P DeWitt, Wiley; Engineering Heat Transfer by N V Suryanarayana, West; [illegible] Heat Transmission by W H McAdams, McGraw-Hill; Conduction of Heat in Solids by Carslaw and Jaeger, Oxford University Press; Thermal Radiation Heat Transfer by R Siegel and J R Howell, Taylor & Francis.

The following references were the basis for generation of thermo-physical data that have been used in the numerous Excel worksheets in this publication; the authors are most grateful to the respective sources: ASHRAE Handbook of Fundamentals (1997), American Society of Heating, Refrigerating and Air-Conditioning Engineers; Fundamentals of Heat and Mass Transfer (2002), F P Incropera and D P DeWitt, Wiley (for properties of saturated water [illegible]); [illegible] steam tables in SI units [illegible] University [illegible] NBS/NRC Steam tables: Thermodynamic and transport properties and computer programs for vapor and liquid states of water in SI units (1984), L Haar, J Gallagher and G S Kell, Hemisphere Publishing Corporation; and CRC Handbook of tables for applied engineering science (1973), R E Bolz and G L Tuve, CRC Press.

Disclaimer and copying policy

Contents of companion CD-ROM

There are two sub-folders within the main 'HT_MKG' folder: 'HT_Workbooks' and 'Allprops'. The former contains Excel workbooks that enable the user to go through the examples given in the hardcopy as well as solve the problems given at the end of each chapter. The latter folder 'Allprops' contains a special software tool for thermodynamic properties, provided by the University of Idaho.

Microsoft Excel Workbooks on CD-ROM sub-folder 'HT_Workbooks'

List of Workbooks
(unless mentioned otherwise, all files have a '.xls' extension

Ex02-04-01	Sequential computation for pressure drop calculation
Ex02-05-02	Lookup tables
Ex02-06-01	Three point quadratic interpolation for equally spaced data points
Ex02-06-02	Four point cubic interpolation for equally spaced data points
Ex02-07-01	Dynamic graphs
Ex02-08-01	Two dimensional lookup table for steam/water substance
Ex02-09-01	Solution of a non-linear algebraic equation via Goal Seek
Ex02-10-01	Solution of a non-linear algebraic equation via False Position method
Ex02-12-01	Matrix operations
Ex02-16-01	Fitting linear models via graphical means
Ex02-17-01	Multi-variable function minimisation via Solver
Ex02-20-01	Outlier analysis
Ex02-21-01	Weighted averages for a given data set
Ex03-05-01	Composite wall with three sections
Ex03-05-02	Complex composite wall
Ex03-07-01	Long composite axi-symmetric cylinder with three sections

Ex08-02-04	Convective loss from an inclined plate with uniform heat flux
Ex08-02-05	Effect of insulation and surface emissivity on the rate of heat loss from a long horizontal cylinder
Ex08-02-06	Heat loss from a hot sphere suspended in oil
Ex08-03-01	Convective heat loss in a vertical channel
Ex08-03-02	Optimum spacing of the plates in an air heater
Ex08-03-03	Convective heat loss from the absorber plate of a solar collector
Ex08-03-04	Convective heat transfer in the annular space between two concentric cylinders
Ex08-04-01	Mixed convection heat transfer from a long horizontal pipe in cross flow
Ex09-02-01	Blackbody radiation
Ex09-02-02	Effect of surface properties on plate temperature
Ex09-03-02	Radiant heat transfer in an enclosure with three gray surfaces
Ex09-04-01	Radiant heat exchange between two rectangular black surfaces
Ex09-04-02	Effect of insulation on radiation heat transfer between surfaces
Ex09-05-01	Radiant heat exchange in an enclosure with two diffuse surfaces
Ex09-05-02	Radiant heat exchange in a diffuse three surface enclosure with an adiabatic wall
Ex10-01-01	Simultaneous convection and radiation from a heated plate
Ex10-02-01	Simultaneous forced-convection, and radiation
Ex10-03-01	Simultaneous free-convection, and radiation
Ex10-04-01	Simultaneous conduction, forced-convection, and radiation
Ex10-05-01	Simultaneous conduction, free-convection, and radiation
Ex10-06-01	Simultaneous conduction, free-convection, and radiation within enclosures
Ex10-07-01	Simultaneous conduction and radiation, with and without free- or forced-convection

Additional data files

Data02-20-01.txt	Non-dimensionalised data derived from a heat transfer study (use with Ex02-20-01.xls)
H20_TP	Thermal conductivity, viscosity, and Pandtl number data for superheated steam
Steamchart1	T–s chart for water/steam substance

Workbooks available from website

In due course in response to readers' request further material may be added by the authors by means of uploading files on this book's websites: http://www.napierengineering.org.uk/research/heatxfer/heatxfer.html, http://www.soe.napier.ac.uk/research/heatxfer/heatxfer.html

Presently, the following MS-Excel workbooks are available from the above website free of charge:

Ex03-07-02.xls	Critical thickness of thermal insulation
Ex04-02-01.xls	Rectangular duct with a coarse grid
Ex04-02-02.xls	Rectangular duct with a finer grid
Ex07-02-06.xls	Hot wire anemometer: Measurement of fluid velocity
Ex08-02-01.xls	Free-convection – vertical plate
Ex10-02-01.xls	Simultaneous forced-convection and radiation
Steamchart1.xls	Temperature-entropy chart for water/steam substance

Highlights of the book and text related to the above sample heat transfer problems are provided in 'Web_examples.doc' file.

Notation

A	area
Bi	finite difference form of the Biot number
C	Celsius
c	specific heat/speed of light in a medium (in a vacuum $c_0 = 2.988 \times 10^8$ m/s).
C_f	friction coefficient
c_p	specific heat capacity at constant pressure
c_v	specific heat capacity at constant volume
D	diameter
D_h	hydraulic mean diameter
E	internal thermal energy/total emissive power of a gray surface
f	friction factor/similarity variable/frequency
F_{ij}	view factor between surface area elements i and j
F_D	drag force
Fo	finite difference form of the Fourier number
G	irradiation
Gr	Grashof number
Gr*	Grashof number for constant heat flux
h	surface heat transfer coefficient/Planks constant/enthalpy
h_{fg}	enthalpy of vaporisation
I	spectral intensity
J	surface radiosity
K	Kelvin
k	thermal conductivity/Boltzmann constant
L	length/mean beam length/thickness
m	mass flow rate
$\dot{m}$	mass flow rate
N_L	number of tubes in the longitudinal plane in a tube bank
N_T	number of tubes in the transverse plane in a tube bank
Nu	Nusselt number

P	perimeter/power/length of chord between two surfaces
p	pressure
Δp	pressure differential
Pe	Peclet number
Pr	Prandtl number
q	heat flux
Q	heat transfer rate
q_g	rate of heat generation per unit volume
r	radius or radial coordinate
R	thermal resistance, gas constant
Ra	Rayleigh number
Re	Reynolds number
$R_{t,T}$	total thermal resistance
S	surface area
S_D	diagonal pitch of a tube bank
S_L	longitudinal pitch of a tube bank
St	Stanton number
S_T	transverse pitch of a tube bank
T	temperature
t	time or thickness
U	overall heat transfer coefficient
V	velocity/volume/voltage
$\dot{V}$	volume flow rate
W	width
X	distance from leading edge of a plate
x	coordinate
y	coordinate
z	coordinate
Δr	space interval
Δt	time interval
Δx	space interval
Δy	space interval
Δz	space interval

Greek letters

α	thermal diffusivity/absorptivity
β	volumetric thermal expansion coefficient
δ	hydrodynamic boundary layer thickness
δ_T	thermal boundary layer thickness
ε	emissivity, pipe roughness
η	fin efficiency

η	similarity variable
θ	longitude or excess temperature/angle
λ	wavelength
μ	viscosity or dynamic viscosity
ν	kinematic viscosity
ξ	unheated starting length of a plate
ρ	density/reflectivity
σ	Stefan–Boltzmann constant
τ	shear stress/transmissivity
ϕ	spherical coordinate or fin effectiveness

Subscripts

∞	free stream/infinity or surrounding medium
λ	spectral
$1, a, i$	relating to material/surface 1
$2, b, j$	relating to material/surface 2
A	section A
AV	average
b	blackbody
black	pertaining to a black surface
c	cross-sectional
cond	referring to conduction
conv	referring to convection
CR	critical
D	based on characteristic diameter
e	entry length/exit
f	relating to fluid film properties/frontal area
fd	fully developed
gray	pertaining to a gray surface
h	hydrodynamic
I	inlet/inner
L	based on characteristic length
lam	laminar
lm	log mean condition
max	maximum
min	minimum
o	outlet/outer
rad	radiation
S	surface
SP	sphere

sur	surroundings
th	thermal
w	water
x	characteristic length

Superscripts

Overbar ($^{-}$)	surface average conditions

1 Introduction to heat transfer

Heat transfer is a naturally occurring phenomenon. Its occurrence takes place by the virtue of temperature difference. The applicability of laws that govern heat transfer are universal and, as shall be demonstrated later in this chapter, even casual observation will show that the heat exchanges between heavenly as well as terrestrial bodies are incessant. The application of the principles of heat transfer is truly universal.

Heat transfer is a highly developed discipline and its processes are based on sound physical laws. There is a great deal of empiricism within the field of study of convection (heat transfer by fluid motion), but, even so, the basic analysis is still derived from well-established laws. The *second law of thermodynamics* states that 'it is impossible to transfer heat from a body at a lower temperature to a body at higher temperature without the use of an external agency'. Furthermore, all three modes of heat transfer – conduction, convection, and radiation are governed by the respective laws due to Fourier, Newton, and Stefan–Boltzmann.

From a very early age we experience the everyday occurrence of heat transfer. The phrase 'getting one's fingers burnt' has indeed become synonymous with bad experience. We learn very early on that woollen clothing provides better insulation than cotton fabric, soaking in the sun provides heat or blowing over our hands while rubbing them on a cold winter morning keeps them warm. Even subconsciously we develop the 'knack' of shivering of limbs to provide warmth to our bodies.

In the following sections, more elaborate examples are provided to demonstrate the universality of heat transfer processes and the applicability of its laws. For obvious reasons, also included herein, are examples taken from engineering practise.

1.1 The cooling down of the universe and galaxies

Barrow and Tipler (1986) have proposed an interesting 'Anthropological cosmological principle' for the Universe. Their postulation is based on the premise that 'observers will reside only in places where conditions are conducive to their evolution and existence'. In their above referred book Barrow and Tipler argue that in order to create and build the life-forming elements – carbon, nitrogen, and oxygen – the simpler elements, hydrogen, and helium were *thermally treated* in the interior of stars following the Big Bang event. The authors argue that the requirement that enough time pass for cosmic expansion to cool off via virtue of radiant heat transfer for the above processes to take place is 10 billion years. Thus, it has been shown that the rate at which heat transfer has taken place during the cooling of Universe has a direct bearing on the time of evolution of carbon-based life.

1.2 Radiating black holes

The laws of Newtonian mechanics state that, provided the mass and its distribution remains unchanged, the gravity at the surface of any astronomical body is inversely proportional to the square of its radius. The escape velocity from any given astronomical body increases with its gravity and consequently, inversely with its radius. Black holes are created when the size of a fuel-exhausted star begins to shrink. The gravity on its surface becomes stronger with the reduction in its size and hence the escape velocity increases. This process continues until a stage is reached where the escape velocity at its surface equals the speed of light. Any light entering the black hole will never escape. As a matter of fact light from other distant stars passing within the vicinity of a black hole may curl around a black hole several times before escaping or falling in.

The conventional wisdom was that a black hole never gets smaller and that nothing can come out of a black hole. However, in 1974 the physicist Hawking presented his theory of Hawking Radiation (Hawking, 1977; Ferguson, 1991). In what was at that time a controversial idea, he suggested that black holes had temperature and emitted radiation.

Based on the principles of quantum mechanics Hawking argued that pairs of particles – pairs of photons and gravitons – continually appear at the *event horizon* of a black hole. For the purpose of this text, and without indulging into the details of astronomical science, we may loosely consider the *event horizon* to be the radius at which the escape velocity is the speed of light. Two particles in a pair that start out together may then move apart. After an interval of time they come together and annihilate one another. Some of the pairs will be pairs of matter particles, one of the pair being an antiparticle. Hawking argued that particle pairs appear at the event horizon. However, before the pair meet again and annihilate each other, the one with the negative energy crosses the event horizon into the black hole. The particle with positive energy, now freed of its partnership, may now escape. To an observer at a distance it appears to come out of the black hole and this is known as Hawking radiation.

The particle with negative energy that falls in the black hole robs it of its positive energy, thus reducing its mass and size. Hawking has thus argued that radiating black holes might continually get smaller and eventually evaporate.

Hawking's theory has been used to ascertain the entropy, temperature, and radiation emission levels of black holes. He has shown that the larger the mass of a black hole, the larger its event horizon and entropy. The greater the entropy, the lower the surface temperature and the rate of emission. He has also proposed the existence of tiny black holes that are the size of the nucleus of an atom. Such tiny black holes he argues would crackle with radiation, indeed it would be more appropriate to call them *white-hot*!

1.3 The Sun and its radiation

1.3.1 *Solar radiation*

The Sun is a medium-sized star. It emits electromagnetic energy, including X-rays, ultraviolet, and infrared radiation as well as visible light at a rate of 3.8×10^{23} kW. However, its attendant planets and their satellites receive only about 1 part in 120,000,000. The small part of this energy that is intercepted by the Earth, on average $1.367\,\text{kW/m}^2$, is of enormous importance to life and to the maintenance of natural processes on the Earth's surface. The energy output of the Sun has its peak at a wavelength of $0.47\,\mu\text{m}$. Solar radiation emerges at the photosphere. Detailed spectral studies reveal that the composition of this region is about 90% hydrogen, 9.9% helium, and a small admixture of heavy elements (Encyclopaedia Britannica, 1997).

1.3.2 *Sun's corona*

The temperature at the Sun's surface, that is, the photosphere, is around 6000°C. However, moving away from the photosphere does not lower the temperature. In the 3000-km zone between the photosphere and the corona, the temperature jumps from 6000°C to over a million celsius. In apparent defiance of common sense, the further away you go from the Sun's surface into the solar atmosphere, the warmer the gas becomes! There have been many theories postulated that try to explain the extraordinary warmth of the Sun's corona. The more popular of these fall into three categories, namely miniature solar flares, atmospheric waves, and electrical dissipation. A detailed discussion of these theories is beyond the scope of the present text and the reader is therefore directed to the NASA (1999) website.

1.4 The Earth's thermal environment

1.4.1 *Terrestrial greenhouse effect*

Energy from the Sun drives the Earth's weather and climate, and in turn heats the Earth's surface. The Earth radiates energy back into space. Atmospheric

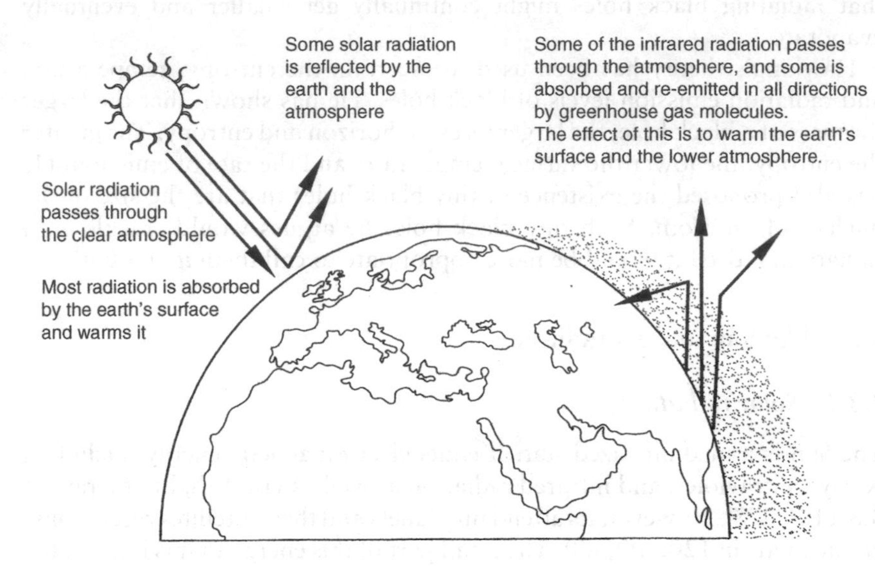

Figure 1.4.1 The terrestrial greenhouse effect.

greenhouse gases (water vapour, carbon dioxide, and other gases) trap some of the outgoing energy, retaining heat somewhat like the glass panels of a greenhouse, as shown in Fig. 1.4.1.

Without this natural 'greenhouse effect', temperatures would be much lower than they are now, and life as known today would not be possible. However, problems may arise when the atmospheric concentration of greenhouse gases increases. Since the beginning of the industrial revolution, atmospheric concentrations of carbon dioxide have increased by nearly 30%, methane concentrations have more than doubled, and nitrous oxide concentrations have risen by about 15%. These increases have enhanced the heat-trapping capability of the Earth's atmosphere. Sulphate aerosols, a common air pollutant, cool the atmosphere by reflecting light back into space. However, sulphates are short-lived in the atmosphere and vary regionally.

As a result of the increase in greenhouse gas concentrations, global mean surface temperatures have an increasing trend. The twentieth century's 10 warmest years all occurred in the last 15 years of the century. Of these, 1998 was the warmest year on record. The snow cover in the Northern Hemisphere and the area of floating ice in the Arctic Ocean have both decreased. Increasing concentrations of greenhouse gases are likely to accelerate the rate of climate change. Scientists expect that the average global surface temperature could rise by 0.6–2.5°C in the next 50 years, and 1.4–5.8°C in the next century.

Most greenhouse gases occur naturally in the atmosphere, while others are produced as a result of human activities. Naturally occurring greenhouse gases include water vapour, carbon dioxide, methane, nitrous oxide, and ozone. Certain human activities, however, add to the levels of most of these naturally occurring gases: *Methane* is emitted during the production and transport of coal, natural gas, and oil. It is also produced as a result of decomposition of organic wastes in municipal solid waste landfills, and the raising of livestock. *Nitrous oxide* is emitted during agricultural and industrial activities, as well as during combustion of solid waste and fossil fuels. The more potent greenhouse gases, which do not occur naturally, include *chlorofluorocarbons* (CFCs), *hydrofluorocarbons* (HFCs), *perfluorocarbons* (PFCs), and *sulphur hexafluoride* (SF6), are generated in a variety of industrial processes. The impact of each gas is represented by its global warming potential (GWP) value.

For further details the reader is referred to Environmental Protection Agency (EPA, 2002) website.

1.4.2 *Terrestrial temperature stratification*

The terrestrial temperature variation with height follows a most peculiar path. Contrary to a 'casual' expectation, the atmospheric temperature actually increases with altitude. The lowest region of our atmosphere is called the troposphere. It contains about half the total mass of Earth's atmosphere and is the layer where weather changes take place. In the troposphere the air becomes colder as one ascends to higher altitudes. The primary source of heat for the troposphere is the Sun-warmed surface of the Earth, which is why the troposphere is warmest near the surface. Just above the troposphere is the stratosphere, where the temperature begins to increase again. The relatively high temperature of the stratosphere is caused by the presence of an ozone layer near an altitude of 25 km. The ozone molecules absorb high-energy ultraviolet rays from the Sun, which warm the atmosphere at that level. Figure 1.4.2 depicts the anomalous temperature stratification under discussion. For further details the reader is once again referred to NASA (1999) website.

1.4.3 *Impact on organisms*

The ability of individual organisms to tolerate or regulate their temperature within the varied conditions experienced in the Earth's thermal environment defines the distribution of life on the planet. Most organisms cannot live in conditions in which the temperature is out with the 0–45°C band because they are not able to maintain a body temperature that is significantly different from that of the environment. Sessile organisms, such as plants and fungi, and animals that cannot move great distances, must therefore be able to withstand the full range of temperatures sustained by their habitat. In contrast, many mobile animals employ behavioural mechanisms to avoid extreme conditions in the short term. Such behaviours vary from

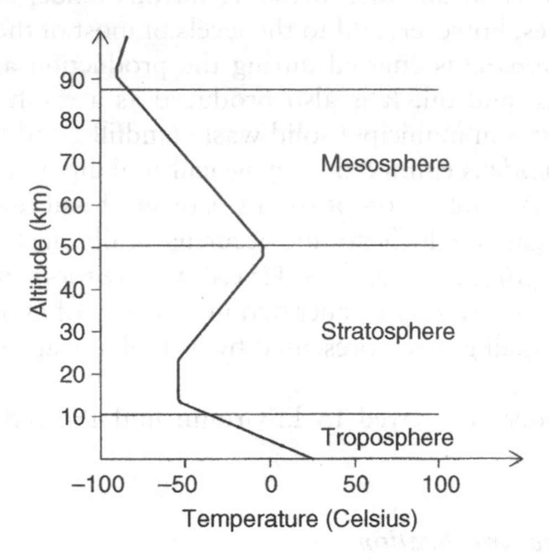

Figure 1.4.2 The atmospheric temperature stratification.

simply moving short distances out of the sun or an icy wind to large-scale migrations.

In respect of plants, it is noted that the thick waxy cuticle of slow growing cacti allows them to survive in hot arid conditions. In more temperate zones, photosynthesising leaves control the rate of gaseous and vapour exchange with their environment by altering the aperture formed between the stomatal (small opening) guard cells on the leaf's underside. If the leaf begins to overheat, the above-mentioned aperture expands, releasing water vapour, which evaporates on the underside, thus cooling the leaf.

Many types of animals employ physiological mechanisms to maintain a constant body temperature, with two categories commonly used to distinguish them. The term 'cold-blooded' refers to reptiles and invertebrates while 'warm-blooded' is generally applied to mammals and birds. In describing the thermal capabilities of these animals, these terms are, however, imprecise. They may be described more accurately as, ectotherm for cold-blooded, and endotherm for warm-blooded. Ectotherms rely on external sources of heat to regulate their body temperatures, and endotherms thermoregulate by generating heat internally.

Endotherms maintain body temperature independently of the environment by the metabolic production of heat. They generate heat internally and control passive heat loss by varying the quality of their insulation or by repositioning themselves to alter their effective surface area. If heat loss exceeds heat generation, metabolism increases to make up the loss. If heat generation exceeds the rate of loss, mechanisms to increase heat loss by evaporation occur. Smaller mammals, having a larger surface area to volume ratio, are more readily affected by local changes in temperature. To compensate, they generally have a relatively higher metabolic rate. The metabolism is the body's engine and the amount of energy released by the metabolism is dependent on the amount of muscular activity. Normally all muscular activity is converted to heat in the body. This topic shall be explored further in Section 1.5.

The thermal resistance of the skin of weasel and caribou are, respectively, 0.23 and 0.85 m^2K/W (Campbell, 1986). An alternate, more convenient unit for expressing the insulation of clothing is a Clo. The latter scale is designed so that a naked person has a Clo value of 0.0 and that of someone wearing a business suit is 1.0. Note that 1 Clo = 0.155 m^2K/W.

Other non-migratory animals have also evolved to tolerate large seasonal variations in their thermal environment. While some larger mammals, being able to build up a sufficient fat store to last the winter, hibernate to overcome the seasonal shortage of food, the common hare of Britain and northern Europe, and many breeds of higher latitude horse, grow thick winter coats, which are shed in spring. In addition, similar species have also evolved, and more recently been selectively bred, to adapt to different environments. For example, the average short wave absorptivity of Zulu cattle ($\alpha = 0.49$) is much lower than that of Aberdeen Angus cattle ($\alpha = 0.89$), allowing them to tolerate much higher levels of incident solar radiation (Campbell, 1986).

The more commonly noted theories for the evolution of stripes on zebras are that they may help to camouflage the animal by breaking its body outline, or, that, as each animal's print is slightly different, they may enable young to recognise parents in the herd. An additional hypothesis is that they may serve to help maintain a stable body temperature through enhancing free convective flow over the body surface. As the surface of the black stripes will be hotter than that of the white, therefore producing higher free convective flow rates, the interaction of the respective flows will serve as to promote turbulence, and hence increase heat loss. Furthermore, the three extant species of zebra, namely, the imperial, common, and mountain, have notably different stripe patterns and stripe widths. The common zebra, found mainly in the African plains, living in a hotter climate, has twenty-four broad stripes as compared to the fifty-five stripes of the mountain species. It may be that the broader stripes have a more marked effect on heat transfer and therefore help the plains zebra avoid overheating.

As they cannot regulate their body temperature by internal mechanisms, when cold, ectotherms are generally slower moving, and may be at a greater risk from predation. Many species have therefore developed strategies to help overcome this problem. Both lizards, and their main predators, snakes, may often be found warming themselves in the morning sun. Both try to gain advantage by increasing their rate of activity through thermal gain. This parallel thermal evolution of both predator and prey has also been observed in different species of early Synapsid. Within this group, Pelycosaurs were mammal like reptiles that evolved from dinosaurs in the early permian period (295–250 million years ago). Both Dimetradon, a reputedly fierce predator, and its prey, the herbivore, Edaphosaurus, had large vertical skin sails along the length of their backs. Palaeontologists believe that the primary function of these sails was for thermoregulation as discussed by Lambert *et al.* (2001). As such these sails (heat transfer fins) may have helped the animals stay cool in hot weather, or more active early in the morning.

Throughout nature, past and present, we see that many of Earth's organisms have developed strategies that have enabled them to adapt to varied thermal environments and colonise the globe. Species have evolved mechanisms whereby they can tolerate both extremes of temperature, and also a large variance in thermal conditions for a given location. Humans, through their ability to adapt to their thermal environment, are one of the worlds widest spread species. Ranging from arid equatorial to higher latitude locations, the air temperature difference, not considering the increased convective loss (chill factor) induced by stronger high latitude winds, may be as much as 40°C in some cases.

Birds have a higher body temperature (~41°C) than other animal species. Seabirds have been interesting subjects for thermoregulation studies, as they live in temperature extremes ranging from the arctic to the equatorial belt. Their nesting sites are often exposed to high winds and intense solar radiation, making the effective temperatures severe and thermoregulation most vital. In general terms birds can tolerate a decrease in heat more easily than an increase (Markum, 1997).

Generally, smaller birds have a greater surface area to volume ratio and as such they need to replace lost heat at a greater rate than larger birds. A larger size for bird is also advantageous in both hot and cold climates as the larger mass translates to a higher heat capacity that allows for higher heat storage and less heat stress in hot thermal environments. Insulation, in the form of feathers and fat, also influences heat conservation. In nesting birds, fat is lost during the incubation of eggs and a bird's dependence on feathers for insulation may change. The extent to which feathers are erect and directed away from the skin, the length of the feathers, and the distribution of the plumage contribute to effective insulation (Lustick, 1984). Fluffier plumage has proportionally more insulating capabilities than flat plumage. Longer, stiffer feathers provide an effective barrier to wind and heat loss thus creating a warmer layer of air adjacent to skin. This is well suited for the bird's heat conservation and survival in colder climates. The distribution of feathers can also vary from species to species. In the herring gull the breast plumage has twice the insulating capabilities of the dorsal coverage, allowing it more variation in controlling its temperature. The wettability of feathers is also an important factor for diving birds for whom it is essential that water does not penetrate to their skin. Owing to its lower thermal conductivity fat is also a barrier for heat loss. Birds have fat distributed generously and evenly over their bodies relative to mammals, except for an increased layer on their breast, which may compensate for the insulation the wings provide to other parts of the body. The colour of the plumage, or its reflectivity, is also a morphological control over heat absorption. Black plumage absorbs more sunlight than white plumage does, but as wind speed increases, the difference between their heat absorption is minimized. This effect becomes more pronounced when feathers are erected, and therefore more insulating.

Adaptations in orientation and posture allow birds to vary the total surface area exposed to the sun and to the wind, and birds without uniform plumage can use orientation to adjust their absorption of solar radiation. Some birds face the Sun to reduce the exposed area and take advantage of a white, reflective breast to reduce heat gain. When the Sun is directly overhead the orientation of the bird becomes insignificant and the erection of the scapular feathers becomes the primary cooling mechanism as it shields the body from direct Sun.

1.5 Heat transfer from humans

From an early age we all learn that different materials in the same environment feel hotter or cooler. In 1757, Benjamin Franklin wrote: 'My desk on which I now write, and the lock of my desk, are both exposed to the same temperature of the air, and have therefore the same degree of heat or cold; yet if I lay my hand successively on the wood and on the metal, the latter feels the coldest, not that it is really so, but being a better conductor, it more readily than the wood takes away and draws into itself the fire that was in my skin', as quoted by Seeger (1973). Here, Franklin was eloquently describing the difference in the rate of heat transfer between his hand and the two above-mentioned materials. Our bodies continuously interact with our thermal environment, and have developed numerous sophisticated mechanisms that attempt to maintain an optimum temperature for our continued survival.

Speaking in a purely physical sense, we note that human beings are heat exchangers. The food that we eat is digested by our bodies, with some of the energy released used to maintain our body temperature. Our thermoregulatory system is quite sophisticated and any change in the temperature level of the various body organs is quickly detected by our numerous sensors. The metabolic energy production within our bodies is dissipated through skin by means of convection, thermal radiation, and secretion of sweat through minute body glands. Any difficulty in achieving this process, resulting from, for example, extremes in environmental temperature, or relative humidity being excessively high, is detected by the body thermostats.

Traditionally, metabolism is measured in Met (1 Met = $58.15\,W/m^2$ of body surface). A normal adult has a surface area of $1.7\,m^2$, and a person in thermal comfort with an activity level of 1 Met will thus have a heat loss of approximately 100 W. Table 1.5.1 gives data for the rate of metabolic heat production for humans for a range of activities. It can be seen that between sleeping and heavy activity there is a 12-fold increase in the amount of heat produced. As we begin to do physical work and our bodies demand more oxygen; we begin to breathe more heavily. Initially, as a result of the

Table 1.5.1 Metabolic heat production in humans

Activity	*Metabolic heat production* (W/m^2)
Sleeping	50
Awake – resting	60
Standing	90
Seated – driving	95
Light work	120
Moderate walking on the flat (4 km/h)	180
Walking on the flat (5.5 km/h) or moderate hard work	250
Walking on the flat (5.5 km/h) with 20 kg pack, or sustained hard work	350
Short periods of strenuous activity, sports, sprinting, or very hard work	600

increased air flow velocities, body temperature is maintained through the increase in heat loss from the respiratory tract. Above a certain workload however, this mechanism is no longer sufficient, and sweat is secreted. As this sweat evaporates on our skin, our bodies are cooled. As a result of the specific nature of the lesion, some tetrapleagics are unable to sweat. External means of cooling, such as fine sprays may be necessary if the subject is in a hot climate (Campbell, 1986).

We enjoy basking under winter Sun, thus exchanging heat from a distant star. We also find pleasure in swimming provided the water is at the right temperature. Our body sensors can easily detect the more rapid increase in the heat transfer coefficient experienced in cooler water as opposed to that experienced under a cool envelope of air.

We also, both consciously and unconsciously, change our posture in response to the sensation of coldness or warmth. Changes in posture alter the convective flow patterns around our bodies, and hence the overall heat transfer coefficient. When we start to feel cold we hunch our shoulders and bring our limbs closer to our bodies. This has the effect of reducing our surface area and hence heat loss. Clark and Toy (1975) found that the variance in flow resulting from altered posture was most marked around the head. The maximum air velocities over the face were found to be of the order of 0.04–0.05 m/s with the subject in the supine (lying down) position, as compared to 0.3–0.5 m/s when standing. As will be seen in Chapter 8, there is a marked difference between the heat transfer coefficient for horizontal and vertical cylinders.

Tregear (1966) and Clark and Edholm (1985) state that while black skin reflects approximately 16% of short wavelength radiation, 30–40% is reflected by white skin. However, as black skin contains more melanin, which affects the skins molecular structure, this skin allows less penetration of solar radiation, thus compensating for its reflection advantage.

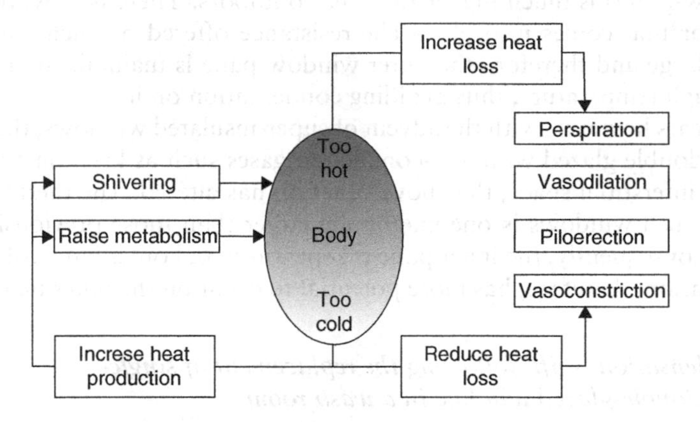

Figure 1.5.1 The human thermoregulatory system.

Furthermore, as more UV is absorbed in the black skins pigmented layer, it is thought that there is less likelihood of the subject becoming sunburnt.

In addition to postural response to temperature, the human body also possesses a number of additional mechanisms for reducing heat loss, and also for generating heat internally. In response to cold, and the movement of cool air over our bodies, the erector pili muscles raise the hairs on our bodies to trap air, providing an insulating envelope and therefore reducing heat loss. Shivering is generally initiated in response to a drop in skin temperature. If the body core temperature also begins to fall, shivering will become more violent as the body tries to raise its internal temperature. In terms of heat generation, shivering is more effective than voluntary muscular contraction. In shivering, as the limbs do not effectively move or do work, all the energy used is dissipated as heat within the muscle as opposed to the 60–70% conversion gained through voluntary contraction (Clark and Edholm, 1985). A schematic diagram of the major human thermoregulatory systems is shown in Fig. 1.5.1.

1.6 Heat transfer mechanisms at work within buildings

1.6.1 Condensation occurrence on single-, double-, or triple-glazed windows

In northern countries (or to be more precise in locations at high latitude) where winters are long and cold, condensation occurrence on windows is quite commonplace. However, the pattern of this condensation occurrence may change depending on the design of the installed window. Traditionally, single-glazed windows, and more recently air-filled double-glazing, have been used in dwellings and other buildings. The condensation occurrence on such glazings is usually on the inside as the interior humidity ratio (and hence the dew-point) is much higher than that outdoors. There is however another factor that comes into play. The resistance offered by such windows is not large and therefore the outer window pane is maintained at a reasonably high temperature, thus avoiding condensation on it.

In recent years however, with the advent of super-insulated windows, that is, triple- or double-glazed with low-conduction gases such as krypton and xenon in the interstitial space, the above situation has altered. The thermal resistance of such windows is one magnitude larger than those previously mentioned. Consequently, the inner pane is kept warm and outer pane cold. The condensation, therefore, has more potential to occur on the outer pane.

1.6.2 Condensation 'shift' following the replacement of single- with double-glazed window in a wash room

In domestic wash rooms that are not normally ventilated, condensation on window pane is a common experience provided the window is constructed using a single pane. The condensation in a wash room is more likely to occur due to the relatively higher levels of moisture. However, if the single-glazing is replaced with a double-pane window the point of initiation of condensation occurrence *shifts* to a colder surface, such as the water flush tank. In this case the condensed water drips on the underside of the latter tank without easily being noticed.

1.6.3 Condensation occurrence on the lower edge of double-glazed windows

It is a common experience that condensation occurrence on the inner pane of a double-glazing initiates at the bottom edge, spreading more thinly upward and towards the side edges. If the moisture level within the room is excessive then the condensation will occur at the top edge, followed by a general spread along the entire glazing. A detailed study on such condensation occurrence was undertaken by Abodahab (1998) and those results have been included in the monograph by Muneer *et al.* (2000).

The above phenomenon is explained thus. All multiple-glazed windows have one or more cavities. Each cavity contains a gas such as air, argon, krypton, xenon, or sulphur hexafluoride. When the gas is subjected to a temperature difference across the enclosure boundaries, it starts to circulate within the cavity. The infill gas ascends along the hot pane of glass and descends along the cold pane. This circulating flow is categorised as free convection within an enclosure or a cavity. Computational Fluid Dynamic (CFD) analysis enables 'flow visualisation' of such phenomena. A series of such CFD temperature and velocity raster plots for free convection flows within window cavities are shown in Fig. 1.6.1.

The circulation of gas within the enclosure is the dominant factor in producing a temperature gradient within the cavity. The metallic edge spacer

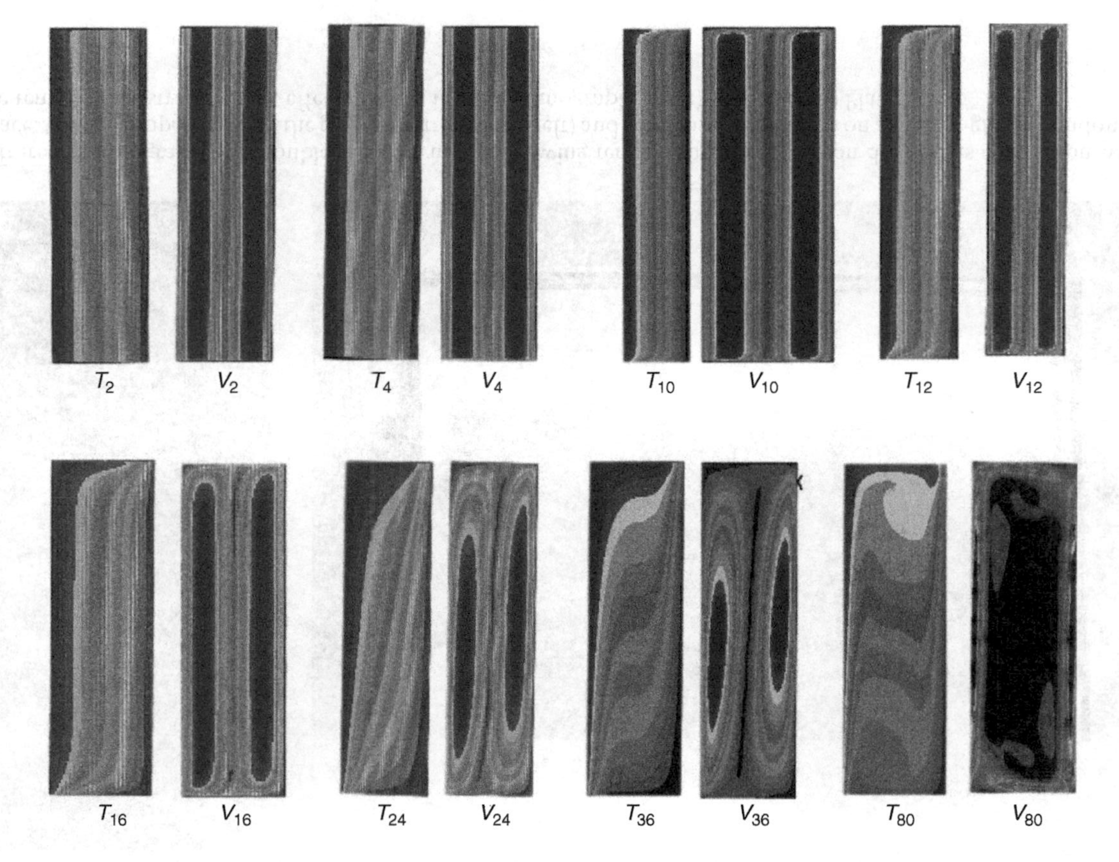

Figure 1.6.1 CFD temperature (T) and velocity (V) raster plots for free convection flows within window cavities. Numbers indicate cavity width in millimeter. (See Colour Plate I.)

Figure 1.6.2 Infrared thermograph of a double-glazed window showing temperature stratification due to gas circulation within the cavity space. Colder temperatures at the glazing bottom edge (left) and condensation pattern on a double-glazed window that reinforces the temperature stratification effects shown in the thermograph (right). (See Colour Plate II.)

also acts as a thermal bridge between the warmer inner and colder outer panes. These two factors result in a two-dimensional temperature distribution with a colder-most bottom edge, followed by colder side-edge and a much warmer top edge. Figure 1.6.2 compares infrared photographs of ordinary double-glazed and super-insulated windows. Notice the much warmer top-edge for the ordinary double-glazing. A photograph of the pattern of condensation occurrence described above is also included.

1.7 Heat transfer mechanisms at work within automobiles

1.7.1 Selective occurrence of frost on automobile windows

We note that on most cold, winter mornings the front windshield of an automobile is worst affected with frost. The disparate occurrence of frost on the front windshield, as opposed to side or back windows of the automobile is so remarkable that whereas the front windshield may have accumulated several millimetres of frost there may be no need to scrape or de-ice the other windows.

There are two main factors that account for this phenomenon, namely:

(1) Larger view factor to sky. On a clear night the sky acts as a near-perfect receiver of thermal radiation. If we take the tilt angles of the front windshield, rear and side windows to typically have values of 30°, 60°, and 90°, respectively, then their respective view factors to the sky would be 0.93, 0.75 and 0.5. This means that while of the proportion of radiant energy exchange between the front windshield and the night sky will be 93%, only 50% of such exchange will take place between the side window and sky. Thus, there will be nearly a double rate of heat loss from the front windshield as opposed to side windows.

    The concept of view factors finds its use in thermal radiation heat transfer and these are discussed more elaborately in Chapter 9.

(2) More front windshield radiation exchange with sky rather than building surfaces. Owing to its low angle with the horizontal the front windshield has low likelihood of seeing a building surface compared with a vertical window. On most housing estates, where cars may be parked on streets the side windows of the automobile will exchange energy from exterior surfaces of heated dwellings and this is bound to raise their temperature even further. Those vertical windows closer to these building exterior surfaces will also have a higher view factor and will therefore receive more radiant energy.

1.7.2 Frost occurrence on the top edge of front windshield

The phenomenon of selective occurrence of frost on automobile windows was discussed in Section 1.7.1. It may also be of note that on mornings that follow a mild cool night, the occurrence of frost on the front windshield may be selective, that is, only the top part of the glass pane may experience frost formation. The reason may be as follows. The convection thermal boundary layer ensues from the bottom edge, which turns to turbulence towards the top edge. Hence higher local convection transfer coefficients are experienced at the top, rather than the bottom edge. This results in a cooler top-part of the windshield, leading to a higher potential for frost formation.

1.8 Refrigeration equipment

Since the start of the process of phasing out of chlorofluoro- and hydrogenated chlorofluorohydrocarbons (cfcs and hcfcs) from refrigeration equipment, the industry has strived to find alternate refrigerants with matching properties. It was felt necessary to phase out the above refrigerants in view of their strong GWP. In this respect ammonia has come back as a suitable candidate.

Ammonia has properties that are quite different to conventional cfc and hcfc refrigerants. First, ammonia is flammable in air at medium concentrations and explosions have occurred after accidental releases from industrial plants. It also poses toxicity problems even at low concentration levels. It has a high adiabatic index (ratio of constant pressure- to constant volume-specific heats) and hence causes a higher temperature rise during compression. However, the above demerits are matched by some of the good heat transfer properties that ammonia possesses.

As well as having a high enthalpy of evaporation, Ammonia has low viscosity and high liquid thermal conductivity and this result in much higher heat transfer coefficients (Table 1.8.1). Thus, a large number of demerits are ignored in favour of good heat transfer properties and its low GWP (Butler, 2002).

1.9 Multipliers for SI units

Throughout this book multipliers of SI units have been used frequently. It is therefore appropriate to introduce the full range of these multipliers at an early stage and these are provided in Table 1.9.1.

Table 1.8.1 Heat transfer coefficients for common refrigerants (Jansson, 1994)

	Ammonia (W/m^{2}K)	*R22* (W/m^{2}K)
Condensation along the outside of tubes	3500–7000	1200–2000
Condensation inside of tubes	2500–6000	1000–1800
Evaporation along the outside of tubes	600–6000	300–3500
Evaporation inside of tubes	1000–6000	450–1800

Table 1.9.1 Multiplier factors for SI Units

Exponent of base 10	*Prefix*	*Abbreviation*
18	Exa	E
15	Peta	P
12	Tera	T
9	Giga	G
6	Mega	M
3	kilo	k
2	hecto	h
1	deca	D
−1	deci	d
−2	centi	c
−3	milli	m
−6	micro	μ
−9	nano	n
−12	pico	p
−15	femto	f
−18	atto	a

References

Abodahab, N. (1998) *Temperature distribution models for double-glazed windows and their use in assessing condensation occurrence*. PhD thesis, Napier University, Edinburgh, UK.

Barrow, J. D. and Tipler, F. J. (1986) *The Anthropic Cosmological Principle*, Oxford University Press, Oxford, England.

Butler, D. (2002) Reasonable alternative. *Building Services Journal*, January, 32–37 (Chartered Institution of Building Services Engineers, London).

Campbell, G. S. (1986) *An Introduction to Environmental Biophysics*, Springer-Verlag, New York.

Clark, R. P. and Edholm, O. G. (1985) *Man and His Thermal Environment*, Edward Arnold, London.

Clark, R. P. and Toy, N. (1975) Natural convection around the human head. *Journal of Physiology*, 244, 283–293.

Encyclopaedia Britannica (1997) Encyclopaedia Britannica Incorporated, USA.

Environmental Protection Agency, EPA (2002) http://www.epa.gov/globalwarming/impacts/index.html

Ferguson, K. (1991) *Stephen Hawking*, Bantam, New York.

Hawking, S. W. (1977) The quantum mechanics of black holes. *Scientific American*, January, 34–40.

Jansson, L. (1994) Ammonia liquid chillers with low charges. *Ammonia Refrigeration Today* (Institute of Refrigeration, London).

Lambert, D., Naish, D., and Wyse, E. (2001) *Encyclopedia of Dinosaurs and Prehistoric Life*, Dorling-Kindersley, London, ISBN 0-7513-0955-9.

Lustick, S. (1984) Thermoregulation in adult seabirds. *Seabird Energetics*, Plenum Press, New York.

Markum, M. E. (1997) *Encyclopaedia Britannica*, Encyclopaedia Britannica Incorporated, USA.

Muneer, T., Abodahab, N., Weir, G., and Kubie, J. (2000) *Windows in Buildings*, Reed-Elsevier, Oxford.
NASA (1999) http://science.nasa.gov/newhome/headlines/ast02sep99_1.htm
Seeger, R. J. (1973) Benjamin Franklin. *New World Physicist*, Pergamon Press, New York.
Tregear, R. T. (1966) *Physical Functions of the Skin*, Academic Press, London.

2 Numerical and statistical analysis using Microsoft Excel

2.1 Introduction

As stated in the preface, throughout this book the Microsoft Excel spreadsheet software has been used to demonstrate its applicability to solve complex engineering problems such as those encountered in heat transfer. The purpose of this chapter is to introduce to the reader the many varied functions and procedures that are available within Excel for this purpose. The authors would recommend a close study of this chapter as this will enable the understanding of examples and solution of unsolved problems that are presented in the remainder of the book. However, if a reader is fully conversant with the Excel environment then they may directly proceed to the chapter of their choice.

Modern engineering design and analysis demand for increased productivity, reduced costs, faster time to work completion, improved quality, and worldwide access to information. In just 10 years, Information Technology (IT) advances have been a catalyst to huge changes in the way design practises are run. Undoubtedly the key drivers to this change have been the explosive breakthrough in the design and development of integrated circuit chips and the emergence of computer aided design (CAD) software. Computer processor technology is one of the most important areas of growth. The London based Institution of Mechanical Engineering predicts that by the year 2005, complete systems, with the storage and processing power of a personal computer, will be available on a single chip. Chips with the capacity of one terabit memory will be commonly available by the year 2010. This will no doubt transform the way in which engineering design and analysis is undertaken. It would be possible to use palm-sized computers and large numerical databases for undertaking numeric-intensive calculations. The real, and perhaps only, barrier to the use of computer hardware would be the software requiring a long training period.

While larger, more complex engineering application packages such as those used for undertaking a finite element analysis (FEA) or computational fluid dynamic (CFD) work represent the top end of the more specialised software market, many people would argue that it is the less complex software applications that have had the most immediate effect on their work environments. The advantage of using a spreadsheet-based computing environment such as Microsoft Excel is that the training times are of the order of, at most, a few hours compared to the several weeks that may be required for some of the above mentioned applications. Further, the learning curve for Excel is very gentle and easy going. Help is available in the form of well-written texts, study guides, and training videos. There are many such texts available and those readers who have not had the opportunity of using Excel in the past may refer to Hallberg *et al.* (1997) and Liengme (2000). The former text provides basic drilling in the use of Excel whereas the latter provides more focussed applications specifically designed for engineering applications.

In general terms, there are some very compelling reasons for switching from manual (calculator-based) to computer-based engineering design. These are:

1. *Cheaper*: Very significant savings can be made in the man-hours required for basic design calculations and optimisation.
2. *Faster*: The results are available immediately. Any new models made available in the scientific journals can be rigorously tested and applied.
3. *Ease of analysis*: The ability to perform 'what-if-analyses' is significantly improved.
4. *Use of graphics*: Information can be displayed, copied, and printed in an easy to understand manner, thus making it easier to communicate results to clients who may not have a scientific background.
5. *Dissemination of information*: Design information may be transmitted through the worldwide web easily and cheaply. Discussion can then take place for design scenarios.
6. *Ease of updating*: Newer databases can easily be incorporated, thus allowing the design process to be kept abreast of the rapid technological changes.
7. *Mainstream activity*: Although some organisations take pride in being 'early adopters', most organisations prefer to use only the mainstream technology. Now with the wider adaptability of computer-based applications in the industry and academia, organisations not using this technology would fall into the 'obsolete' class.

2.2 Excel – a user-friendly environment for number crunching

The history of the development of computer spreadsheets may be traced to VisiCalc which, in the early 1970s, created a significant impact in the corporate field. Prior to the launch of spreadsheets, scientists and engineers had not considered the use of personal computers (PCs) worthwhile. However, VisiCalc changed that image. Some journals claim that without the spreadsheet, the PC would not have developed so rapidly in the early 1980s.

The chances are that most people who perform any work with numbers do have recourse to a spreadsheet. VisiCalc was rapidly replaced by Lotus 1-2-3 in the 1980s, which was faster, more comprehensive, and easier to use. However, Lotus were slow in transferring their product to the Microsoft Windows environment and consequentially lost ground to their nearest rival, Microsoft Excel. Today, with over 100 million users worldwide the sales of Excel are second only to Web browser software. Excel comes with a range of templates and on-screen training tips. For most science and engineering students, as previously outlined, it requires only a few hours training to master the software at an intermediate level. It is also the corporate standard for spreadsheets.

It is likely that the average Excel user uses only a small fraction of this software's capabilities, which now extend to well beyond simply calculating lists of figures or making colourful charts. The power of Excel for engineering use is exploitable through its multitude of functions, that is, macros (to perform repeatedly a chain of commands or operations), **lookup** tables (for performing interpolations on large property tables and data querying), **Goal Seek** (for solving non-linear equations), **Solver** (for solution of a given set of non-linear equations and optimisation), and inter-linked graphs (for performing what-if and other analyses).

As the above mentioned capabilities have only recently been made available, new books are presently emerging which highlight this software's computing potential. Further development is bound to take place in the near future. Owing to their very nature, engineering subjects easily lend themselves to the use of spreadsheets. In this respect packages such as Mathcad, Mathematica or Matlab face strong competition from spreadsheets, which, in contrast, requires only a fraction of the training time.

2.3 The functionality of Excel – an example-based tour

The following sections give readers a focussed introduction to the potential of Excel as a CAD tool. Only those **functions, tools,** and **procedures** which are of direct relevance to the present text are included. All examples and computational tools provided in the companion CD use Excel as the problem-solving environment. It was therefore felt that a brief tour of the most essential features of Excel would be of benefit to the reader. As stated above it is assumed that the reader has a basic familiarity with the Excel package. Those readers who would like to gain the basic skills for using Excel are referred to the previously mentioned texts. In addition to introducing the relevant Excel facilities, an overview of the appropriate mathematical techniques is also given.

2.4 Sequential computation chain

Sequential computation is the most fundamental, and perhaps one of the most useful properties of Excel. Demonstration of this feature is provided in the form of the following example using Ex02-04-01.xls workbook contained in the CD accompanying this book.

Example 2.4.1 Pressure drop associated with fluid flows within closed conduits.

An appropriate introduction to the effects of viscosity for fluid flows within closed conduits and external to a solid surface shall be presented in Chapters 6 and 7. Note that Chapter 3 provides a brief introduction to those thermophysical properties that define the thermal and dynamic behaviour of materials. The present example considers the flow of air within a conduit, and the subsequent loss of energy due to the effects of fluid viscosity. When a fluid moves past a solid interface, a boundary layer is formed adjacent to the interface. The viscosity property of the fluid dictates a variation of velocity normal to the interface from a nil value at the wall (adhesion) to a maximum value at the outer edge of the boundary layer. This variation in the fluid velocity represents a loss in momentum, and hence a resistance to the flow (drag force). The associated pressure loss (Δp) may be shown to be directly related to the length (L) of the conduit, the fluid density (ρ), the square of the fluid velocity (V), and inversely related to the conduit diameter (D). Thus,

$$\Delta p = f\rho(L/D)(V^2/2) \tag{2.4.1}$$

The variable 'f' in the above equation represents the Darcy–Weisbach friction factor. Consider the flow of air under standard conditions, through a 4.0 mm diameter drawn tubing (roughness, ε = 0.0015 mm) with an average velocity of 50 m/s. Determine the pressure drop in a 0.1 m section of the tube.

Solution

- Open the workbook Ex02-04-01.xls. This workbook consists of one worksheet, namely, *Compute* which includes given data as well as the computations performed by Excel.
- Read the example text carefully and insert the above given data in cells **(B4:B7)**. Under standard temperature and pressure, the density of air (ρ) = 1.23 kg/m^3, and its viscosity (μ) = 1.79×10^{-5} Ns/m^2. Insert these values in **cells (B8 and B9)** respectively.
- The solution is shown in **cell B18**. Excel performs all calculations as given below.
  - The Reynolds number, Re = ($\rho VD/\mu$) = 1.37×10^4 **cell B15**. Thus, the flow is turbulent (Re > 3000).
  - The relative roughness (ε/D) = 3.75×10^{-4} **cell B16**.

- Using the Swamee and Jain equation, the friction factor is obtained:

$$f = \frac{1.325}{\left[\ln\left((\varepsilon/D)/3.7\right) + \left(5.74/\mathrm{Re}^{0.9}\right)\right]^2} = 0.0292 \text{ cell } \mathbf{B17}$$

  Note that the friction factor can also be found by using other models such as Colebrook equation (see Example 2.9.1) or from the Moody chart (Moody, 1944) which gives ($f = 0.0298$).
- The pressure drop

$$\Delta p = f\frac{l}{D}\frac{1}{2}\rho V^2 = 1122\,\text{Pa cell } \mathbf{B18}$$

  Note that you can explore the effect of the pipe diameter, length, roughness, or the flow velocity on the pressure drop by changing the input data in **cells (B4:B9)**.

A detailed discussion on boundary layer is available in Schmidt *et al.* (1993) and Incropera and DeWitt (2002).

2.5 Linear interpolation and lookup tables

2.5.1 Linear interpolation

Interpolation will not be a new concept for the reader as linear interpolation is often used to find intermediate values in tables of mathematical functions. In effect, the tabulated function is approximated by passing a straight line through the two tabulated points on either side of the missing point. The equation of this line is then solved for the desired ordinate, y.

Referring to Fig. 2.5.1, the equation of the straight line connecting the two tabulated points P_1 and P_0 on the plot of $y = f(x)$ is:

$$\frac{y - y_0}{x - x_0} = \frac{y_1 - y_0}{x_1 - x_0} \tag{2.5.1}$$

If we let s be the fractional part of the interval between the two given values of x so that

$$s = \frac{x - x_0}{x_1 - x_0} \tag{2.5.2}$$

we get

$$y = y_0 + s(y_1 - y_0) \tag{2.5.3}$$

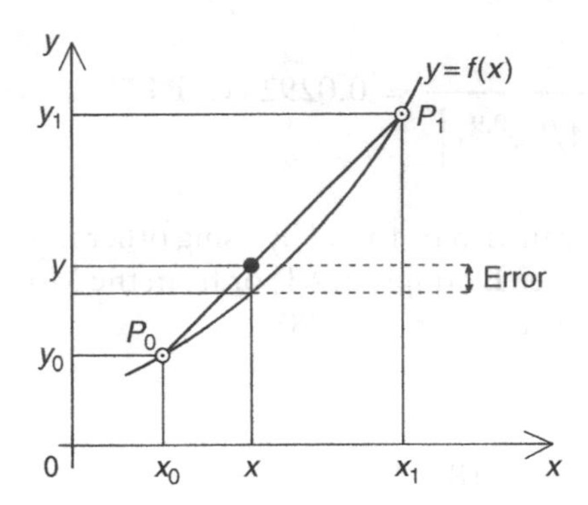

Figure 2.5.1 Linear interpolation.

Table 2.5.1 Enthalpy data for saturated water vapour at various temperatures

Temperature (°C)	*Enthalpy* (J/kg)
200	2791.3
220	2800.2
240	2802.3
260	2796.0
280	2779.4
300	2749.1

Example 2.5.1 Linear interpolation of thermophysical data.

Table 2.5.1 presents enthalpy data for saturated water vapour at various temperatures. Using linear interpolation, find the enthalpy corresponding to a temperature of 270°C. Note that the *exact* value of enthalpy, given in steam tables (Haar *et al.*, 1984), is 2789.1 J/kg.

The fractional part of the interval is

$$s = (270 - 260)/(280 - 260) = 0.5$$

and therefore,

$$\text{enthalpy} = 2796.0 + 0.5(2779.4 - 2796.0) = 2787.7\,\text{J/kg}$$

Thus, the error involved is 0.05%.

2.5.2 Lookup *tables*

Excel provides a very useful facility for creating **lookup** tables. Excel's **lookup** function is based on the linear interpolation method as described above.

Through the **lookup** facility the user may find one piece of information that is based on another piece of information. A **lookup** table consists of a column or row of ascending values, called 'compare values', and corresponding data for each compare value. This is demonstrated in worksheet *Properties* of the workbook Ex02-05-02.xls, which gives the thermophysical properties for air. In this illustration, the first column (temperature of the gas) contains data for the compare values. The corresponding data are the thermophysical property values.

Tabulated data, in **lookup** form, are often used by design engineers. For example, a large number of heat exchanger selection tables are available from most manufacturers. In the present context, demonstration of the Excel **lookup** function is given through application of this function to a **lookup** table for thermophysical properties for air. Most heat transfer problems require an estimation of the thermophysical properties of a fluid at a certain temperature. Inevitably, this requires interpolations for each of the desired thermophysical properties.

Example 2.5.2 Calculation of hydrodynamic and thermal boundary layer thickness.

In convection heat transfer one encounters the concept of hydrodynamic and thermal boundary layers. The thermal boundary layer may be defined as the region of influence within which the temperature of the fluid flowing over a solid interface changes from the interface temperature to the free stream temperature of the fluid.

If δ and δ_T represent the respective thickness of the hydrodynamic and thermal boundary layers, then their values may be obtained from the following formulae,

$$\text{For } \mathrm{Re}_x < \mathrm{Re}_{cr} \qquad \delta = \frac{5x}{\mathrm{Re}_x^{0.5}} \quad \text{and} \quad \delta_T = \frac{\delta}{\mathrm{Pr}^{1/3}} \tag{2.5.4}$$

$$\text{For } \mathrm{Re}_x > \mathrm{Re}_{cr} \qquad \delta = \frac{0.37x}{\mathrm{Re}_x^{0.2}} \quad \text{and} \quad \delta_T = \delta \tag{2.5.5}$$

The Reynolds number, $\mathrm{Re}_x = Ux/\nu$, where U is the uniform velocity, x the distance from the leading edge and ν is the dynamic viscosity. Prandtl number, Pr is a fluid property, where $\mathrm{Pr} = C_p\mu/k$.

Consider the flow of air at a temperature of 40°C, its flow being parallel to a flat plate. The uniform velocity of air is given to be 3 m/s. The plate is maintained at a uniform temperature of 60°C. Calculate the hydrodynamic and thermal boundary layer thickness (δ and δ_T, respectively) at a distance of 1 m from the leading edge of the plate.

Solution

- Open the workbook Ex02-05-02.xls. This workbook consists of two worksheets, namely, *Compute* and *Properties*.
- Activate the worksheet *Compute* by clicking on its tab. This worksheet includes the computations performed by Excel and a schematic diagram for the example.
- Read the example text carefully and insert the given data in **cells (B4:B7)**.
- The required δ and δ_T are shown in **cell B16** and **cell B17**, respectively. Excel performs all calculations as given below.
  - The film temperature, $T_f = (T_s + T_\infty)/2 = 50°C$ **cell B14**.
  - The table of thermophysical properties for air is provided in **cells (A11:H45)** of *Properties* worksheet. This worksheet is accessed by clicking on its name tab. The air temperatures are given in the far-left column, **cells (A11:A45)**, in Kelvin. Therefore, T_f should be $(50 + 273.15 = 323.15\,K)$ **cell B6**. This temperature (323.15 K) does not exist in the temperature **cells (A11:A45)**. In this routine, the temperatures immediately below and above T_f, termed T_{down} and T_{up}, are first located in **cells (A11:A45)**. Second, the properties at T_{down} and T_{up} are determined. Finally, the required properties at T_f are found by interpolating between the corresponding properties at T_{down} and T_{up}. This cumbersome process of calculations can be automated by using the Excel **lookup** facility. Thus, the required properties are determined automatically each time the value of T_f is changed. The **lookup** facility is explained below.

T_{down} **cell B7** is found by using the following form of **Vlookup** function:

T_{down} = **Vlookup (lookup-value, table-array, column-index)**

where

- **lookup-value** is the value to be found in the first column of the table, that is, the value of **cell B6** = 323.15 K;
- **table-array** is the table to be searched, that is, **cells (A11:H45)** (values of **cells (A11:A45)** must be in ascending order);
- **column-index** is the column number in the table from which the matching value must be returned, for example, **1** returns the value in the first column (the temperature column).

Vlookup searches for a value in the leftmost column of a table, and returns a value in the same row from a column you specify in the table. If **Vlookup** cannot find the **lookup-value**, it returns the largest value in the table that is less than or equal to the **lookup-value**. If the **lookup-value** is smaller than the smallest value in the first column, **Vlookup** returns the #N/A error value. In this example, **Vlookup** returns the value of 300 K for T_{down} **cell B7**.

The air properties at T_{down} are found in the same way. The temperature 300 K **cell B7** is the **lookup-value**. The **column-index** should be changed according to the number of the column of the required property, for example, 2 for ρ, 3 for c_p, 4 for μ, and so on. For example to find the value of ρ at 300 K, **cell C7** = 1.161 kg/m^3, use **Vlookup (B7,A11:A45,2)**.

To find T_{up}, the table array (**A11:H45**) must be in a descending order as shown in **cells (A47:H81)**.

Three Excel functions are then used to find the value of T_{up}. These functions are shown in **cells (J11:J13)**.

1 **cell J11=Match (B6,A47:A81,−1)**: This function gets the relative position of the smallest value in the far-left column of the array table (**A47:H81**) that is greater than or equal to the **lookup-value** T_f in **cell B6**. In this case, the **Match** function gives a value of 30. This means that T_{up} is located in row number 30 in the array table (**A41:H75**).
2 **cell J12=Address (J11+46,1)**: This function converts the relative position of a cell to an absolute position. In this case, it gives the absolute position for T_{up} in the worksheet *Properties* as **cell A76**.
3 **cell J13=Indirect (J12)**: This function gives the value stored in a cell whose address is known. In this case, it gives the value stored in **cell A76** which is the required T_{up}.

The properties of air at T_{up} are found by using the **Vlookup** function as explained above in the case of the properties for T_{down}. For example, ρ at 350 K, **cell C5** = 0.995 kg/m^3 [**Vlookup (B5,A11:A45,2)**].

Finally, the properties of air at T_f are calculated by performing a linear interpolation between the corresponding values at T_{down} and T_{up}, for example, the value of ρ at 323.15 K,

cell C6 = If(C5=C7,C5,(B6-B7)*(C5-C7)/(B5-B7)+C7)

This expression takes into account the possibility that the **lookup-value** T_f may exist in the lookup-table, that is, $T_{down} = T_{up}$.

Therefore, the properties at T_f are shown in **cells (C6:I6)**. Only the following properties are needed in this example: $\nu = 1.82 \times 10^{-5}$ m^2/s **cell F6** and Pr = 0.704 **cell I6**.

- The Reynolds number, $\mathrm{Re}_x = (Vx/\nu) = 1.65 \times 10^5$ (sheet *Compute*, **cell B15**). For forced convection over a flat plate, the critical Reynolds number above which the boundary layer becomes turbulent is $\mathrm{Re}_{cr} = 5 \times 10^5$. Therefore, the boundary layer at a distance of 1 m from the leading edge is laminar.
- δ and δ_T are calculated from Eqs (2.5.4) and (2.5.5), respectively. Note that a conditional test should be conducted on Re_x in order to determine δ and δ_T. For this purpose the Excel **IF** function is used which has the

following form,

If(logical_test, value_if_true,value_if_false)

where **logical_test** is any value or expression that can be evaluated to TRUE or FALSE, **value_if_true** is the value that is returned if **logical_test** is TRUE, and **value_if_false** is the value that is returned if **logical_test** is FALSE.

Cell B16 and **cell B17** show how the above equations are used to calculate δ and δ_T with the aid of **IF** function. The required δ and δ_T are 12.3 and 13.9 mm, respectively.

2.6 Polynomial interpolation

If higher accuracy than can be obtained by linear interpolation is desired, an equation may be written connecting *any* number of points in the region of interest; the more points used, the better (usually) the accuracy. The type of function chosen is usually a polynomial, although other functions such as a Fourier series could also be used.

One widely used interpolating polynomial is the *Lagrange interpolation polynomial*. This formula will be discussed in the following two sections for the general case of unequally spaced intervals of the independent variable x and then for the more usual case of equally spaced intervals.

2.6.1 Lagrange's interpolation polynomial for unequal intervals

Given a set of $n+1$ tabulated points

$$(x_0, y_0), (x_1, y_1), (x_2, y_2), \ldots, (x_n, y_n) \tag{2.6.1}$$

we seek an equation of a curve that will pass through all these points. One such equation is the Lagrange polynomial:

$$y = L_0(x)y_0 + L_1(x)y_1 + L_2(x)y_3 + \cdots + L_n(x)y_n \tag{2.6.2}$$

where

$$\begin{aligned}
L_0(x) &= \frac{(x-x_1)(x-x_2)(x-x_3)\cdots(x-x_n)}{(x_0-x_1)(x_0-x_2)(x_0-x_3)\cdots(x_0-x_n)} \\
L_1(x) &= \frac{(x-x_0)(x-x_2)(x-x_3)\cdots(x-x_n)}{(x_1-x_0)(x_1-x_2)(x_1-x_3)\cdots(x_1-x_n)} \\
L_2(x) &= \frac{(x-x_0)(x-x_1)(x-x_3)\cdots(x-x_n)}{(x_2-x_0)(x_2-x_1)(x_2-x_3)\cdots(x_2-x_n)} \\
&\ \ \vdots \\
L_n(x) &= \frac{(x-x_0)(x-x_1)(x-x_2)\cdots(x-x_{n-1})}{(x_n-x_0)(x_n-x_1)(x_n-x_2)\cdots(x_n-x_{n-1})}
\end{aligned} \tag{2.6.3}$$

We may convince ourselves that this polynomial passes through each of our data points by evaluating it at the point (x_r, y_r). Substituting x_r for x in the first of Eqs (2.6.3), we get:

$$L_0(x_r) = \frac{(x_r - x_1)(x_r - x_2)\cdots(x_r - x_r)\cdots(x_r - x_n)}{(x_0 - x_1)(x_0 - x_2)\cdots(x_0 - x_r)\cdots(x_0 - x_n)} = 0 \qquad (2.6.4)$$

In a similar way, all the other *L*s will vanish, except

$$L_r(x_r) = \frac{(x_r - x_0)(x_r - x_1)\cdots(x_r - x_n)}{(x_r - x_0)(x_r - x_1)\cdots(x_r - x_n)} = 1 \qquad (2.6.5)$$

Therefore, Eq. (2.6.2) reduces to

$$y = L_r(x_r)y_r = y_r \qquad (2.6.6)$$

Showing that the point (x_r, y_r) does in fact satisfy the equation.

2.6.2 *Lagrange's polynomial method simplified for three-point interpolation*

The Lagrange polynomial is rather forbidding in the general form of Eqs. (2.6.2) and (2.6.3) and is also inconvenient to program in that form. Let us simplify these equations for the case of three unequally spaced data points (x_0, y_0), (x_1, y_1), and (x_2, y_2). Interpolation over three data points is sometimes called *quadratic interpolation* because the interpolating polynomial will turn out to be of second degree. Let us define '*R*' variables as

$$\begin{aligned} R_0 &= \frac{y_0}{(x_0 - x_1)(x_0 - x_2)} \\ R_1 &= \frac{y_1}{(x_1 - x_0)(x_1 - x_2)} \\ R_2 &= \frac{y_2}{(x_2 - x_0)(x_2 - x_1)} \end{aligned} \qquad (2.6.7)$$

Then, from Eqs (2.6.3)

$$\begin{aligned} L_0(x)y_0 &= R_0(x - x_1)(x - x_2) \\ L_1(x)y_1 &= R_1(x - x_0)(x - x_2) \\ L_2(x)y_2 &= R_2(x - x_0)(x - x_1) \end{aligned} \qquad (2.6.8)$$

Equation (2.6.6) then becomes

$$\begin{aligned} y &= R_0(x - x_1)(x - x_2) + R_1(x - x_0)(x - x_2) + R_3(x - x_0)(x - x_1) \\ &= R_1\left[x^2 - (x_1 + x_2)x + x_1x_2\right] + R_1\left[x^2 - (x_0 + x_2)x + x_0x_2\right] \\ &\quad + R_2\left[x^2 - (x_0 + x_1)x + x_0x_1\right] \end{aligned}$$

Collecting all the terms we get

$$y = C_0 + C_1x + C_2x^2 \tag{2.6.9}$$

where

$$\begin{aligned} C_0 &= R_0x_1x_2 + R_1x_0x_2 + R_2x_0x_1 \\ C_1 &= -R_0(x_1 + x_2) - R_1(x_0 + x_2) - R_2(x_0 + x_1) \\ C_2 &= R_0 + R_1 + R_2 \end{aligned} \tag{2.6.10}$$

Equation (2.6.9) enables us to write an algorithm for three-point interpolation. It finds the second-degree equation passing through the three points (x_0y_0), (x_1y_1), and (x_2y_2) and computes the ordinate Y for some intermediate value of X falling anywhere between x_0 and x_2. Remember that x_0, x_1, and x_2 do not have to be equally spaced.

For the usual case where the xs are *equally spaced*, with interval h, the equation simplifies to,

$$\begin{aligned} C_0 &= \frac{y_0x_1x_2 - 2y_1x_0x_2 + y_2x_0x_1}{2h^2} \\ C_1 &= \frac{-y_0(x_1 + x_2) + 2y_1(x_0 + x_2) - y_2(x_0 + x_1)}{2h^2} \\ C_2 &= \frac{y_0 - 2y_1 + y_2}{2h^2} \end{aligned} \tag{2.6.11}$$

Example 2.6.1 Quadratic interpolation of thermophysical data.

Redo Example 2.5.1 using quadratic interpolation, taking data for one lower temperature (260°C), and two higher values (280 and 300°C) from Table 2.5.1.

Solution

- Open the workbook Ex02-06-01.xls. This workbook consists of one worksheet, namely, *Compute*.
- Insert the 'X' data from that given in Table 2.5.1 in **cells (B4:B6)**.
- Insert the respective 'Y' data given in Table 2.5.1 in **cells (D4:D6)**.
- Insert the *required* X value in **cell B8**.
- Simultaneously press **Alt + F8** followed by **Alt + R**. Excel performs the necessary calculations, and returns the values of 1765.1, 8.4175, and −0.017125 for C_0, C_1, and C_2, in **cells (B16:B18)**, respectively.

The required Y value, enthalpy for the specified *candidate* X, is calculated as 2789.2 J/kg, **cell B21**.

Note that in contrast to linear interpolation, the error involved in this case has been reduced to 0.01%.

Three or four-point interpolation is advantageous only where data points are precisely known, as in published tables of mathematical functions or data from carefully performed experiments. The method of four point, or cubic, interpolation is now presented.

2.6.3 *Four-point interpolation*

Following the same procedure as used in the previous section, we get the equations:

$$y = C_0 + C_1 x + C_2 x^2 + C_3 x^3 \tag{2.6.12}$$

where

$$\begin{aligned} C_0 &= -R_0 x_1 x_2 x_3 - R_1 x_0 x_2 x_3 - R_2 x_0 x_1 x_3 - R_3 x_0 x_1 x_2 \\ C_1 &= R_0(x_1 x_2 + x_1 x_3 + x_2 x_3) + R_1(x_0 x_2 + x_2 x_3 + x_0 x_3) \\ &\quad + R_2(x_0 x_1 + x_1 x_3 + x_0 x_3) + R_3(x_0 x_1 + x_1 x_2 + x_0 x_2) \\ C_2 &= -R_0(x_1 + x_2 + x_3) - R_1(x_0 + x_2 + x_3) \\ &\quad - R_2(x_0 + x_1 + x_3) - R_3(x_0 + x_1 + x_2) \\ C_3 &= R_0 + R_1 + R_2 + R_3 \end{aligned} \tag{2.6.13}$$

$$\begin{aligned} R_0 &= \frac{y_0}{(x_0 - x_1)(x_0 - x_2)(x_0 - x_3)} \\ R_1 &= \frac{y_1}{(x_1 - x_0)(x_1 - x_2)(x_1 - x_3)} \\ R_2 &= \frac{y_2}{(x_2 - x_0)(x_2 - x_1)(x_2 - x_3)} \\ R_3 &= \frac{y_3}{(x_3 - x_0)(x_3 - x_1)(x_3 - x_2)} \end{aligned} \tag{2.6.14}$$

As with the three-point formula, simplification is possible when the data points are equally spaced.

Example 2.6.2 Cubic interpolation of thermophysical data.

Redo Example 2.5.1 using cubic interpolation taking two lower and two higher data points from Table 2.5.1.

Solution

- Open the workbook Ex02-06-02.xls. This workbook consists of one worksheet, namely, *Compute*.
- Insert the ‘X’ data given in Table 2.5.1 in **cells (B4:B7)**.
- Insert the respective ‘Y’ data given in Table 2.5.1 in **cells (D4:D7)**.

- Insert the *required* X value in **cell B8**.
- Simultaneously press **Alt + F8** followed by **Alt + R**. Excel performs the necessary calculations, and returns the values of 3312.1, −8.2142, 0.04238, and -7.08×10^{-5} for C_0, C_1, C_2, and C_3 in **cells (B16:B19)**, respectively.
- The required Y value, enthalpy for the specified X, is calculated as 2789.2 J/kg, **cell B21**.

Note that in contrast to both linear and quadratic interpolation, the error involved in this case has been further reduced to 0.004%.

2.7 What-if analysis using interactive graphs

An interesting feature of Excel is that it provides a dynamic link between numeric and graphic environments. While spreadsheet tables are essential for querying information expressed in the numerical format, often, graphical outputs are also sought by designers to provide an insight into the overall picture. The following example will demonstrate this feature of Excel.

Example 2.7.1 Boundary layer thickness plotted as a function of the distance from the plate leading edge.

Using data from Example 2.5.2, plot the hydrodynamic (δ) and thermal boundary (δ_T) layer thickness as a function of the distance from the leading edge.

Solution

- Open the workbook Ex02-07-01.xls. This workbook consists of two worksheets, namely, *Compute* and *Properties*.
- Activate the worksheet *Compute* by clicking on its tab. This worksheet includes the computations performed by Excel and the required plot.
- Note that using Eqs (2.5.4) and (2.5.5), δ and δ_T are calculated for a range of 'x' values [see **cells (B14:K14)**]. The reader may wish to select any other suitable range for x.
- At $x = 0.1$ m **cell B14**, the calculated Re_x, δ, and δ_T are 1.65×10^4, 3.9, and 4.4 mm, respectively [see **cells (B15:B17)**]. To facilitate the task of writing the formulae for Re_x, δ, and δ_T for each value of x, these formulae are to be keyed-in only in **cells (B15:B17)** and then copied to the **cells (C15:K17)**.

Note the imbedded chart is dynamically linked, that is, any change in the input values alters the graph.

2.8 Advanced (two-dimensional) lookup tables

In Section 2.5 the **lookup** function was introduced. That particular facility enables linear interpolation of data and search in one-dimensional tables, that is, only one independent variable can be handled. In engineering practice however, particularly in thermofluid design, two-dimensional tables are commonly used. For example, in heat transfer calculations related to steam boiler design and analysis, one encounters thermodynamic and transport properties that are functions of two independent variables, namely, temperature and pressure. Three such property matrices are given in the file Ex02-08-01.xls, provided on the companion CD. These matrices include tables for density, enthalpy, and entropy for a steam/water mixture for a common range of temperature and pressure.

Presently, the Visual Basic for Applications (VBA) facility within Excel has been used to develop a two-dimensional search and interpolation routine. Its use is demonstrated via Example 2.8.1.

Example 2.8.1 Two-dimensional lookup table for thermophysical properties.

Use the VBA program provided within Ex02-08-01.xls to explore interpolation for density of steam/water mixture. Compute density for the range of temperature and pressure values given within the *Calculations* worksheet of the above mentioned Excel file.

By simultaneously pressing the '**Alt**' + '**F11**' keys, display the VBA routine. The function '**InterpolateD(T, P)**' developed by the present authors is displayed. This function will linearly interpolate density values for any given temperature and pressure. The formula given in **cell C5** may be copied down column C to obtain density values for other temperature and pressure combinations. Interpolated values for enthalpy and entropy may be obtained in a similar manner by copying down their respective equations given in **cells (D5:E5)**.

It is only possible to interpolate within one given phase of the substance, that is, the user may interpolate for compressed liquid or superheated phases. The user may wish to improve the accuracy of computations by employing the quadratic or cubic interpolation techniques presented in Section 2.6.

2.9 Solution of a non-linear algebraic equation by Newton's method

Design engineers often encounter a situation where the result of a formula, or the final result at the end of a series of computations is known (or desired), but not the input value. To solve such a problem the Excel **Goal Seek** facility may be used. Excel varies the value in the specified cell until the result in the target cell is obtained. Example 2.9.1 will adequately demonstrate this procedure.

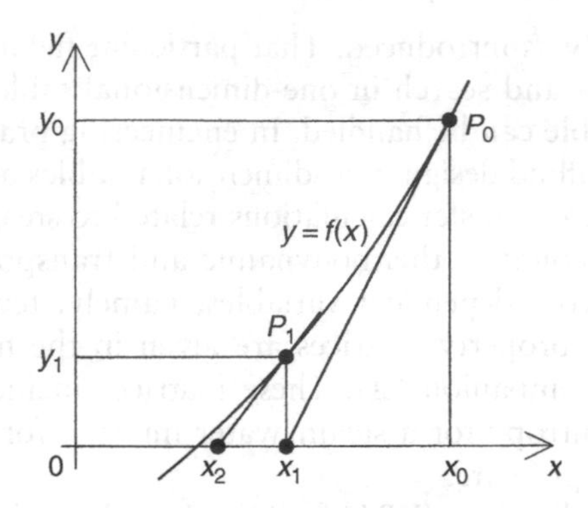

Figure 2.9.1 The Newton–Raphson method.

2.9.1 *Newton's method* (Goal Seek)

The *Newton*, or *Newton–Raphson*, method of approximating roots is probably the most widely used method because of its rapid convergence and ease of programming. It requires only one first guess which, if not sufficiently close to the root, may sometimes cause divergence, or convergence on the wrong root. To use this method it is necessary to know the first derivative of the given function.

Referring to Fig. 2.9.1, we make the initial guess X_0, from which we compute the ordinate Y_0. Evaluating the first derivative of our function at X_0, we obtain the slope m of the tangent to the curve at P_0, which crosses the x axis at X_1. We take X_1 as our second approximation to the root. Since

$$f'(X_0) = \frac{(Y_0 - 0)}{(X_0 - X_1)} \tag{2.9.1}$$

we get

$$X_1 = X_0 - \frac{Y_1}{f'(X_0)} \tag{2.9.2}$$

Our new approximation X_1 is now used to compute Y_1 and the slope at P_1 from which a third approximation X_2 is found, and so on until the required accuracy is achieved.

Excel's **Goal Seek** function is based on the Newton–Raphson method, and this is demonstrated by the following example.

Example 2.9.1 Solution of Colebrook's equation for fluid friction factor via Newton's method.

Redo Example 2.4.1 using the Colebrook's (1939) model given below and Newton's method for solving the non-linear equation.

Colebrook's model for fluid friction factor

$$\frac{1}{\sqrt{f}} = -2\log_{10}\left(\frac{\varepsilon/D}{3.7} + \frac{2.51}{\mathrm{Re}\sqrt{f}}\right) \tag{2.9.3}$$

Solution

Open the workbook Ex02-09-01.xls and follow the same steps as in Example 2.4.1.

- Colebrook's non-linear equation requires a trial and error solution for friction factor. Therefore, assume a value for 'f' and insert it in **cell B17**. Convert Colebrook's equation to the form $F(f) = 0$ and insert it in **cell B18**. Use the **Goal Seek** function to find the value of 'f', **cell B17**, that makes the value of **cell B18** = 0. **Goal Seek** is based on Newton's method and this works by starting with an initial value in the target cell, then zeroing in on a solution ($f = 0.0291$).
- The pressure drop

  $$\Delta p = f\frac{l}{D}\frac{1}{2}\rho V^2 = 1119\,\text{Pa cell B19}$$

  is in good agreement with the answer of Example 2.4.1 which gave a result of 1122 Pa.

The use of the **Goal Seek** procedure is now explained in a step-by-step manner.

1. From the **Tools** menu select **Goal Seek.** If for any reason the **Goal Seek** function was not originally installed with Excel software then run the set-up program to install it.
2. In the **Set cell** box, enter the name of the target (or results) cell. In this particular instance the cell address is **B18** and its value is 4.57. This quantity corresponds to the assumed value of friction factor, $f = 0.01$.
3. In the **To value** box, enter the value of zero.
4. In the **By changing cell** box, enter the reference of the cell which contains the value for the friction factor, that is, **cell B17**.

Click on the **OK** button to initiate the **Goal Seek** process. If **Goal Seek** is not able to find a solution, it displays an error message. If it finds a solution, it shows the **Goal Seek Status** box which gives the **Current Value** to which the solution converges, 6.39×10^{-6} in the present case. If this value is acceptable, click **OK**.

2.10 Solution of a non-linear algebra equation via False Position method

This method, also called *regula falsi* and the *secant method*, is similar to the method of interval halving in that it requires two initial approximations. The method will sometimes work if both initial guesses are on the same side of the root, but to guarantee convergence, they should bracket the root.

From Fig. 2.10.1, the two initial approximations are X_1 and X_2. The line connecting P_1 and P_2 crosses the x axis at X_3, which is taken as the next approximation to the root.

The slope of this line is

$$m = \frac{Y_2 - Y_1}{X_2 - X_1} = \frac{Y_2 - 0}{X_2 - X_3} \tag{2.10.1}$$

from which

$$X_3 = X_2 - Y_2 \frac{X_2 - X_1}{Y_2 - Y_1} \tag{2.10.2}$$

Example 2.10.1 Solution of Colebrook's equation for fluid friction factor via False Position method.

Redo Example 2.4.1 using the Colebrook's (1939) model, Eq. (2.9.3), and the **False Position** method for solving the non-linear equation.

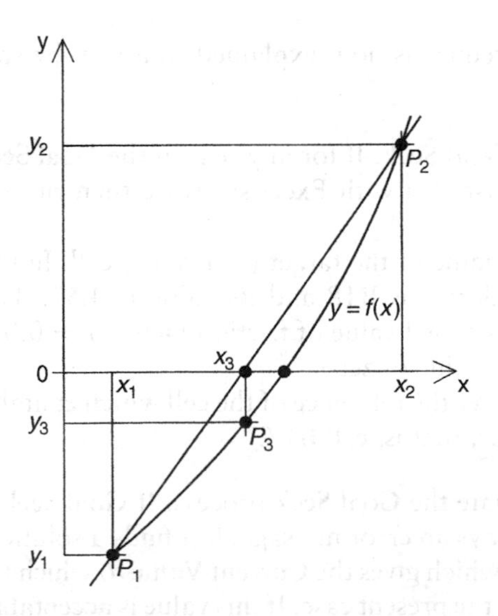

Figure 2.10.1 The False Position method.

Solution

Refer to the Excel workbook Ex02-10-01.xls in the companion CD. Open the workbook and then launch the VBA Macro by simultaneously pressing **Alt + F11** keys. A routine for **False Position** method is furnished. Study the code carefully then try out the solution for 'f' by changing diameter, relative roughness and flow velocity. In each case the solution is gained by running the VBA sub-routine. This routine is executed by simultaneously pressing the **Alt + F8** keys and then either clicking the **Run** button or pressing **Alt + R** keys.

As for the **Goal Seek** method presented in Section 2.9, the above method also returns a value of 0.0291 for f. The macro has been designed to also give the number of iterations required to obtain this value.

2.11 Conversion of partial differential equations to algebraic equations

2.11.1 The nodal network applied to the heat conduction equation

In contrast to an analytical solution, which allows for temperature determination at *any* point of interest in a medium, a numerical solution enables determination of the temperature at *discrete* points only. The first step in any numerical analysis must therefore be to select these points. Referring to Fig. 2.11.1, this is done by subdividing the medium of interest into a number of small regions and assigning to each a reference point that is at its centre. The reference point is frequently termed a *nodal point* (or simply *node*), and the aggregate of points is termed a *nodal network*, *grid*, or *mesh*. The nodal points are designated by a numbering scheme that, for a two-dimensional

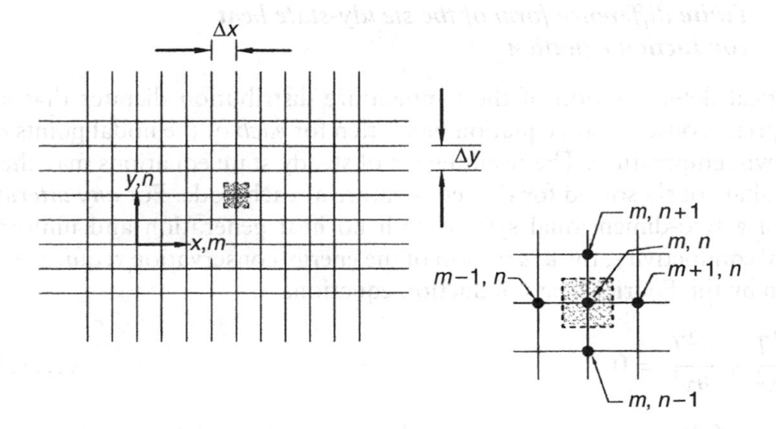

Figure 2.11.1 Two-dimensional conduction heat transfer nodal network.

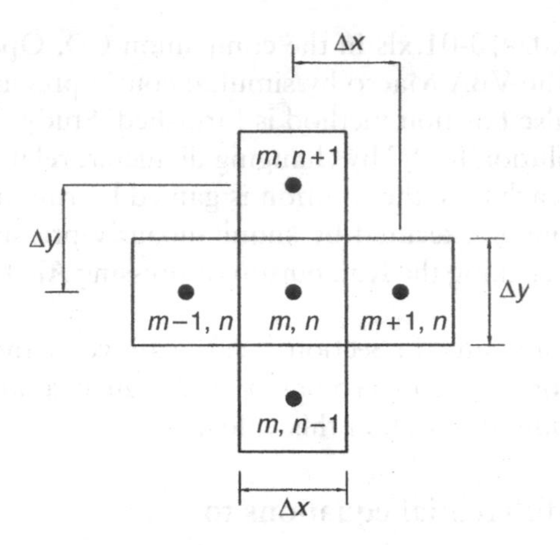

Figure 2.11.2 Two-dimensional conduction heat transfer with respect to an interior node.

system, may take the form shown in Fig. 2.11.1. The x and y locations are designated by the m and n indices, respectively.

It is important to note that each node represents a certain region, and its temperature is a measure of the *average* temperature of the region. For example, the temperature of the node m, n of Fig. 2.11.2 may be viewed as the average temperature of the surrounding shaded area. The numerical accuracy of calculations that are performed depends strongly on the number of designated nodal points. If this number is small the accuracy will be limited. However, a large number of points may be chosen to obtain a finer mesh and thus obtain higher accuracy.

2.11.2 *Finite difference form of the steady-state heat conduction equation*

Numerical determination of the temperature distribution dictates that an appropriate conservation equation be written for *each* of the nodal points of unknown temperature. The resulting set of steady-state equations may then be simultaneously solved for the temperature at each node. For *any interior* node of a two-dimensional system with no heat generation and uniform thermal conductivity, the *exact* form of the energy conservation requirement is given by the Fourier heat conduction equation,

$$\frac{\partial^2 T}{\partial x^2} + \frac{\partial^2 T}{\partial y^2} = 0 \qquad (2.11.1)$$

However, if the system is characterised in terms of a nodal network, it may be necessary to work with a *finite-difference* form of the above equation.

A finite-difference equation which is suitable for the interior nodes of a two-dimensional system may be inferred directly from Eq. (2.11.1). Consider the second derivative, $\partial^2 T/\partial x^2$. From Fig. (2.11.2), the value of this derivative at the m, n nodal point may be approximated as,

$$\left.\frac{\partial^2 T}{\partial x^2}\right|_{m,n} \approx \frac{\partial T/\partial x|_{m+1/2,n} - \partial T/\partial x|_{m-1/2,n}}{\Delta x} \quad (2.11.2)$$

The temperature gradients may in turn be expressed as a function of the nodal temperatures. That is,

$$\left.\frac{\partial T}{\partial x}\right|_{m+1/2,n} \approx \frac{T_{m+1,n} - T_{m,n}}{\Delta x} \quad (2.11.3)$$

$$\left.\frac{\partial T}{\partial x}\right|_{m-1/2,n} \approx \frac{T_{m,n} - T_{m-1,n}}{\Delta x} \quad (2.11.4)$$

Substituting Eqs (2.11.3) and (2.11.4) into Eq. (2.11.2), we obtain

$$\left.\frac{\partial^2 T}{\partial x^2}\right|_{m,n} \approx \frac{T_{m+1,n} + T_{m-1,n} - 2T_{m,n}}{(\Delta x)^2} \quad (2.11.5)$$

Proceeding in a similar fashion, it is readily shown that

$$\left.\frac{\partial^2 T}{\partial y^2}\right|_{m,n} \approx \frac{\partial T/\partial y|_{m,n+1/2} - \partial T/\partial y|_{m,n-1/2}}{\Delta y}$$

$$\approx \frac{T_{m,n+1} + T_{m,n-1} - 2T_{m,n}}{(\Delta y)^2} \quad (2.11.6)$$

Using a network for which $\Delta x = \Delta y$ and substituting Eqs (2.11.5) and (2.11.6) into Eq. (2.11.1) we obtain

$$T_{m,n+1} + T_{m,n-1} + T_{m+1,n} + T_{m-1,n} - 4T_{m,n} = 0 \quad (2.11.7)$$

Hence for the m, n node the heat conduction equation, which is an exact differential equation, is reduced to an approximate algebraic equation. This approximate, finite-difference form of the heat equation may be applied to any interior node that is equidistant from its four neighbouring nodes.

2.12 Solution of a linear system of equations (matrix algebra)

2.12.1 The matrix inversion method

Consider a system of N finite-difference equations corresponding to N unknown temperatures. In writing these equations, it is convenient to identify the nodes by a single integer subscript, rather than by the double subscript

(m,n) previously used. The procedure for setting up a matrix inversion begins by expressing the equation as

$$\begin{aligned}
&a_{11}T_1 + a_{12}T_2 + a_{13}T_3 + \cdots + a_{1N}T_N = C_1\\
&a_{21}T_1 + a_{22}T_2 + a_{23}T_3 + \cdots + a_{2N}T_N = C_2\\
&\qquad\qquad\vdots\\
&a_{N1}T_1 + a_{N2}T_2 + a_{N3}T_3 + \cdots + a_{NN}T_N = C_N
\end{aligned} \tag{2.12.1}$$

where the quantities $a_{11}, a_{12}, \ldots, C_1, \ldots$ are known coefficients and constants involving quantities such as $\Delta x, k, h$, and T_∞. Using matrix notation, these equations may be expressed as

$$[A][T] = [C] \tag{2.12.2}$$

where

$$A \equiv \begin{bmatrix} a_{11} & a_{12} & \cdots & a_{1N} \\ a_{21} & a_{22} & \cdots & a_{2N} \\ \vdots & \vdots & \cdots & \vdots \\ a_{N1} & a_{N2} & \cdots & a_{NN} \end{bmatrix}, \quad T \equiv \begin{bmatrix} T_1 \\ T_2 \\ \vdots \\ T_N \end{bmatrix}, \quad C \equiv \begin{bmatrix} C_1 \\ C_2 \\ \vdots \\ C_N \end{bmatrix}$$

The square *coefficient matrix* $[A]$ *elements* are designated by a double subscript notation, for which the first and second subscripts refer to rows and columns, respectively. The matrices $[T]$ and $[C]$ have a single column and are known as *column vectors*. If the matrix multiplication implied by the left-hand side of Eq. (2.12.1) is performed, Eq. (2.12.2) is obtained.

The solution vector may now be expressed as

$$[T] = [A]^{-1}[C] \tag{2.12.3}$$

where $[A]^{-1}$, the inverse of $[A]$, and $[A]$ are inter-related by the following identity,

$$[A]^{-1} \cdot [A] \equiv \begin{bmatrix} 1 & 0 & \ldots & 0 \\ 0 & 1 & \ldots & 0 \\ 0 & \ldots & 1 & 0 \\ 0 & \ldots & 0 & 1 \end{bmatrix} \tag{2.12.4}$$

Note that the unit matrix shown on the right-hand side of Eq. (2.12.4) contains a diagonal of unit numbers with all other elements being zeroes.

Example 2.12.1 Solution of a linear system of equations using matrix inversion.

While solving problems in heat transfer one often encounters a set of simultaneous equations with several unknowns. These equations may or may not be linear. While the solution of non-linear problems shall be dealt with in Section 2.13, presently, the solution of a linear system of equations using matrix operations is demonstrated.

Consider the following system of algebraic equations,

$$x + y + z = 6$$
$$3x + 2y + 2z = 13$$
$$2x + 2y + z = 9$$

These equations may now be written in matrix form as,

$$\begin{bmatrix} 1 & 1 & 1 \\ 3 & 2 & 2 \\ 2 & 2 & 1 \end{bmatrix} \begin{bmatrix} x \\ y \\ z \end{bmatrix} = \begin{bmatrix} 6 \\ 13 \\ 9 \end{bmatrix}$$

The solution to the unknowns may be obtained from,

$$\begin{bmatrix} x \\ y \\ z \end{bmatrix} = \begin{bmatrix} 1 & 1 & 1 \\ 3 & 2 & 2 \\ 2 & 2 & 1 \end{bmatrix}^{-1} \begin{bmatrix} 6 \\ 13 \\ 9 \end{bmatrix}$$

The Microsoft Excel environment may be used to obtain solution to the above equations quite effectively. Open workbook Ex02-12-01.xls which has just one worksheet. The coefficient matrix and its inversion are respectively shown in yellow and green colours. Note that the inverse of any given matrix is obtained by first highlighting a blank array of an appropriate size, then entering the function, **Minverse (A5:C7)**. The 'Ctrl', 'Shift', and 'Enter' keys should be pressed simultaneously after entering the above function to fill the inverse matrix array.

2.12.2 *Gauss–Seidel iteration*

The Gauss–Seidel method is a powerful and extremely popular iterative technique. Application to the system of equations represented by Eq. (2.12.1) is facilitated by the following procedure.

1 To whatever extent possible, the equations should be reordered to provide diagonal elements whose magnitudes are larger than those of other elements in the same row. That is, it is desirable to sequence the equations such that

$$|a_{11}| > |a_{12}|, |a_{13}|, \ldots, |a_{1N}|; \; |a_{22}| > |a_{21}|, |a_{23}|, \ldots, |a_{2N}|;$$

and so on.

2 After re-ordering, each of the N equations should be written in explicit form for the temperature associated with its diagonal element. Each

temperature in the solution vector would then be of the form:

$$T_i^{(k)} = \frac{C_i}{a_{ii}} - \sum_{j=1}^{i-1} \frac{a_{ij}}{a_{ii}} T_j^{(k)} - \sum_{j=i+1}^{N} \frac{a_{ij}}{a_{ii}} T_j^{(k-1)} \tag{2.12.5}$$

where $i = 1, 2, \ldots, N$. The superscript k refers to the level of the iteration.

3 An initial ($k = 0$) value is assumed for each temperature T_i. Subsequent computations may be reduced by selecting values based on rational estimates.
4 New values of T_i are then calculated by substituting assumed ($k = 0$) or new ($k = 1$) values of T_j into the right-hand side of Eq. (2.12.5). This step is the first iteration ($k = 1$).
5 Using Eq. (2.12.5), the iteration procedure is continued by calculating new values of $T_i^{(k)}$ from the $T_j^{(k)}$ values of the current iteration, where $1 \le j \le i - 1$, and the $T_i^{(k-1)}$ values of the previous iteration, where $i + 1 \le j \le N$.
6 The iteration is terminated when a prescribed *convergence criterion* is satisfied. The criterion my be expressed as

$$\text{Max}\left|T_i^{(k)} - T_i^{(k-1)}\right| \le \varepsilon \tag{2.12.6}$$

where ε represents an acceptable degree of uncertainty in the calculated temperature.

2.13 Solution of non-linear system of equations

2.13.1 Introduction to optimisation

This section is primarily concerned with stating and describing the so-called classical optimisation procedures. These derive, after suitable embellishments to make them useful, from the application of the calculus to the basic problem of finding the maximum or minimum of a continuous function. These techniques have an intrinsic utility in that they can sometimes be used to solve problems that are not too complex and do not involve more than a few variables. More important, however, is the theoretical significance of the results of classical optimisation theory. This theory is of importance in the development of solution methods to non-linear programming problems. The theory underlies any discussion of the development of algorithms for solving any complex optimisation problem.

2.13.2 Functions of one variable: mathematical background

We shall here summarise some definitions and results that will be required in subsequent sections.

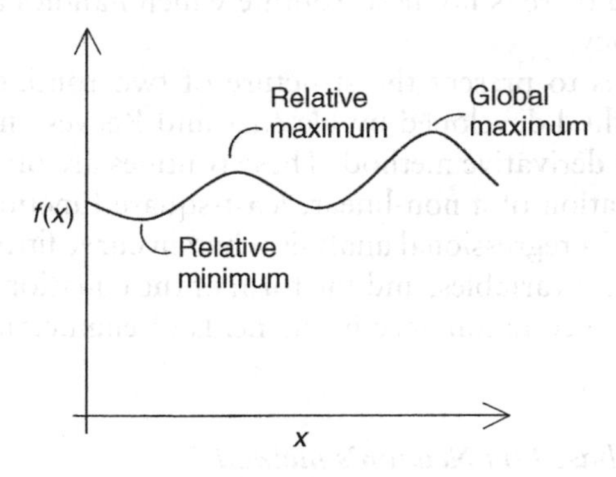

Figure 2.13.1 Relative maxima and minima, and the global maximum of a function $f(x)$.

If a function $f(x)$ takes on its *absolute maximum* at a point x^* if $f(x) \leq f(x^*)$ for *all* x. The absolute maximum is often called the *global maximum*. The definition of *global maximum* can be obtained from the preceding definition by simply reversing the inequality between $f(x)$ and $f(x^*)$. If the initial inequality holds true for a value of x within a limited range of x values, the point is termed a *relative*, or *local maximum*.

In Fig. 2.13.1 we see several examples of relative maxima and minima, as well as a global maximum designated. In heat transfer problems there may be situations where one may also encounter weak relative maxima or minima. One such example is the optimum cavity spacing for minimisation of heat loss from a double-glazing. As shall be demonstrated in Chapter 8, there is no *critical* cavity spacing where a *sharp* minimum value of heat transfer occurs.

In addition to the above definitions, the literature of optimisation often makes reference to a stationary point. A *stationary point* is any point at which $f'(x) = 0$.

In dealing with optimisation problems we encounter objective functions which are to be either maximised or minimised. The techniques used for solving such problems are iterative in nature and are therefore suitable for use on computers.

Computer optimisation routines may be classified into three main categories, namely: direct search, first-derivative, and second-derivative types. During the latter part of the twentieth century a number of routines, which fall into the three above categories were developed. For fuller details, the reader is directed to the literature (Fletcher and Powell, 1963; Fletcher and Reeves, 1964; Powell, 1964; Greig, 1980; and Muneer, 1988). These routines can be efficiently used for minimisation of a function while also finding application in such diverse fields as economics, operations research, mathematics, engineering, statistics, and regression analysis. Each routine has certain merits and demerits and there is no 'best' routine which handles all problems with the same efficiency.

The purpose of this section is to present the structure of two routines, namely, the first derivative method developed by Fletcher and Reeves and, second, Newton's second order derivative method. These routines are often used in relation to the minimisation of a non-linear, least-square function. The problem is essentially one of a regressional analysis wherein curve fitting is to be attempted between several variables, and the form of the equation is non-linear. Such applications are commonplace in the fields of engineering heat transfer.

2.13.3 Optimisation routine based on Newton's method

This method, introduced in Section 2.9 for a single non-linear equation, requires second derivatives of the function being minimised. The algorithm is based on Newton's formula which states that if X_r is an approximation to the root of the equation; then a better approximation is in general given by X_{r+1} where

$$X_{r+1} = X_r - \frac{f(X_r)}{f'(X_r)} \tag{2.13.1}$$

For a general function which has to be minimised, the algorithm reduces to:

$$X_{r+1} = X_r - B_r^{-1} g_r \tag{2.13.2}$$

where g_r and B_r are, respectively, the gradient vector and Hessian matrix at the iteration.

$$g_r = \left(\frac{\partial f}{\partial x_1}, \frac{\partial f}{\partial x_2}, \ldots, \frac{\partial f}{\partial x_n} \right)$$

$$B_r = \left\| \frac{\partial^2 f}{\partial x_i \partial x_j} \right\|$$

2.13.4 Optimisation routine based on Fletcher–Reeves method

This algorithm, also known as the method of conjugate gradients, uses the first derivatives of the function which may be either evaluated numerically or provided to the routine by manual differentiation. It uses a set of *P*s based on the negative gradients $-g_r$ at successive points. The algorithm can be summarised by the following relations:

(i) $P_1 = g_1$
(ii) X_{r-1} is the minimum of 'f' along the line $X_r + \alpha P_r$
(iii) $P_{r+1} = -g_{r+1} + \beta P_r$
(iv) $\beta = (g_{r+1}^{\mathrm{T}} g_{r+1})/(g_r^{\mathrm{T}} g_r)$

For a general function, cycles of n points are calculated, restarted each time with current gradients. It may be noted that in step (ii) the multi-dimensional problem transforms into a one-dimensional optimisation problem. Further details to the above two routines are provided in Muneer (1988).

2.14 Linear regression using least squares method

2.14.1 Fitting a straight line to a set of data points

Often in heat transfer experimentation the data need to be fitted to a straight line. This section describes how to find the equation of the straight line that is the best fit for a particular set of data. Many sets of data can be fitted with a straight line. More importantly, a large number of other equations can be rewritten in the form of a straight line, and can thus be fitted to data by means of the simple methods presented in this section.

2.14.2 Graphical method

In this method, also called the *method of selected points*, we simply plot the data points, draw a straight line with a straight edge, and read the slope m and y-intercept b of the drawn line. The equation of the line, in slope-intercept form, is then

$$y = mx + b \tag{2.14.1}$$

If the line does not intercept the y-axis and is too far from it to be extended, we read the slope of the line and the coordinates (x_1, y_1) of any point *on the line*. Then, using the point-slope form of the equation of a straight line, we get,

$$m = \frac{y - y_1}{x - x_1} \tag{2.14.2}$$

which can be re-arranged into slope-intercept form as

$$y = mx + (y_1 - mx_1) \tag{2.14.3}$$

The graphical method is fast and simple and can give quite good results if done carefully. It is useful for an approximate check on the more accurate method of least squares.

2.14.3 Method of least squares

A second means of determining the constants m and b is the *method of least squares* which requires that we *minimise* the sum of the squares of the residuals, or

$$\sum r^2 = \sum [y - f(x)]^2 = \text{minimum} \tag{2.14.4}$$

From calculus, we know that in order to find the minimum of some function we must take the first derivative and set it equal to zero. Taking (partial) derivatives with respect to b,

$$\begin{aligned}
\frac{\partial}{\partial b}\sum r^2 &= \frac{\partial}{\partial b}\sum [y - f(x)]^2 \\
&= \frac{\partial}{\partial b}\left[\sum (y - mx - b)^2\right] \\
&= \frac{\partial}{\partial b}\Big[(y_1 - mx_1 - b)^2 + (y_2 - mx_2 - b)^2 \\
&\qquad + \cdots + (y_n - mx_n - b)^2\Big] \\
&= -2(y_1 - mx_1 - b) - 2(y_2 - mx_2 - b) \\
&\qquad - \cdots - 2(y_n - mx_n - b) \\
&= -2\sum y + 2m\sum x + 2nb
\end{aligned} \tag{2.14.5}$$

where n is the number of data points. Setting the right-hand side of Eq. (2.14.5) equal to zero, we get

$$m\sum x + nb = \sum y \tag{2.14.6}$$

Now taking partial derivatives with respect to m,

$$\begin{aligned}
&\frac{\partial}{\partial m}\left[\sum (y - mx - b)^2\right] \\
&\quad = -2(y_1 - mx_1 - b)x_1 \\
&\qquad - 2(y_2 - mx_2 - b)x_2 - \cdots - 2(y_n - mx_n - b)x_n \\
&\quad = -2\sum xy + 2m\sum x^2 + 2b\sum x
\end{aligned} \tag{2.14.7}$$

and setting this also to zero, we get

$$m\sum x^2 + b\sum x = \sum xy \tag{2.14.8}$$

Equations (2.14.6) and (2.14.8) are called the normal equations, and their simultaneous solution by means of Eqs (2.14.9) and (2.14.10) will yield the values of m, slope and b, intercept:

$$m = \frac{n\sum xy - \sum x \sum y}{n\sum x^2 - \left(\sum x\right)^2} \tag{2.14.9}$$

$$b = \frac{\sum x^2 \sum y - \sum x \sum xy}{n\sum x^2 - \left(\sum x\right)^2} \tag{2.14.10}$$

The use of these equations will be made clearer by a numerical example.

Example 2.14.1 Linear regression – fitting data to a straight line.

Find the equation of the straight line that will best fit the eight data points given in the first two columns of Table 2.14.1, as shown in Fig. 2.14.1, using the least squares method.

Table 2.14.1 Data for the eight points shown in Fig. 2.14.1

x	y	x^2	xy
1	3.249	1	3.249
2	3.522	2	7.044
3	4.026	9	12.08
4	4.332	16	17.33
5	4.900	25	24.50
6	5.121	36	30.73
7	5.601	49	39.21
8	5.898	64	47.18
$\sum x = 36$	$\sum y = 36.649$	$\sum x^2 = 204$	$\sum xy = 181.3$

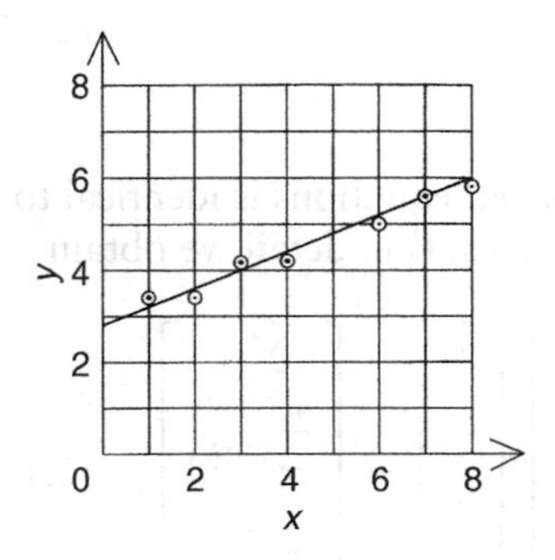

Figure 2.14.1 Fitting data to a straight line: example of linear regression.

Solution

From Table 2.14.1, the summed values for the respective columns may be inserted directly into Eqs (2.14.9) and (2.14.10) to give:

$$\text{slope, } m = \frac{8(181.3) - 36(36.5)}{8(204) - (36)^2} = 0.3904$$

$$\text{intercept, } b = \frac{204(36.65) - 36(181.3)}{8(204) - (36)^2} = 2.824$$

The equation of our best fitting line is therefore:

$$y = 0.3904x + 2.824$$

2.15 The matrix approach to multiple linear regression

In fitting a multiple regression model, it is much more convenient to express the mathematical operations using matrix notation. Suppose that there are κ regressor variables and n observations $(x_{i1}, x_{i2}, \ldots, x_{i\kappa}, y_i)$, $i = 1, 2, \ldots, n$ and that the model relating the regressors to the response is

$$y_i = \beta_0 + \beta_1 x_{i1} + \beta_2 x_{i2} + \cdots + \beta_\kappa x_{i\kappa} + \varepsilon_i, \quad i = 1, 2, \ldots, n \tag{2.15.1}$$

This model is a system of n equations that can be expressed in matrix notation as

$$y = X\beta + \varepsilon \tag{2.15.2}$$

where

$$y = \begin{bmatrix} y_1 \\ y_2 \\ \vdots \\ y_n \end{bmatrix} \quad X = \begin{bmatrix} 1 & x_{11} & x_{12} & \cdots & x_{1\kappa} \\ 1 & x_{21} & x_{22} & \cdots & x_{2\kappa} \\ \vdots & \vdots & \vdots & \vdots & \vdots \\ 1 & x_{n1} & x_{n2} & \cdots & x_{n\kappa} \end{bmatrix} \tag{2.15.3}$$

$$\beta = \begin{bmatrix} \beta_0 \\ \beta_1 \\ \vdots \\ \beta_\kappa \end{bmatrix} \quad \text{and} \quad \varepsilon = \begin{bmatrix} \varepsilon_1 \\ \varepsilon_2 \\ \vdots \\ \varepsilon_n \end{bmatrix} \tag{2.15.4}$$

One can see that the matrix form of the normal equations is identical to the scalar form. Writing out Eqs (2.15.3) and (2.15.4) in detail we obtain

$$\begin{bmatrix} n & \sum_{i=1}^{n} x_{i1} & \sum_{i=1}^{n} x_{12} & \cdots & \sum_{i=1}^{n} x_{i\kappa} \\ \sum_{i=1}^{n} x_{i1} & \sum_{i=1}^{n} x_{i1}^2 & \sum_{i=1}^{n} x_{i1}x_{i2} & \cdots & \sum_{i=1}^{n} x_{i1}x_{i\kappa} \\ \vdots & \vdots & \vdots & \vdots & \vdots \\ \sum_{i=1}^{n} x_{i\kappa} & \sum_{i=1}^{n} x_{i\kappa}x_{i1} & \sum_{i=1}^{n} x_{i\kappa}x_{i2} & \cdots & \sum_{i=1}^{n} x_{i\kappa}^2 \end{bmatrix} \begin{bmatrix} \hat{\beta}_0 \\ \hat{\beta}_1 \\ \vdots \\ \hat{\beta}_\kappa \end{bmatrix} = \begin{bmatrix} \sum_{i=1}^{n} y_i \\ \sum_{i=1}^{n} x_{i1}y_1 \\ \vdots \\ \sum_{i=1}^{n} x_{i\kappa}y_i \end{bmatrix} \tag{2.15.5}$$

If the indicated matrix multiplication is performed, the scalar form of the normal equations (i.e. Eq. (2.15.1)) will result. In this form it is easy to see that X^1X is a $(p \times p)$ symmetric matrix and X^1y is a $(p \times 1)$ column vector. Note the special structure of the X^1X matrix. The diagonal elements of X^1X are the sums of squares of the elements in the columns of X, and the off-diagonal elements are the sums of cross-products of the elements in the columns of X. Furthermore, note that the elements of X and y are the sum of cross-products of the columns of X and the observations (y_i).

The fitted regression model is

$$\hat{y}_i = \hat{\beta}_0 + \sum_{j=1}^{K} \hat{\beta}_j x_{ij}, \quad i = 1, 2, \ldots, n \tag{2.15.6}$$

2.16 Fitting linear models via graphical means – Excel-based facility

Example 2.16.1 Linear regression using Excel.

File Ex02-16-01.xls and the its companion plot (provided within the aforementioned file) present the results of measurements carried out to develop a non-dimensional mathematical relationship representing convection heat transfer from a cylinder (y-axis) and the velocity of a given gas in cross flow (x-axis). Use procedures available within the Excel environment to fit a quadratic relationship between the 'y' and the 'x' variable.

Solution

Open the file Ex02-16-01 and left click on the embedded chart shown to highlight it. From the top line menu list select 'Chart', followed by 'Add Trendline'.

- From the 'Type' menu select 'polynomial' and order '2'. Click on the 'Options' tab and select 'Display equation on chart', followed by OK. Excel automatically fits a second-degree polynomial function to the chart data and displays its equation on the chart.

2.17 Non-linear regression (Solver)

In Section 2.15, the basic mathematics required for non-linear regression was presented. In the study of thermal science, which relies a great deal on empiricism, non-linear regression is commonplace, and this application is now introduced.

2.17.1 The objective function

The objective function chosen in this section is related to a thermodynamic equation of state. The equation of state is an equation relating the pressure, density, and temperature of a particular substance. In this section, an equation of state for steam was employed using thermodynamic property data from steam tables. The equation of this form is known as the Beattie–Bridgeman equation and has five constants to be determined by regression. The equation is written in its full form as:

$$P = RT\rho + \left(RTB - A - \frac{RC}{T^2}\right)\rho^2 + \left(aA - RTBb - \frac{RBC}{T^2}\right)\rho^3 + \frac{RBbc}{T^2}\rho^4 \tag{2.17.1}$$

where P is the pressure in kPa, ρ the density of superheated steam in kg/m^3, and T the temperature of the steam in Kelvin. A, B, a, b, and c are the constants to be determined. The equation was fit to a 72-point data grid which covered a range of pressure from 100 kPa to 20 MPa, while the temperature range was from 200 to 1300°C.

The objective function (S) is the sum of the squares of the residual. Thus

$$S = \sum (P_{\text{calc}} - P_{\text{obs}})^2 \tag{2.17.2}$$

which reduces to

$$S = \sum \left\{ (RT_\rho - P) + \left(RTB - A - \frac{RC}{T^2}\right)\rho^2 + \left(Aa - RTBb - \frac{RBC}{T^2}\right)\rho^3 + \frac{RBbc}{T^2}\rho^4 \right\}^2 \tag{2.17.3}$$

The problem now becomes that of optimisation wherein the objective function S is to be minimised. Table 2.17.1 shows a comparison of the results obtained using Newton's method, and the Fletcher–Reeves algorithm (see Section 2.13). It is clear from the coefficients of correlation given that the Fletcher–Reeves algorithm is the better in performance. This superiority stems from many factors. First, the Fletcher–Reeves algorithm does not require evaluation of the second-order derivatives and construction and inversion of the Hessian matrix in every iteration, as it is in Newton's method. Thus for a general function being minimised for n variables, Newton's routine requires the evaluation of n^2 second-order derivatives and inversion of $n \times x$ Hessian matrices. This additional computing tasks also result in longer computer time being required by Newton's routine. Section 2.19 presents information on the coefficient of correlation, and other statistical methods that are used in the evaluation of regression models.

Table 2.17.1 Comparison of the performance of Newton's method and the Fletcher–Reeves algorithm

Method	*Iterations required*	*Total number of evaluations*			*Optimum value of function*	*Coefficient of correlation*	*Approximate computer time for each iteration*
		Function	*First derivatives*	*Second derivatives*			
Newton's	23	—	115	575	298641	0.995	35
Fletcher–Reeves	5	61	305	—	41630	0.999	17

2.17.2 *Overview of the* Solver *function*

The Excel **Solver** function is based on well-established numeric methods for equation solving and optimisation. These methods supply relevant numeric inputs to the candidate model and through an iterative procedure generates the results. In Section 2.9, Excel's **Goal Seek** facility was demonstrated. **Solver**, while similar to **Goal Seek**, also incorporates the facility for solving a family (or set) of linear, or non-linear, equations. In addition, as mentioned above, multi-dimensional optimisation problems may be solved for function maximisation or minimisation. In the authors' experience this facility of Excel is probably the most powerful feature, as shall be demonstrated via Example 2.17.1.

Solver can handle up to 200 independent variables with upper or lower bounds, along with an additional 100 constraints. However, **Solver** only works 'locally'. It may therefore not give all the possible solutions.

To use the **Solver** facility (provided in the **Tools** menu) the user defines the problem that needs to be solved by firstly identifying a **target** cell, whose value is to be maximised, minimised, or made to reach a specified value, then the **changing** cells (containing the numerical values of the variables). Lastly, the **constraints** imposed upon the optimisation routine are included in the **Solver Parameters** dialog box. Full control of the solution process is possible via the use of the **Solver Options** dialog box. It is possible for the user to select the solution time and the desired number of iterations, precision of constraints, integer tolerance, automatic (independent-variable) scaling, and the solution method used by the **Solver**.

Within the choice of solution method three options are available, namely; the **Estimates**, **Derivatives**, and **Search** procedures. The **Estimates** option enables the selection of a **Tangent** option suitable for near-linear problems or the **Quadratic** option which is more suitable for problems with a high degree of non-linearity. The **Derivatives** option enables a choice between **Forward differencing** (faster and yet approximate computation) and **Central differencing** (slower but precise computation). Finally, the **Search** option makes it possible to experiment with the **Newton** (steepest descent routine which requires less processor work) and **Conjugate** gradient (requires fewer iterations to obtain convergence) methods. Further information on the **Solver** facility is provided in Microsoft (1992) and Person (1992).

Example 2.17.1 Non-linear regression using Excel's **Solver** facility.

Computational Fluid Dynamics (CFD) is the use of computers for modelling and evaluating the behaviour of fluids in given situations, for example, the natural convection of infill gases within cavities of double-glazed windows. Table 2.17.2 shows a sample of the values of convective heat transfer coefficients (h, W/m^{2}K) obtained by Muneer and Han (1996) for different window configurations by using CFD. Research work has shown that the window aspect ratio (H/L) being the ratio of the window height to cavity width, also has a significant bearing on h.

Table 2.17.2 Internal convective heat transfer coefficients, h (W/m^2 K) for different double-glazed window configurations

Configuration and gas filler	*Enclosure gap* (mm)							
	4	8	12	16	20	24	28	50
0.6 m high, $t_I = 20°C$, $t_O = 0°C$, Air	6.25	3.16	2.20	1.90	1.90	1.99	2.07	2.13
0.6 m high, $t_I = 20°C$, $t_O = 0°C$, Argon	4.05	2.05	1.47	1.31	1.34	1.38	1.40	1.38
0.6 m high, $t_I = 20°C$, $t_O = 0°C$, Krypton	2.15	1.14	0.96	1.00	1.03	1.04	1.03	1.01
0.6 m high, $t_I = 20°C$, $t_O = 0°C$, Xenon	1.21	0.74	0.73	0.76	0.77	0.76	0.75	0.75
1 m high, $t_I = 20°C$, $t_O = 0°C$, Air	6.25	3.14	2.16	1.76	1.64	1.67	1.75	1.80
1 m high, $t_I = 20°C$, $t_O = 5°C$, Air	6.25	3.14	2.14	1.71	1.53	1.52	1.57	1.66
1 m high, $t_I = 20°C$, $t_O = -5°C$, Air	6.15	3.10	2.15	1.79	1.73	1.79	1.84	1.85
1 m high, $t_I = 25°C$, $t_O = 5°C$, Air	6.30	3.17	2.17	1.76	1.63	1.65	1.70	1.77
1 m high, $t_I = 15°C$, $t_O = 5°C$, Air	6.15	3.11	2.12	1.74	1.64	1.68	1.73	1.75
2.4 m high, $t_I = 20°C$, $t_O = 0°C$, Air	—	—	—	1.68	1.49	—	—	—
2.4 m high, $t_I = 20°C$, $t_O = -10°C$, Air	—	—	—	—	1.57	—	—	—
2.4 m high, $t_I = 30°C$, $t_O = -10°C$, Air	—	—	—	—	1.19	—	—	—
2.4 m high, $t_I = 30°C$, $t_O = -10°C$, Air	—	—	—	—	0.94	—	—	—

Using the above mentioned table, develop a model for the Nusselt number, $\mathrm{Nu_L}$ in the form,

$$\mathrm{Nu_L} = c\mathrm{Ra_L^m}\left(\frac{H}{L}\right)^n \tag{2.17.4}$$

where $\mathrm{Ra_L}$ is the Rayleigh number and c, m, and n are numerical constants.

Solution

- Open the workbook Ex02-17-01.xls. This workbook consists of one worksheet, namely, *Main* which includes the computations performed by Excel.
- For each of the cases given in **cells (A8:A44)**, the aspect ratio H/L is calculated by using the data from the above table [see **cells (B8:B44)**].
- $\mathrm{Ra_L}$ and $\mathrm{Nu_L}$, measured values are given in **cells (C8:C44)** and **cells (D8:D44)** respectively. The data given in the above table were used in obtaining $\mathrm{Ra_L}$ and $\mathrm{Nu_L}$. This will be explained in more detail in Chapter 8.
- Assume values for c, m, and n and insert them in **cells (C3, C4, and C5)**, respectively, for example, $c = 0.3, m = 0.2$, and $n = -0.4$.
- For each case, obtain a computed value of $\mathrm{Nu_{L,computed}} = c\mathrm{Ra_L^m}(H/L)^n$ using the above values of c, m, and n [see **cells (E8:E44)**].
- Calculate the square of difference between $\mathrm{Nu_{L,computed}}$ and $\mathrm{Nu_{L,measured}}$ for each given case [see **cells (F8:F44)**].
- Calculate the sum of the squares of differences **cell F45**.
- To develop a model for $\mathrm{Nu_L}$ in the form $Nu_L = c\mathrm{Ra_L^m}(H/L)^n$, the values of c, m, and n that will minimise the sum of the squares of differences, **cell F45**, are needed. This can be done via **Solver** facility as follows.

1. From the Tools menu select '**Solver**'. If for any reason the **Solver** function was not originally installed with Excel software then run the set-up program to install it.
2. In the **Set target cell** box, enter the name of the target (or results) cell. In this example the cell address is **F45** and the content of this cell shows that the sum of the squares of differences is 1.05×10^2. This quantity corresponds to the assumed values of $c = 0.3$, $m = 0.2$, and $n = -0.4$.
3. Select the **Min** option.
4. In the **By changing cells** box, enter the references of the cells which contain the values for c, m, and n, that is, **C3, C4, C5**.
5. Choose the **Solve** button. If the 'time limit exceeded' message is displayed, click the 'continue' button.

Upon successful convergence the user has a choice of retaining the **Solver** solution or reverting to the original settings. Answer reports can also be obtained. The answer report for the current example is shown in

Table 2.17.3 'Solver' answer report
Microsoft Excel 8.0e Answer Report
Worksheet: [Ex02-17-01.xls] Main

Target cell (Min)			
Cell	**Name**	**Original value**	**Final value**
F45	{NuLmeasured – NuL, computed}2	1.05E + 02	1.94E + 00
Adjustable cells			
Cell	**Name**	**Original value**	**Final value**
C3	*c*	0.30	0.59
C4	*m*	0.20	0.21
C5	*n*	–0.40	–0.32

Constraints
NONE

Table 2.17.3. The **Solver** facility minimises the sum of the squares of differences to a value of 1.94, resulting in values of c, m, and n as 0.59, 0.21, and -0.32, respectively.

2.18 Measures of deviation

In heat transfer experimentation a large number of data need to be processed statistically. Towards that end this section will introduce some of the basic statistical parameters.

2.18.1 *Standard deviation*

When a given data population is finite and consists of N values, we may define the population variance, or standard deviation, as

$$\sigma^2 = \frac{\sum_{i-1}^{N}(x_i - \mu)^2}{N} \tag{2.18.1}$$

2.18.2 *The coefficient of variation*

Occasionally, it is desirable to express variation as a fraction of the mean. A dimensionless measure of relative variation called the sample coefficient of variation is used to accomplish this. The sample coefficient of variation is defined as

$$cv = \frac{\sigma}{\bar{x}} \tag{2.18.2}$$

The coefficient of variation is useful when comparing the variability of two or more data sets that differ considerably in the magnitude of the observations. For example, the coefficient of variation might be useful in comparing the variability of daily space heating usage within samples of single-family residences in London, England, and Edinburgh, Scotland during January.

Example 2.18.1 Use of coefficient of variation to compare the variability of multiple data sets.

Measurements made with a thermocouple of the temperature of a cup of heated oil have a mean of 40°C and a standard deviation of 0.1°C, whereas measurements made with a second thermocouple of the temperature of an air stream have a mean of 28°C and a standard deviation of 0.075°C. The coefficients of variation are

$$cv_{\text{oil}} = \frac{0.1}{40} = 0.0025$$

and

$$cv_{\text{air}} = \frac{0.075}{28} = 0.0027$$

respectively. Therefore, the measurements made with the first thermocouple exhibit relatively less variability than those made with the second one.

2.18.3 Mean bias error (MBE) and root mean square error (RMSE)

To enable further insight in the performance evaluation of any given convection heat transfer model, MBEs and RMSEs may be obtained. These parameters are defined as

$$\text{MBE} = \sum (Y_c - Y_o)/n \tag{2.18.3}$$

$$\text{RMSE} = \left[\sum (Y_c - Y_o)^2/n\right]^{1/2} \tag{2.18.4}$$

These formulae provide MBE and RMSE which have the same physical units as the dependent variable, Y. In some instances non-dimensional MBE (NDMBE) and RMSE (NDRMSE) are required. These are obtained as

$$\text{NDMBE} = \sum [(Y_c - Y_o)/Y_o]/n \tag{2.18.5}$$

$$\text{NDRMSE} = \left\{\sum [(Y_c - Y_o)/Y_o]^2/n\right\}^{1/2} \tag{2.18.6}$$

2.19 Coefficient of determination and correlation coefficient

2.19.1 Coefficient of determination, r^2

The ratio of explained variation $[\sum (Y_c - Y_m)^2]$ to the total variation $[\sum (Y_o - Y_m)^2]$ is called the coefficient of determination. Y_m is the mean of the observed Y values. The ratio lies between zero and one. A high value of r^2 is desirable as this shows a lower unexplained variation.

2.19.2 *Coefficient of correlation, r*

The square root of the coefficient of determination is defined as the coefficient of correlation, r. It is a measure of the relationship between variables based on a scale ranging between +1 and −1. Whether r is positive or negative depends on the inter-relationship between x and y, that is, whether they are directly proportional (y increases as x increases) or vice versa. Once r has been estimated for any fitted model its numerical value may be interpreted as follows. Let us assume that for a given regression model $r = 0.9$. This means $r^2 = 0.81$. It may be concluded that 81% of the variation in Y has been explained (removed) by the model under discussion, leaving 19% to be explained by other factors.

2.19.3 *Student's t-distribution*

Often the modeller is faced with the question as to what quantitative measure is to be used to evaluate the value of r^2 obtained for any given model. Clearly, r^2 would depend on the size of the data population. For example, a lower value of r^2 obtained for a model fitted against a large database may or may not be better than another model which used a smaller population. In such situations the student's t-test may be used for comparing the above two models. Example 2.19.1 demonstrates the use of this test of significance for r^2.

Example 2.19.1 Use of Student's t-test for evaluating the significance for r^2.

For a given regression model for convection between dimensionless wind speed and dimensionless heat transfer from a heated surface, $r^2 = 0.64$ for 12 pairs of data points. Using Student's t-test investigate the significance of r. The test statistic

$$t = (n - 2)^{0.5}\{r/(1 - r^2)\}$$

where n is the number of data points and $(n - 2)$ is the degrees of freedom (df). Thus, test statistic

$$t = (12 - 2)^{0.5}\{0.8/(1 - 0.64)\} = 4.216$$

In this example there are 10 df. Thus from Table 2.19.1 the value of $r = 0.8$ is significant at 99.8% but not at 99.9% (note that for $df = 10, t = 4.216$ lies between 4.144 and 4.587 corresponding to columns for 0.998 and 0.999, respectively).

In layman terms this means that using the above regression model, the heat transfer component may be estimated with a 99.8% confidence.

Table 2.19.1 Student's *t*-distribution

	P				
df	0.95	0.98	0.99	0.998	0.999
1	12.706	31.821	63.657	318.310	636.620
2	4.303	6.965	9.925	22.327	31.598
3	3.182	4.541	5.841	10.214	12.924
4	2.776	3.747	4.604	7.173	8.610
5	2.571	3.365	4.032	5.893	6.869
6	2.447	3.143	3.707	5.208	5.959
7	2.365	2.998	3.499	4.785	5.408
8	2.306	2.896	3.355	4.501	5.041
9	2.262	2.821	3.250	4.297	4.781
10	2.228	2.764	3.169	4.144	4.587
15	2.131	2.602	2.947	3.733	4.073
20	2.086	2.528	2.845	3.552	3.850
25	2.060	2.485	2.787	3.450	3.725
30	2.042	2.457	2.750	3.385	3.646
40	2.021	2.423	2.704	3.307	3.551
60	2.000	2.390	2.660	3.232	3.460
120	1.980	2.358	2.617	3.160	3.373
200	1.972	2.345	2.601	3.131	3.340
500	1.965	2.334	2.586	3.107	3.310
1000	1.962	2.330	2.581	3.098	3.300
∞	1.960	2.326	2.576	3.090	3.291

2.20 Outlier analysis

In heat transfer experimentation one may encounter data that lie unusually far removed from the bulk of the data population. Such data are called 'outliers'. One definition of an outlier is that it lies three or four standard deviations or more from the mean of the data population. The outlier indicates peculiarity and suggests that the datum is not typical of the rest of the data. As a rule, an outlier should be subjected to particularly careful examination to see whether any logical explanation may be provided for its peculiar behaviour. Automatic rejection of outliers is not always very wise. Sometimes an outlier may provide information that arises from unusual conditions. Outliers may however be rejected if the associated errors may be traced to observations or any defects in the experimental apparatus.

Statistically a 'near-outlier' is an observation that lies outside 1.5 times the inter-quartile range. The inter-quartile is the interval from the 1st quartile to the 3rd quartile. The *near-outlier* limits are mathematically defined by:

$$\text{Lower outlier limit} = \text{1st quartile} - 1.5 * (\text{3rd quartile} - \text{1st quartile}) \tag{2.20.1}$$

$$\text{Upper outlier limit} = \text{3rd quartile} + 1.5 * (\text{3rd quartile} - \text{1st quartile}) \tag{2.20.2}$$

Likewise, *far-outliers* are defined as the data whose limits are defined below:

$$\text{Lower limit} = \text{1st quartile} - 3 * (\text{3rd quartile} - \text{1st quartile}) \quad (2.20.3)$$

$$\text{Upper limit} = \text{3rd quartile} + 3 * (\text{3rd quartile} - \text{1st quartile}) \quad (2.20.4)$$

If there are a high number of outliers in the data set it signifies that the observations have a high degree of variability.

Figure 2.20.1 shows a plot of experimental data, obtained from a convection heat transfer study. The data have been non-dimensionalised, so that each of the two variables, x and y, can only have a minimum and maximum value of 0 and 1, respectively. It can be shown that it is possible to use all of the given data to obtain a functional relationship for the dependent and independent quantities shown in the above mentioned figure. However, in applying the simple least-squares method, as previously defined, the influence of the outliers identified as points '*A*' and '*B*' would be significant. An outlier analysis identifies whether or not a given set of observations lie within the statistically acceptable limits.

Example 2.20.1 Identification of outliers in a given data set.

Use the procedure presented above to establish whether the data points '*A*' and '*B*' in Fig. 2.20.1 are indeed outliers and whether they should be included in defining a functional relationship between the plotted parameters.

Solution

The data presented in Fig. 2.20.1 are saved within file Ex02-20-01.xls. Open this Excel file and note that the data are arranged in an ascending order in column *A*.

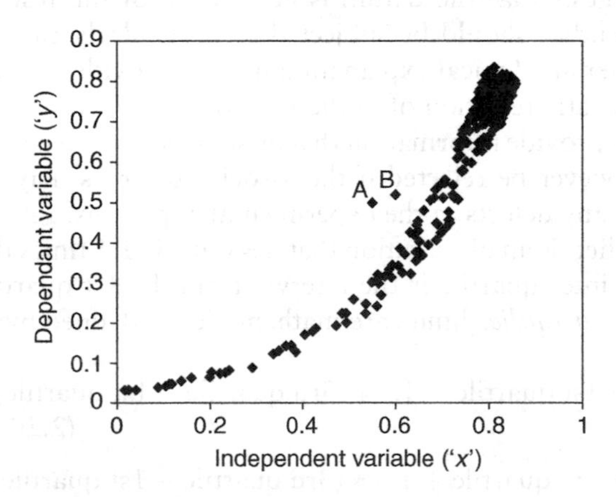

Figure 2.20.1 Relationship between two parameters for a heat transfer study.

To undertake outlier analysis, data falling within a narrow band of the independent variable, encompassing the data points to be queried, are first selected. A judicious choice between bandwidth and the number of local data points is required to provide an accurate definition of the statistical measures required for outlier analysis. Points *A* and *B* in the above mentioned figure have coordinate values of (0.55, 0.5) and (0.6, 0.52), respectively. In consideration of the size of their respective local populations, a bandwidth of 0.1 unit (for the independent variable) will be appropriate for both points, such that:

$$\text{For point 'A'}, 0.5 < x < 0.6; \text{ and, for point 'B'}, 0.55 < x < 0.65$$

The solution is now presented in detail for point '*A*', and it is up to the reader to verify the conclusion drawn for point '*B*'.

Cells **(B29:B42)** contain data range that are relevant to analysis for point, *A*. Highlight **cell D42** and insert the following Excel formula for calculating the first quartile, Q1

= **Quartile(B29:B42,1)**

Likewise, **Q3** may be calculated in **cell E42** by keying-in the following formula

= **Quartile(B29:B42,3)**

In this manner, for point '*A*', values of 0.2625 and 0.32 are obtained for quartiles Q1 and Q3, respectively. Applying Eqs (2.20.1)–(2.20.4) reveals that the near and far outlier upper limits for the data population adjacent to point '*A*' are 0.349 and 0.435. As the dependent variable value for point '*A*' is 0.50, it is clear that this datum is outwith the upper limit, and may therefore be excluded in defining the above mentioned functional relationship.

Similarly, point '*B*' is also found to lie outwith the defined limits.

The linear regressions obtained with, and without, the given outliers are:

$$y = 1.5474x^2 - 0.4173x + 0.0677, \quad r^2 = 0.9429 \text{ (with outliers)}$$

$$y = 1.6024x^2 - 0.4686x + 0.0732, \quad r^2 = 0.9470 \text{ (without outliers)}$$

Thus, the regression model, as well as the coefficient of determination may change as a result of filtering out the suspect datum.

For a more rigorous discussion on outlier analysis the reader is referred to Draper and Smith (1998) and Montgomery and Peck (1992).

2.21 Weighted-averages

In the above section a procedure for the identification of outliers was presented. It was also shown that the outliers may exert a significant influence on the quality and functionality of the regression model. As automatic rejection of outliers should be avoided, a procedure that uses a weighted-least squares method for regression model building may be used. Using such a method reduces the influence of an outlier on the mathematical relationship derived, thus obviating the tedium of outlier identification, analysis, and, removal, where deemed necessary. While detailed description of such procedures is outwith the scope of the present text, a full review is presented by Draper and Smith (1998). Presently, however, a procedure is presented that obtains weighted-averages of data.

The weighting formula used is a matter of choice. In Example 2.21.1 weights that are in inverse proportion to the deviation of any given datum from the mean of the population are used. Users may wish to alter this procedure by employing weights that are in inverse square proportion or, indeed, other power functions may be used. Once the weighted-averages have been obtained, a suitable regression model may then be developed in an almost routine manner.

2.21.1 *Weighted least squares*

Suppose that we are fitting the line $Y = B_0 + B_1 x + \epsilon$, but the variance of Y depends on the level of x; that is,

$$V(Y_i|x_i) = \sigma_i^2 = \frac{\sigma^2}{w_i}, \quad i = 1, 2, \ldots, n \tag{2.21.1}$$

where the w_i are constants, often called weights. For an appropriate objective function, the resulting least squares normal equations are

$$\begin{aligned} \hat{\beta}_0 \sum_{i-1}^{n} w_i + \hat{\beta}_1 \sum_{i-1}^{n} w_i x_i &= \sum_{i-1}^{n} w_i y_i \\ \hat{\beta}_0 \sum_{i-1}^{n} w_i x_i + \hat{\beta}_1 \sum_{i-1}^{n} w_i x_i^2 &= \sum_{i-1}^{n} w_i x_i y_i \end{aligned} \tag{2.21.2}$$

Example 2.21.1 Linear regression using weighted least squares.

In Example 2.20.1 data obtained from heat transfer experimentation was presented. Presently, attention is drawn towards the text file 'Data02-20-01.txt' that contains an identical set of data, but in a text, rather than Excel file format. The latter has an upper row limit of 65536. However, with a text-based format there is no upper limit for the amount of data that may be processed.

The data are presented as column vectors of an independent and a dependent variable. A scatter plot of the data, obtained via Excel software, was presented in Fig. 2.20.1. Possible outliers were also identified within the latter figure.

Using a weighing function that is inversely proportional to the deviation of any given datum from the mean of the population obtain a polynomial relationship between '*y*' and '*x*'.

Solution

This sort of analysis is not available as a built-in function within the Microsoft Excel environment. However, the Visual Basic for Application platform may be used very effectively for writing the relevant routine. Such a routine has been developed in file 'Ex02-21-01.xls'.

- Open the file 'Ex02-21-01.xls' after launching Excel and examine the 'weighted_average' routine under discussion. You may view the routine by simultaneously pressing **Alt + F11** keys, then clicking the Modules, then weighted_average tabs.
- Input the 'No. of intervals' in **cell B3** of the *Main* worksheet. In the present example this number has been selected as 10.
- Access and execute the required macro from the macro box by clicking **Alt + F8** keys followed by **Alt + R** keys.
- The macro retrieves data from the 'Data02-20-01.txt' file and performs the necessary calculations. Ensure that the Ex02-21-01.xls and Data02-20-01.txt files are placed within the same folder in your PC.
- The data returned may then be used to derive a more accurate mathematical relationship than would be achieved from a simple best fit approach.

Close examination of the 'arithmetic-' and 'weighted-average' columns shows the presence of outliers, that is, the greater the difference between the above quantities, the greater the probability of outliers.

References

Colebrook, C. F. (1939) Turbulent flow in pipes with particular reference to the transition between the smooth and rough pipe laws. *J. Inst. Civil Eng.* 11.

Draper, N. R. and Smith, H. (1998) *Applied regression analysis*, 3 edn, Wiley, New York.

Fletcher, R. and Powell, M. J. D. (1963) *Comput. J.* 6, 163–168.

Fletcher, R. and Reeves, C. M. (1964) *Comput. J.* 7, 149–154.

Greig, D. M. (1980) *Newton's Method for Function Minimization, Optimisation*, Longman, London.

Haar, L., Gallagher, J., and Kell, S. (1984) *NBS/NRC Steam Tables: Thermodynamic and Transport Properties and Computer Programs for Vapor and Liquid States of Water in SI Units*, Hemisphere Publishing Corporation, New York.

Hallberg, B., Kinkoph, S., Ray, B., and Nielsen, J. (1997) Special edition: *Using Microsoft Excel 97*, Que Corporation, Santa Rosa, California.

Incropera, F. P. and DeWitt, D. P. (2002) *Fundamentals of Heat and Mass Transfer*, 5th edn, Wiley, New York.

Liengme, B. V. (2000) *A Guide to Microsoft Excel for Scientists and Engineers*, Wiley, New York.

Microsoft (1992) *Microsoft Excel User's Guide* 2, Microsoft Corporation, Seattle.

Montgomery, D. C. and Peck, E. A. (1992) *Introduction to Linear Regression Analysis*, 2nd edn, Wiley, New York.

Moody, L. (1944) Friction factors for pipe flow, *ASME Trans.* 66, 671–684.

Muneer, T. (1988) Comparison of optimisation methods for non-linear, least squares minimisation. *Int. J. Math. Educ. Sci. Tech.* 19 (1), 192–197.

Muneer, T. and Han, B. (1996) Simplified analysis for free convection in enclosures-application to an industrial problem, *Energy Conversion and Management*, 37, 1463.

Person, R. (1992) *Using Excel 4 for Windows*, Que Corporation, Santa Rosa, CA.

Powell, M. J. D. (1964) *Comput. J.* 7, 155–162.

Schmidt, F. W., Henderson, R. E., and Wolgemuth, C. H. (1993) *Introduction to Thermal Sciences*, Wiley, New York.

3 One-dimensional, steady-state conduction

3.1 Nature and principles of conduction

3.1.1 *Introduction*

As we have already discussed, heat transfer is a process by which heat flows from hot to cold regions until the temperatures between the regions are equalised. If the process is determined purely by atomic and molecular activity at which energy is transferred between adjacent atoms and molecules, but without any bulk motion, the heat transfer process is then known as *conduction*. It is not the purpose of this book to discuss the detailed microscopic mechanism of conduction in different media, and the reader is referred to other textbooks and monographs on this subject (Luikov, 1968; Holman, 1990; Bejan, 1993; Kreith and Bohn, 1993). Instead, we concentrate on the macroscopic description and analysis of this important problem.

3.1.2 *Fourier's law of conduction*

Fourier was aware that Newton had already observed that the rate at which heat is transferred between regions at different temperatures is proportional to the temperature difference. He conducted simple experiments on steady-state one-dimensional heat flow in solids. These experiments are simple in analytical sense, in that the temperature does not vary with *time (steady-state* conditions) and that the temperature changes only in the direction of the heat flow (*one-dimensional flow* condition). However, to achieve such conditions experimentally is extraordinarily difficult, but Fourier was an able experimentalist. Fourier postulated that in these situations the rate of heat transfer across a solid layer, Q is directly proportional to the area perpendicular to the direction of the heat flow, A and the temperature difference, ΔT and indirectly proportional to the thickness, Δx. Since *heat flux* q is defined as

$$q = \frac{Q}{A} \tag{3.1.1}$$

Fourier hypothesis can be expressed by the equation

$$q = -k\frac{\Delta T}{\Delta x} \tag{3.1.2}$$

where the negative sign is introduced because heat transfer is always in the direction of decreasing temperature (or negative temperature gradient). Numerous other experiments in solids and other media have confirmed Eq. (3.1.2), which can be expressed in its differential form, and referred to as Fourier's law of heat conduction, as

$$q_x = -k\frac{\mathrm{d}T}{\mathrm{d}x} \tag{3.1.3}$$

where the constant of proportionality is called the *thermal conductivity* of the material.

Furthermore, it has been shown experimentally that Fourier's conditions of steady-state and one-dimensional heat flow are not necessary and that the Fourier's law of heat conduction has a general validity

$$q_n = -k\frac{\partial T}{\partial n} \tag{3.1.4}$$

that is heat flux q_n in a direction n, normal to an *isotherm* (surfaces of constant temperatures), is proportional to the temperature gradient in the direction perpendicular to the isothermal surface. The thermal conductivity of the material, k, together with other thermophysical properties are discussed in Section 3.4.

3.2 General heat diffusion equation

3.2.1 Derivation in Cartesian coordinate system

The determination of temperature distribution and the associated heat fluxes is the aim of many engineering calculations used in the design of equipment, which is either temperature sensitive or which is used to control the temperature in a particular environment. The temperature distribution, or the *temperature field*, in a body will depend not only on the properties of the material, but also on the initial conditions and the conditions imposed on its boundaries. The temperature distribution is determined by applying the principle of energy conservation. The only restriction used here will be that heat is only transferred by conduction, and hence that convection and radiation can be neglected. This is satisfied in practice for heat transfer in solid media.

Consider an infinitesimally small control volume shown in Cartesian coordinates, as shown in Fig. 3.2.1. The control volume is a rectangular parallelepiped $\mathrm{d}x \times \mathrm{d}y \times \mathrm{d}z$. Let the heat flux across the surface $ABCD$

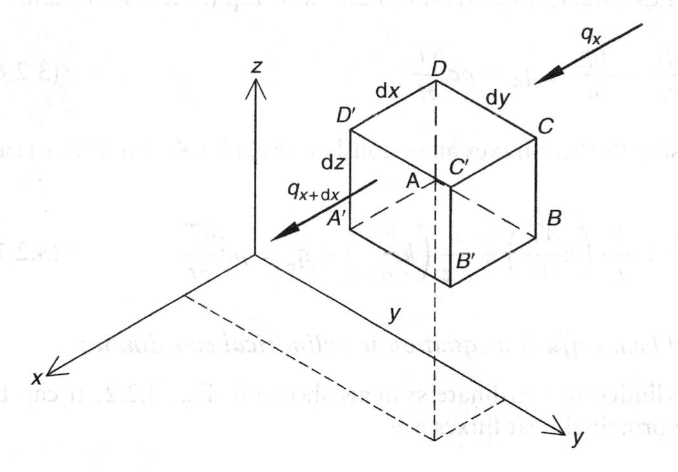

Figure 3.2.1 Control volume in Cartesian coordinates.

be q_x. The heat flux across the surface $A'B'C'D'$ can then be expressed from Taylor series expansion with the higher terms neglected as

$$q_{x+\mathrm{d}x} = q_x + \frac{\partial q_x}{\partial x}\,\mathrm{d}x \tag{3.2.1}$$

and similarly

$$q_{y+\mathrm{d}y} = q_y + \frac{\partial q_y}{\partial y}\,\mathrm{d}y \tag{3.2.2}$$

$$q_{z+\mathrm{d}z} = q_z + \frac{\partial q_z}{\partial z}\,\mathrm{d}z \tag{3.2.3}$$

Furthermore, the medium contains an energy source which generates heat at a rate of q_g per unit volume. Hence the rate of increase of internal thermal energy of the control volume, E is

$$\begin{aligned}(q_x - q_{x+\mathrm{d}x})\,\mathrm{d}y\,\mathrm{d}z + (q_y - q_{y+\mathrm{d}y})\,\mathrm{d}x\,\mathrm{d}z + (q_z - q_{z+\mathrm{d}z})\,\mathrm{d}x\,\mathrm{d}y \\ + q_g\,\mathrm{d}x\,\mathrm{d}y\,\mathrm{d}z = \mathrm{d}E\end{aligned} \tag{3.2.4}$$

Assuming that the material does not undergo a change of phase only sensible, and not latent, effects are considered and the rate of increase of internal thermal energy E can be expressed as

$$\mathrm{d}E = \rho c \frac{\partial T}{\partial t}\,\mathrm{d}x\,\mathrm{d}y\,\mathrm{d}z \tag{3.2.5}$$

where ρ is the density, c is the specific heat of the material, and t is time.

Substituting Eqs (3.2.1)–(3.2.3) and (3.2.5) into Eq. (3.2.4) we obtain

$$-\frac{\partial q_x}{\partial x} - \frac{\partial q_y}{\partial y} - \frac{\partial q_z}{\partial z} + q_g = \rho c \frac{\partial T}{\partial t} \tag{3.2.6}$$

Finally, expressing the heat fluxes analogously to Eq. (3.1.4), Eq. (3.2.6) can be written as

$$\frac{\partial}{\partial x}\left(k\frac{\partial T}{\partial x}\right) + \frac{\partial}{\partial y}\left(k\frac{\partial T}{\partial y}\right) + \frac{\partial}{\partial z}\left(k\frac{\partial T}{\partial z}\right) + q_g = \rho c \frac{\partial T}{\partial t} \tag{3.2.7}$$

3.2.2 General heat diffusion equation in cylindrical coordinates

Consider the cylindrical coordinate systems shown in Fig. 3.2.2. It can be shown that the principal heat fluxes are

$$q_r = -k\frac{\partial T}{\partial r} \tag{3.2.8}$$

$$q_\theta = -\frac{k}{r}\frac{\partial T}{\partial \theta} \tag{3.2.9}$$

$$q_z = -k\frac{\partial T}{\partial z} \tag{3.2.10}$$

where r is the radius, θ is the longitude, and z is the axial coordinate.

It can be shown (Carslaw and Jaeger, 1976) that the heat diffusion equation is then given by

$$\frac{1}{r}\frac{\partial}{\partial r}\left(kr\frac{\partial T}{\partial r}\right) + \frac{1}{r^2}\frac{\partial}{\partial \theta}\left(k\frac{\partial T}{\partial \theta}\right) + \frac{\partial}{\partial z}\left(k\frac{\partial T}{\partial z}\right) + q_g = \rho c \frac{\partial T}{\partial t} \tag{3.2.11}$$

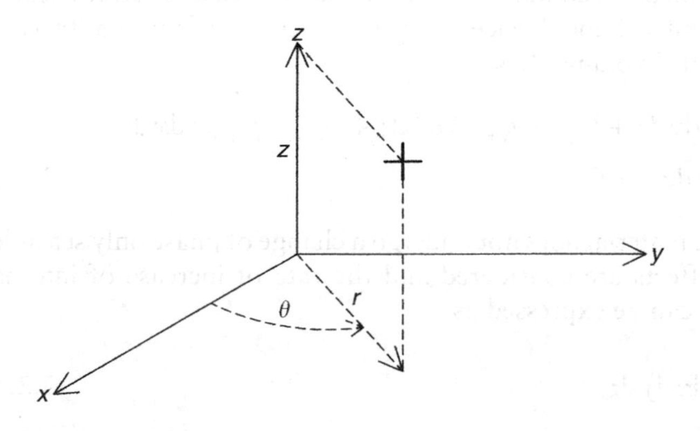

Figure 3.2.2 Cylindrical coordinate system.

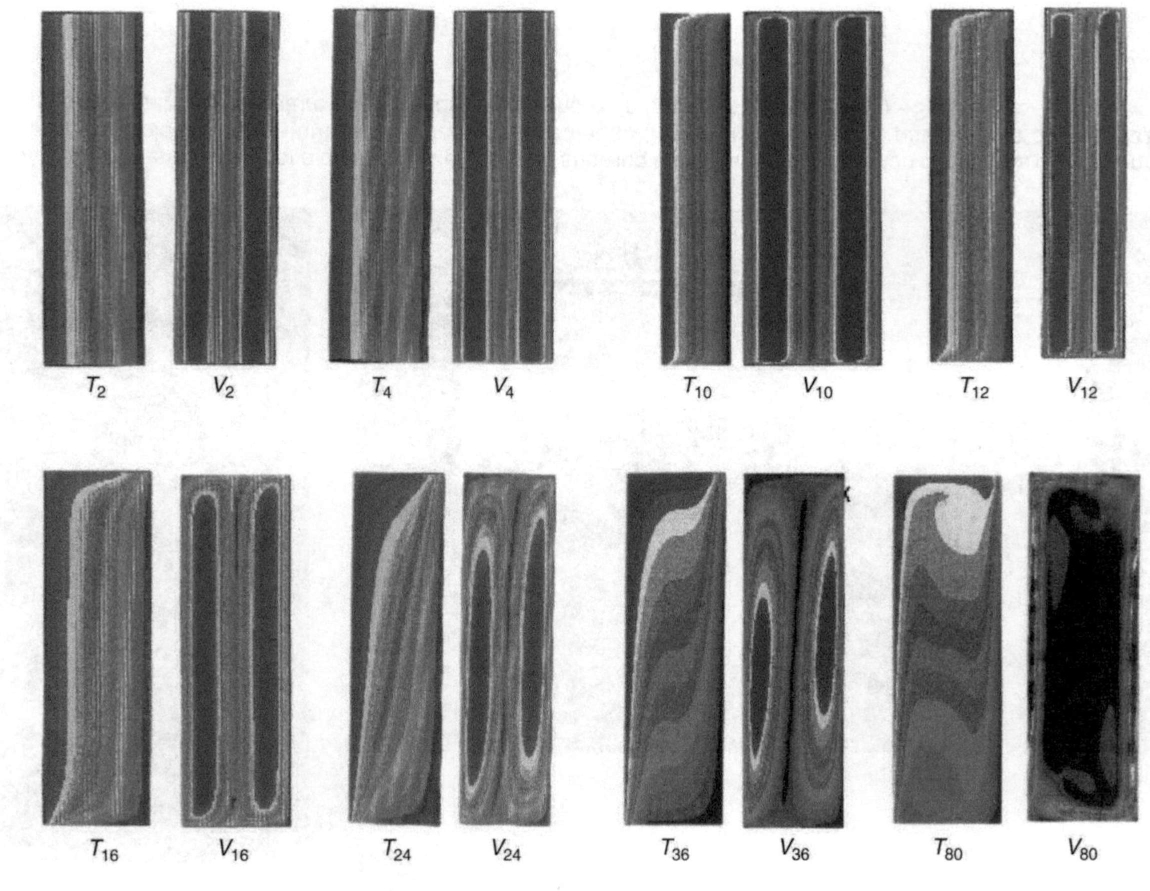

Colour Plate I CFD temperature (T) and velocity (V) raster plots for free convection flows within window cavities. Numbers indicate cavity width in millimeter. (See Figure 1.6.1.)

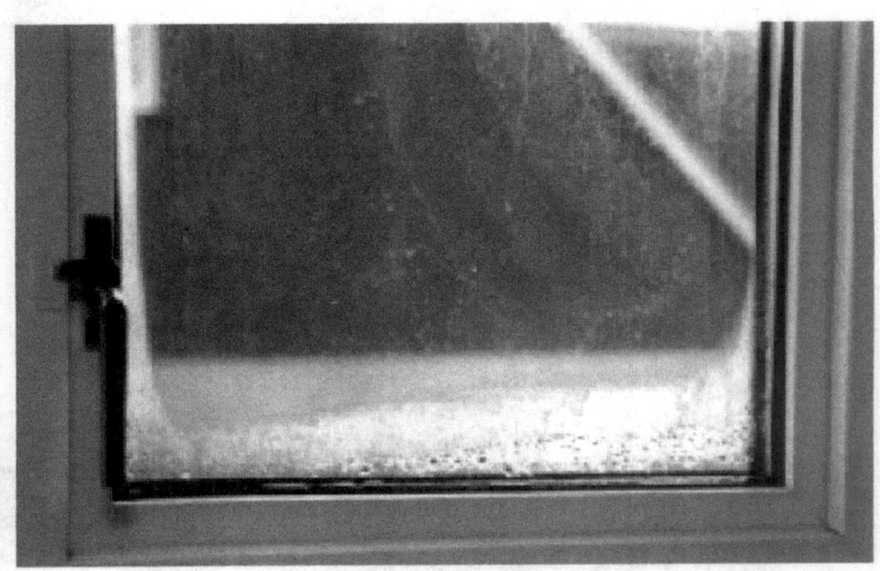

Colour Plate II Infrared thermograph of a double-glazed window showing temperature stratification due to gas circulation within the cavity space. Colder temperatures at the glazing bottom edge (left) and condensation pattern on a double-glazed window that reinforces the temperature stratification effects shown in the thermograph (right). (See Figure 1.6.2.)

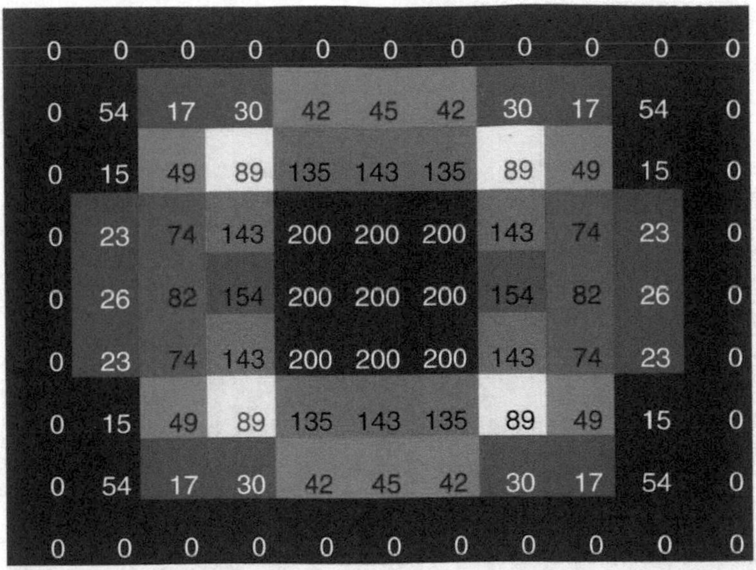

0	0	0	0	0	0	0	0	0	0	0
0	54	17	30	42	45	42	30	17	54	0
0	15	49	89	135	143	135	89	49	15	0
0	23	74	143	200	200	200	143	74	23	0
0	26	82	154	200	200	200	154	82	26	0
0	23	74	143	200	200	200	143	74	23	0
0	15	49	89	135	143	135	89	49	15	0
0	54	17	30	42	45	42	30	17	54	0
0	0	0	0	0	0	0	0	0	0	0

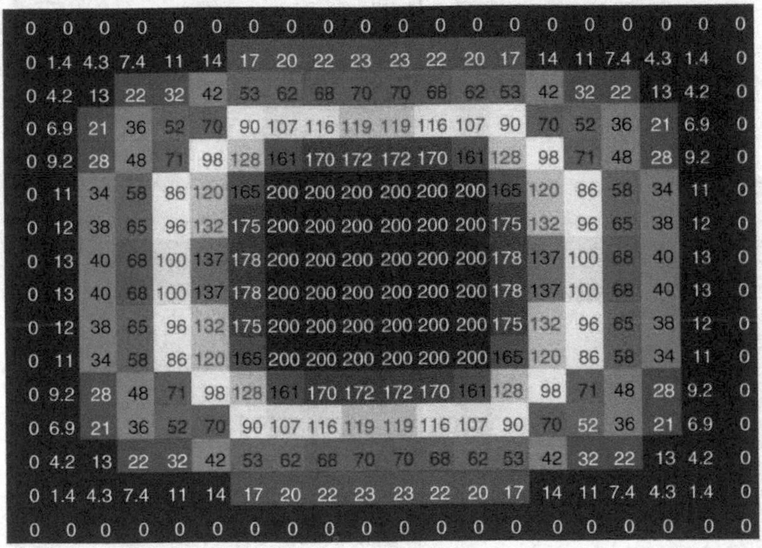

0	0	0	0	0	0	0	0	0	0	0	0	0	0	0	0	0	0	0	0
0	1.4	4.3	7.4	11	14	17	20	22	23	23	22	20	17	14	11	7.4	4.3	1.4	0
0	4.2	13	22	32	42	53	62	68	70	70	68	62	53	42	32	22	13	4.2	0
0	6.9	21	36	52	70	90	107	116	119	119	116	107	90	70	52	36	21	6.9	0
0	9.2	28	48	71	98	128	161	170	172	172	170	161	128	98	71	48	28	9.2	0
0	11	34	58	86	120	165	200	200	200	200	200	200	165	120	86	58	34	11	0
0	12	38	65	96	132	175	200	200	200	200	200	200	175	132	96	65	38	12	0
0	13	40	68	100	137	178	200	200	200	200	200	200	178	137	100	68	40	13	0
0	13	40	68	100	137	178	200	200	200	200	200	200	178	137	100	68	40	13	0
0	12	38	65	96	132	175	200	200	200	200	200	200	175	132	96	65	38	12	0
0	11	34	58	86	120	165	200	200	200	200	200	200	165	120	86	58	34	11	0
0	9.2	28	48	71	98	128	161	170	172	172	170	161	128	98	71	48	28	9.2	0
0	6.9	21	36	52	70	90	107	116	119	119	116	107	90	70	52	36	21	6.9	0
0	4.2	13	22	32	42	53	62	68	70	70	68	62	53	42	32	22	13	4.2	0
0	1.4	4.3	7.4	11	14	17	20	22	23	23	22	20	17	14	11	7.4	4.3	1.4	0
0	0	0	0	0	0	0	0	0	0	0	0	0	0	0	0	0	0	0	0

Colour Plate III Steady state, two-dimensional heat conduction analysis for a rectangular duct: Coarse grid (top) and fine grid (bottom). (See Figure 4.2.5.)

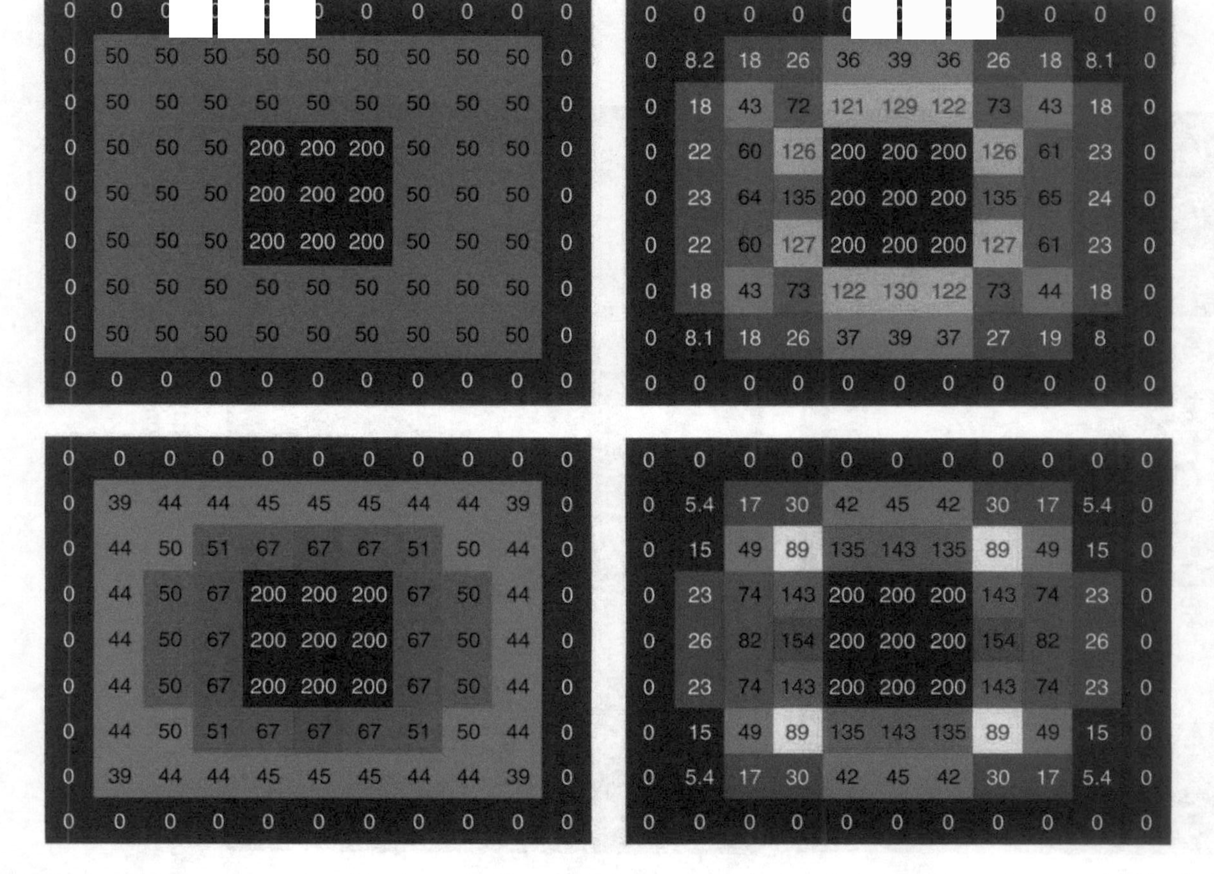

Colour Plate IV Transient, two-dimensional heat conduction analysis for a rectangular duct: Initial condition $t = 0$ s (top-left); $t = 40$ s (bottom-left); $t = 400$ s (top-right); and $t = 4000$ s (bottom-right). Note that at $t = 4000$ s steady-state conditions have been achieved as this plot now compares favourably with that shown in Plate III (top-half). (See Figure 5.5.2.)

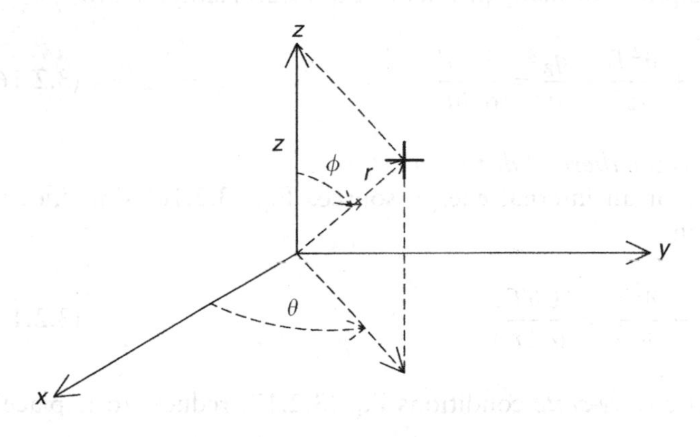

Figure 3.2.3 Spherical coordinate system.

3.2.3 *General heat diffusion equation in spherical coordinates*

Consider the spherical coordinate systems shown in Fig. 3.2.3. It can be shown that the principal heat fluxes are

$$q_r = -k\frac{\partial T}{\partial r} \tag{3.2.12}$$

$$q_\theta = -\frac{k}{r\sin\phi}\frac{\partial T}{\partial \theta} \tag{3.2.13}$$

$$q_\phi = -\frac{k}{r}\frac{\partial T}{\partial \phi} \tag{3.2.14}$$

where the coordinates are shown in Fig. 3.2.3.

It can be shown (Carslaw and Jaeger, 1976) that the heat diffusion equation is then given by

$$\frac{1}{r^2}\frac{\partial}{\partial r}\left(kr^2\frac{\partial T}{\partial r}\right) + \frac{1}{r^2\sin^2\phi}\frac{\partial}{\partial \theta}\left(k\frac{\partial T}{\partial \theta}\right) + \frac{1}{r^2\sin\phi}\frac{\partial}{\partial \phi}\left(k\sin\phi\frac{\partial T}{\partial \phi}\right) + q_g = \rho c\frac{\partial T}{\partial t} \tag{3.2.15}$$

3.2.4 *Various simplifications*

The general equations have been solved for a large number of problems (Luikov, 1968; Carslaw and Jaeger, 1976; Özisik, 1993), but the equations can be simplified to demonstrate the behaviour of these systems. All the simplifications will be demonstrated in Cartesian coordinates using Eq. (3.2.7). First, if it is assumed that the thermal conductivity is constant, which is a reasonable assumption for many problems, Eq. (3.2.7) simplifies to

$$\frac{\partial^2 T}{\partial x^2} + \frac{\partial^2 T}{\partial y^2} + \frac{\partial^2 T}{\partial z^2} + \frac{q_g}{k} = \frac{1}{\alpha}\frac{\partial T}{\partial t} \tag{3.2.16}$$

where $\alpha = k/\rho c$ is the *thermal diffusivity*.

In the absence of an internal energy source, Eq. (3.2.16) simplifies to Fourier's equation:

$$\frac{\partial^2 T}{\partial x^2} + \frac{\partial^2 T}{\partial y^2} + \frac{\partial^2 T}{\partial z^2} = \frac{1}{\alpha}\frac{\partial T}{\partial t} \tag{3.2.17}$$

Moreover, under *steady-state* conditions Eq. (3.2.17) reduces to Laplace's equation:

$$\frac{\partial^2 T}{\partial x^2} + \frac{\partial^2 T}{\partial y^2} + \frac{\partial^2 T}{\partial z^2} = 0 \tag{3.2.18}$$

Further simplifications can be obtained. For example, for *one-dimensional steady-state* heat transfer in a *homogeneous medium*, Eq. (3.2.18) simplifies to

$$\frac{d^2 T}{dx^2} = 0 \tag{3.2.19}$$

3.3 Boundary and initial conditions

As mentioned above the temperature fields in any body depend not only on the properties of the material, but also on the initial conditions and the conditions imposed on its boundaries. Hence, before we can proceed we have to specify the appropriate *initial* and *boundary* conditions. The initial condition is given by specifying the temperature T at time $t = 0$ at all points of the solid body:

$$T_{t=0} = T(x, y, z) \tag{3.3.1}$$

The form of the temperature field obtained by solving the governing equation must be such that as time tends to zero, the condition given by Eq. (3.3.1) is satisfied at all points of the solid body.

Three boundary or surface conditions are commonly used. The first kind of boundary conditions specifies the temperature on the surface of the body. The surface temperature, T_S can be either constant or it can be a function of time. The first condition is closely approximated when the surface is in close contact with another medium which is undergoing a change of phase, such as during the melting of solids or boiling of liquids.

The second kind of boundary conditions specifies the heat flux on the surface of the body

$$-k\left(\frac{\partial T}{\partial n}\right)_S = q_S \tag{3.3.2}$$

where the differentiation is in the direction of the outward normal at the surface. The surface heat flux can be either constant or it can be a function of time. Generally, constant heat flux is difficult to realise in practice, but it can be approximated reasonably well with electric heating of a thin layer next to the surface of the body. A special case of this condition is obtained when the heat flux on the surface is zero. This corresponds to a perfectly insulated surface, also difficult to realise in practice.

Hence, the second kind of boundary conditions is not easily achieved. The actual conditions observed may be better approximated by the third kind of boundary conditions. In this condition the surface heat flux is proportional to the difference between the surface temperature, T_S and the temperature of the surrounding medium, T_∞:

$$-k\left(\frac{\partial T}{\partial n}\right)_S = h(T_S - T_\infty) \tag{3.3.3}$$

where the constant of proportionality h is referred to as a heat transfer coefficient (or, sometimes, as a film coefficient or surface conductance). It should be noted that in practice there are some difficulties in defining precisely the meaning of T_∞, but for external flows past solid bodies it can be closely approximated by the free stream temperature. Additionally, the heat transfer coefficient may not be easily determined and may not necessarily be constant. However, even with these limitations, this is a very versatile boundary condition.

3.4 Thermophysical properties

Several important properties required for the analysis of conduction heat transfer have been already introduced: thermal conductivity k, density ρ, and specific heat c. These properties, together with additional properties discussed below, are collectively known as *thermophysical properties*. Two categories of thermophysical properties are generally considered: (i) *transport properties* and (ii) *thermodynamic properties*.

Transport properties govern the rates of transfer of energy, momentum, and mass. For example, thermal conductivity k controls the diffusion rate of heat transfer and kinematic viscosity ν controls the diffusion rate of momentum transfer.

Thermodynamic properties describe the equilibrium state of a system. For example, density ρ and specific heat c are two such properties.

These and other thermophysical properties are extensively discussed in many available textbooks (Brodkey and Hershey, 1988). The recommended values of the thermophysical properties are also generally available (Rohsenow and Hartnett, 1973; Perry, 1998). Property tables for a selection of materials are provided in appendices to this book.

3.5 The plane wall

3.5.1 *Introduction*

We consider one-dimensional steady-state conduction in a plane wall, as shown in Fig. 3.5.1. The general heat diffusion Eq. (3.2.7) reduces to

$$\frac{d}{dx}\left(k\frac{dT}{dx}\right) + q_g = 0 \tag{3.5.1}$$

and this must be solved, subject to suitable boundary conditions. Equation (3.5.1) has no general analytical solutions (Carslaw and Jaeger, 1976; Özisik, 1993). However, analytical solutions can be obtained if the equation is simplified further. Some of the simplifications approximate very closely the observed conditions, and thus have widespread applications.

First we consider the case of no internal heat generation. This simplifies Eq. (3.5.1) to

$$k\frac{dT}{dx} = A_c \tag{3.5.2}$$

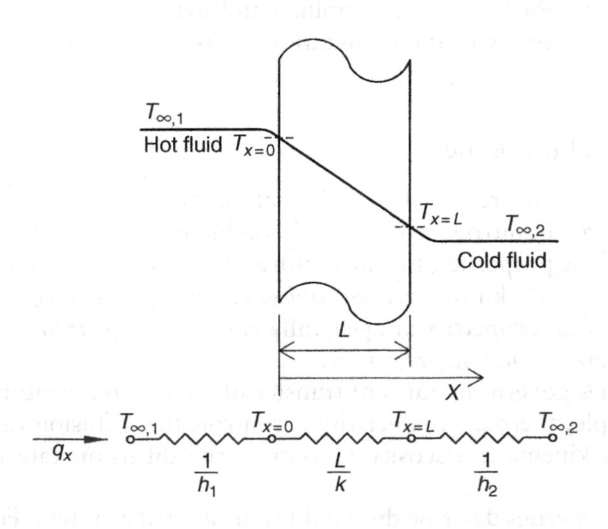

Figure 3.5.1 Steady-state conduction in a plane wall.

where A_c is a constant independent of x. A comparison of Eq. (3.1.3) with Eq. (3.5.2) shows that for one-dimensional steady-state conduction in a plane wall with no heat generation the heat flux q_x is constant. In general the thermal conductivity k is a function of both the temperature and the position

$$k = k(T, x) \tag{3.5.3}$$

and the analytical integration of Eq. (3.5.2) can be difficult. However, further insight can be obtained by further simplification. It is first assumed that the thermal conductivity k is independent of temperature; the general solution of Eq. (3.5.2) then is

$$T = T_{x=0} + A_c \int_0^x \frac{1}{k}\,dx \tag{3.5.4}$$

where $T_{x=0}$ is the surface wall temperature at $x = 0$. Second, it is assumed that the thermal conductivity k is constant; Eq. (3.5.4) can be solved to obtain

$$T = (T_{x=L} - T_{x=0})\frac{x}{L} + T_{x=0} \tag{3.5.5}$$

where L is the thickness of the wall and $T_{x=L}$ is the surface wall temperature at $x = L$. Using Eq. (3.1.3) the heat flux can be then calculated as

$$q_x = \frac{k}{L}(T_{x=0} - T_{x=L}) \tag{3.5.6}$$

Equation (3.5.6) can be used if the surface wall temperatures are known. However, as discussed extensively in the chapters on convection (Chapters 6–8), it is more usual to use the plane wall to separate two fluids of known temperature. In such cases the surface temperatures of the wall are not known. The third boundary condition, given by Eq. (3.3.3) can then be written for the two wall surfaces as

$$q_{x=0} = h_1(T_{\infty,1} - T_{x=0}) \tag{3.5.7}$$

$$q_{x=L} = h_2(T_{x=L} - T_{\infty,2}) \tag{3.5.8}$$

where h_1 and h_2 are the respective heat transfer coefficients and $T_{\infty,1}$ and $T_{\infty,2}$ are the free stream fluid temperatures (or the temperatures some distance from the wall). Noting that the heat fluxes in the above three equations are identical, these equations can be combined to obtain

$$q_x = U(T_{\infty,1} - T_{\infty,2}) \tag{3.5.9}$$

where the overall heat transfer coefficient U is

$$U = \frac{1}{(1/h_1) + (L/k) + (1/h_2)} \tag{3.5.10}$$

Hence the heat flux is given by the overall temperature difference ($T_{\infty,1} - T_{\infty,2}$) and the overall heat transfer coefficient. The reciprocal of the overall heat transfer coefficient U is the total thermal resistance $R_{t,T}$

$$R_{t,T} = \frac{1}{h_1} + \frac{L}{k} + \frac{1}{h_2} \tag{3.5.11}$$

If we, then, define thermal resistance for conduction $R_{t,cond}$ and thermal resistance for convection $R_{t,conv}$ as

$$R_{t,cond} = \frac{L}{k} \tag{3.5.12}$$

$$R_{t,conv} = \frac{1}{h} \tag{3.5.13}$$

and noting that the convection from the hot fluid to the wall, the conduction in the wall, and the convection from the wall to the cold fluid take place in series, we can use the circuit analogy and obtain the overall thermal resistance (or the total thermal resistance) by adding the thermal resistances obtained from separate consideration of each element (as shown in Fig. 3.5.1).

3.5.2 *Composite wall*

We consider first one-dimensional steady-state conduction in a composite plane wall, consisting of plane wall A (thickness L_A and thermal conductivity k_A) and plane wall B (thickness L_B and thermal conductivity k_B), shown in Fig. 3.5.2. Experimental observations show that in many situations there can

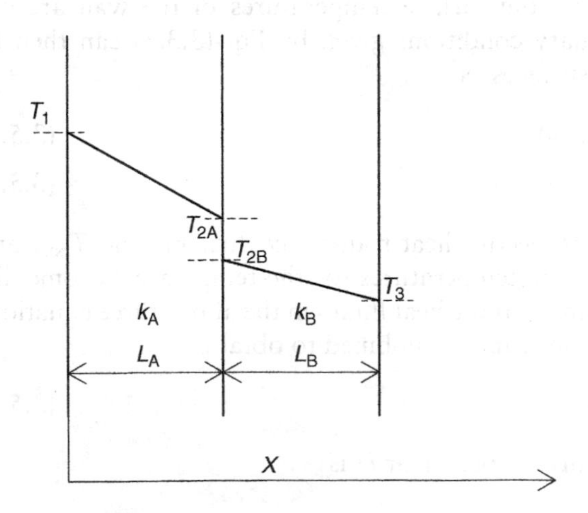

Figure 3.5.2 Plane composite wall with two sections.

be a significant temperature change $(T_{2A} - T_{2B})$ across the interface between the two plane walls. The temperature change is due to the imperfect contact between the two walls, which is significantly influenced, for example, by surface roughness. To account for this phenomena thermal contact resistance $R_{t,cont}$ is introduced, defined analogously to Eq. (3.5.9) as

$$R_{t,cont} = \frac{T_{2A} - T_{2B}}{q_x} \tag{3.5.14}$$

Thermal contact resistance can be appreciable, and its control is important. Further discussion is given in Bejan (1993), Incropera and DeWitt (1996), and Kreith and Bohn (1993).

Next consider the composite wall shown in Fig. 3.5.3. Analogously to above, the heat flux is given by

$$q_x = \frac{T_{\infty,1} - T_{\infty,4}}{R_{t,T}} \tag{3.5.15}$$

where the total thermal resistance is given by

$$R_{t,T} = \frac{1}{h_1} + \frac{L_A}{k_A} + R_{t,cont,AB} + \frac{L_B}{k_B} + R_{t,cont,BC} + \frac{L_C}{k_C} + \frac{1}{h_4} \tag{3.5.16}$$

Equation (3.5.16) can be generalised for a composite wall with any number of sections.

Example 3.5.1 Composite wall with three sections.

A plane composite wall, shown in Fig. 3.5.3, consists of three sections, each with a constant thermal conductivity. The composite wall separates two fluids with known free stream temperatures. The heat transfer coefficients

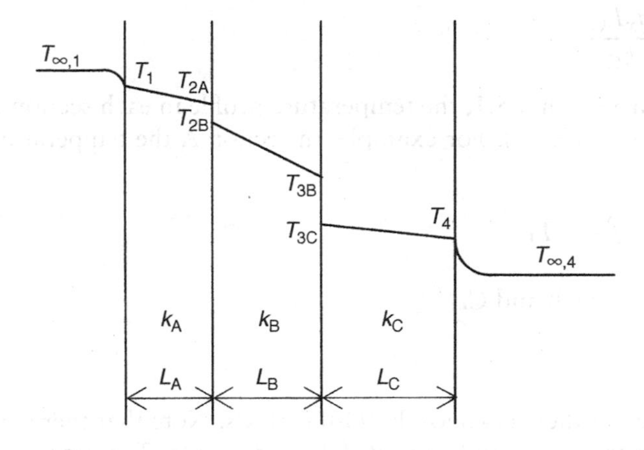

Figure 3.5.3 Plane composite wall with three sections.

between the fluids and the respective external surfaces of the composite wall are also known. Finally, the appropriate contact resistances have also been obtained. There is no heat generation anywhere in the composite wall. Assume that steady-state conditions have been reached, and that the problem can be approximated by one-dimensional conduction. Determine the temperature distribution in the composite wall.

Analysis

Since the problem is based on the analysis presented in Section 3.5.2, the total thermal resistance $R_{t,T}$ can be calculated from Eq. (3.5.16), and the heat flux q_x then obtained from Eq. (3.5.15). However, as discussed above, the heat fluxes across all elements of the system are identical and equal to q_x. Hence

$$\frac{q_x}{h_1} = T_{\infty,1} - T_1$$

and

$$T_1 = T_{\infty,1} - \frac{q_x}{h_1}$$

and similarly

$$T_{2A} = T_1 - \frac{q_x L_A}{k_A}$$

$$T_{2B} = T_{2A} - q_x R_{t,cont,AB}$$

$$T_{3B} = T_{2B} - \frac{q_x L_B}{k_B}$$

$$T_{3C} = T_{3B} - q_x R_{t,cont,BC}$$

$$T_4 = T_{3C} - \frac{q_x L_C}{k_C}$$

Finally, as shown in Section 3.5.1, the temperature profile in each section is linear and given by Eq. (3.5.5). For example, in section A the temperature profile is given by

$$T = (T_{2A} - T_1)\frac{x}{L_A} + T_1$$

and similarly for sections B and C.

Solution

The solution is given in the workbook Ex03-05-01.xls. Note that there are three worksheets in this workbook: *Input data and results, Temperatures,* and *Distribution and graph.*

The first worksheet, *Input data and results*, allows for the input of the data, and presents the numerical results for the surface temperature of the three sections and the overall heat flux, and the temperature distributions in a graphical form. The input data are entered in **cells (B3:B14)**, with the data already entered used for illustration only. The overall heat flux q_x is given in **cell B17**, and the surface temperatures in **cells (B19:B24)**. The graph shows the temperature profiles in the composite wall. It should be noted that the graph and its labels are dynamically linked to the solution, so that the reader can see the influence of the various parameters on the resultant temperature profiles. The presentation of the results is improved by clicking **Full Screen** in **View**.

The second worksheet, *Temperatures*, is used to calculate the total thermal resistance, $R_{t,T}$, the heat flux, q_x, and the surface temperatures of the three sections.

The last worksheet, *Distribution and graph*, is used to scale the temperatures and distances for the preparation of the graph and to provide the appropriate labelling. It should be noted that there is no express need for the user to understand the mechanics of this worksheet.

Discussion

The solution, shown in the worksheet *Input data and results*, can be used to experiment with the input data and to observe their influence. Additionally, the following investigations can be made.

1. As the film heat transfer coefficient h_1 increases, the temperature difference $T_{\infty,1} - T_1$ decreases. Hence for very high values of the heat transfer coefficient the surface temperature approaches the free stream fluid temperature. This is discussed further in Chapter 7.
2. Hence for very high values of both heat transfer coefficients the surface temperatures T_1 and T_2 are virtually identical to the fluid temperatures $T_{\infty,1}$ and $T_{\infty,2}$, respectively. The heat transfer is then determined by conduction through the composite wall, when the outside temperatures are $T_{\infty,1}$ and $T_{\infty,2}$, respectively.
3. A composite wall consisting of only two sections can also be investigated. This can be achieved, for example, by creating section D, which combines sections A and B. Section D is then defined by thermal conductivity k_D, with $k_A = k_B = k_D$ and $L_D = L_A + L_B$, with the contact resistance between section A and section B set to zero ($R_{t,cont,AB} = 0$).
4. A simple plane wall can be similarly modelled.

Other situations can be modelled and the reader is invited to experiment further. Finally, the temperature distribution in the two fluids is considered in Chapters 7 and 8.

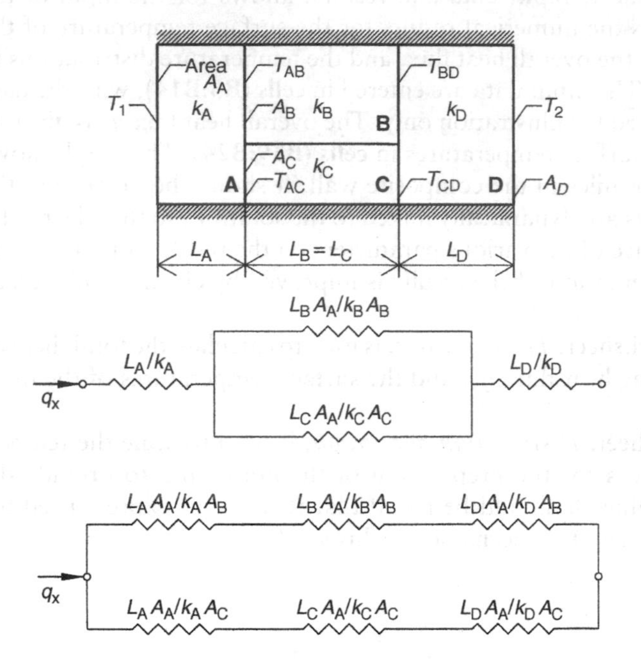

Figure 3.5.4 Complex composite wall.

Example 3.5.2 Complex composite wall.

More complex composite walls can, under some conditions, be also solved using one-dimensional steady-state conduction. Consider, the four section composite wall shown in Fig. 3.5.4 in which the two opposite faces are thermally insulated. Assume that steady-state conditions have been reached, and that the problem can be approximated by one-dimensional conduction. Determine the rate of conduction heat transfer through the wall and the interfacial temperatures. Discuss the validity of the solution.

Analysis

Since the overall heat flow is primarily in one direction, we will assume that we can use the one-dimensional analysis developed in this chapter. We will first assume that the surfaces normal to the x-direction are isothermal. Hence the temperature of the interface between sections A and B, T_{AB} is equal to the temperature of the interface between sections A and C, T_{AC}

$$T_{AB} = T_{AC}$$

and similarly

$$T_{BD} = T_{CD}$$

The rate of heat transfer Q_A across section A is

$$Q_A = \frac{k_A A_A}{L_A}(T_1 - T_{AB})$$

where k_A is the thermal conductivity, L_A is the length, and A_A is the cross-sectional area of section A. Similarly

$$Q_B = \frac{k_B A_B}{L_B}(T_{AB} - T_{BD})$$

$$Q_C = \frac{k_C A_C}{L_C}(T_{AB} - T_{BD})$$

$$Q_D = \frac{k_D A_D}{L_D}(T_{BD} - T_2)$$

Since the total rate of heat transfer Q_x is related to the other parameters by

$$Q_x = Q_A = Q_B + Q_C = Q_D$$

the above equations can be combined

$$Q_x\left(\frac{L_A}{k_A A_A} + \frac{1}{k_B A_B/L_B + k_C A_C/L_C} + \frac{L_D}{k_D A_D}\right) = T_1 - T_2$$

Since

$$q_x A_A = Q_x$$

we finally obtain

$$q_x = \frac{T_1 - T_2}{R}$$

where the thermal resistance R is given as

$$R = \frac{L_A}{k_A} + \frac{1}{(k_B/L_B)(A_B/A_A) + (k_C/L_C)(A_C/A_A)} + \frac{L_D}{k_D}$$

This can be modelled by the first series–parallel representation in Fig. 3.5.4.

However, we can also make another assumption; we can assume that there is no heat transfer between surfaces parallel to the x-direction (assumption of adiabatic parallel surfaces). This, of course implies that

$$T_{AB} \neq T_{AC}$$

$$T_{BD} \neq T_{CD}$$

It can then be shown that the thermal resistance R is given as

$$\frac{1}{R} = \frac{1}{(L_A/k_A)(A_A/A_B) + (L_B/k_B)(A_A/A_B) + (L_D/k_D)(A_A/A_B)} + \frac{1}{(L_A/k_A)(A_A/A_C) + (L_C/k_C)(A_A/A_C) + (L_D/k_D)(A_A/A_C)}$$

This can be modelled by the second series-parallel representation in Fig. 3.5.4.

The series–parallel representation of heat transfer problems is a powerful technique, but it must be used with caution. For example, it is important to note that the two solutions above are not identical and that different results will be obtained for the thermal resistances and hence for the rates of heat transfer. It may be more appropriate to solve such problems numerically, as discussed in Chapter 4.

Solution

The solution is given in the workbook Ex03-05-02.xls. Note that there are three worksheets in this workbook: *Input data and results, Calculations*, and *Figure*.

The first worksheet, *Input data and results*, allows for the input of the data, and presents the numerical results for the thermal resistance, overall heat flux, overall rate of heat transfer and interfacial temperatures T_{AB}, T_{AC}, T_{BD}, and T_{CD}. The numerical results are given for both isothermal and adiabatic approximations. The worksheet also presents a diagram of the interfacial temperatures for the two approximations. The input data are entered in **cells (B3, B4, B7,** and **B8)** and **cells (B10:B16)**, with the data already entered used for illustration only. The results for the thermal resistance are given in **cells (B20** and **C20)** for the isothermal and adiabatic approximations respectively. Similarly, the results for the overall heat flux are given in **cells (B21** and **C21)**, for the overall rate of heat transfer in **cells (B22 and C22)**, for temperature T_{AB} in **cells (B23 and C23)**, for temperature T_{AC} in **cells (B24** and **C24)**, for temperature T_{BD} in **cells (B25 and C25)** and for temperature T_{CD} in **cells (B26 and C26)** respectively. The interfacial temperatures in the two diagrams are dynamically linked to the input data. The presentation of the results is improved by clicking **Full Screen in View**.

The second worksheet, *Calculations*, is used to calculate the results and the last worksheet, *Figure*, is used to prepare the diagrams.

Discussion

The solution, shown in the worksheet *Input data and results*, can be used to experiment with the input data and to observe their influence. The following observations can be made:

1 If the thermal conductivity k_B is set equal to the thermal conductivity k_C the two approximations give identical results. The reason is that in this case the system is truly one-dimensional and hence both assumptions, that is, surfaces normal to the direction of the heat flow being isothermal and surfaces parallel with the direction of the heat flow being adiabatic, are satisfied.
2 The difference between the results of the two approximations increases with increasing difference between the thermal conductivities of sections B and C.
3 The difference between the results of the two approximations can be used to judge the validity of the assumptions used; the smaller the difference the better the approximation and more reliable the results.

Other situations can be modelled and the reader is invited to experiment further.

3.6 Other systems

In many other situations thermal systems exhibit temperature variations in predominantly one direction and insignificant errors are introduced when temperature gradients in other directions are neglected. Such behaviour can be observed in special radial systems, such as cylinders and spheres, and in extended surfaces, which are used to enhance the rates of heat transfer. These systems are discussed in the following sections.

However, it is important to point out that the assumption of one-dimensional behaviour is an approximation, and that estimates, either implicit or explicit, of its validity and the associated errors should always be made. Such estimates must take into account the real multi-dimensional behaviour of the system, and will be discussed in detail in Chapter 4 when two-dimensional steady-state heat conduction is considered.

3.7 Cylindrical systems

3.7.1 *Introduction*

In many cases it is possible to study cylindrical systems with axi-symmetric one-dimensional models with temperature gradients in the radial direction only. For this condition to be satisfied the following conditions must hold:

- the system, including its boundary and initial conditions, must be axi-symmetric (which eliminates the temperature gradient along the longitudinal coordinate, θ);
- the system must extend to infinity along the axial coordinate (which eliminates the temperature gradient along the axial coordinate, z).

We consider steady-state conduction in such a system. The general heat diffusion Eq. (3.2.11) reduces to

$$\frac{1}{r}\frac{\mathrm{d}}{\mathrm{d}r}\left(kr\frac{\mathrm{d}T}{\mathrm{d}r}\right) + q_g = 0 \tag{3.7.1}$$

and this must be solved, subject to suitable boundary conditions.

Equation (3.7.1) has no general analytical solutions, but analytical solutions for a wide range of problems have been obtained (Carslaw and Jaeger, 1976; Özisik, 1993). However, as above, relatively straightforward analytical solutions can be obtained if the equation is simplified further. Once again, many of the simplifications approximate very closely the observed conditions, and thus have widespread applications.

First we consider the case of no internal heat generation. This simplifies Eq. (3.7.1) to

$$k\frac{\mathrm{d}T}{\mathrm{d}r} = \frac{A_c}{r} \tag{3.7.2}$$

where A_c is a constant independent of r. A comparison of Eq. (3.2.8) with Eq. (3.7.2) shows that in this case the radial heat flux q_r is indirectly proportional to the radial coordinate r. In general the thermal conductivity k is a function of both the temperature and the radial position

$$k = k(T, r) \tag{3.7.3}$$

and, as above, the analytical integration of Eq. (3.7.2) can be difficult. However, if we again assume that the thermal conductivity k is independent of temperature, the general solution of Eq. (3.7.2) is

$$T = T_1 + A_c \int_{r_1}^{r} \frac{1}{k\,r}\mathrm{d}r \tag{3.7.4}$$

where T_1 is the temperature on the radial coordinate r_1. Assuming next that the thermal conductivity is constant, Eq. (3.7.4) can be solved to obtain the typical logarithmic temperature distribution.

$$T = T_1 + \frac{T_2 - T_1}{\ln(r_2/r_1)} \ln\left(\frac{r}{r_1}\right) \tag{3.7.5}$$

where T_2 is the temperature on the radial coordinate r_2.

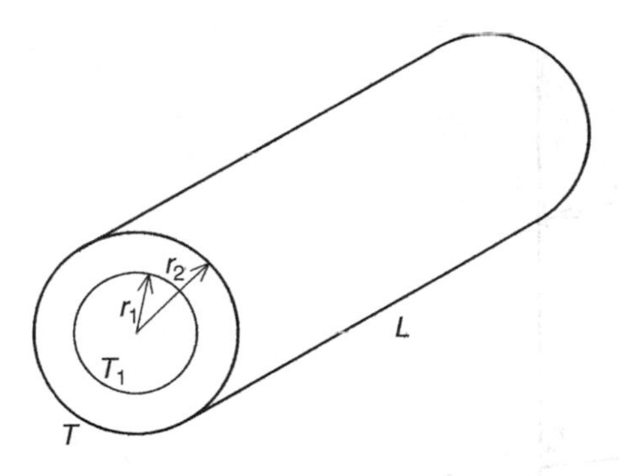

Figure 3.7.1 Long hollow axi-symmetric cylinder.

3.7.2 *Long hollow axi-symmetric cylinder*

Consider a hollow axi-symmetric cylinder of length L, shown in Fig. 3.7.1, whose outer surface, of radius r_2, is maintained at temperature T_2 and whose inner surface, of radius r_1, is maintained at temperature T_1. If the cylinder is sufficiently long (or, more precisely, if the aspect ratio L/r_2 is sufficiently large) the system can be regarded as one-dimensional and the simplifications used above can be used.

The temperature variation within the cylinder is then given by Eq. (3.7.5). The rate of heat transfer Q across any cylindrical surface can then be expressed from Eqs (3.1.1) and (3.2.8) as

$$Q = -kA\frac{\mathrm{d}T}{\mathrm{d}r} \tag{3.7.6}$$

where

$$A = 2\pi rL \tag{3.7.7}$$

Combining Eqs (3.7.6) and (3.7.7) with Eq. (3.7.2) it becomes clear that in this case Q is a constant; in other words, the rate of conduction heat transfer in the radial direction is constant. Substituting from Eq. (3.7.5) we next obtain rate of conduction heat transfer per unit length

$$\frac{Q}{L} = -\frac{2\pi k}{\ln(r_2/r_1)}(T_2 - T_1) \tag{3.7.8}$$

3.7.3 *Long composite axi-symmetric cylinder*

Similarly to Section 3.5.2 we can now analyse a composite cylinder, such as the one shown in Fig. 3.7.2. A general case is considered with the third (or fluid) boundary condition on the inner and the outer surfaces and contact

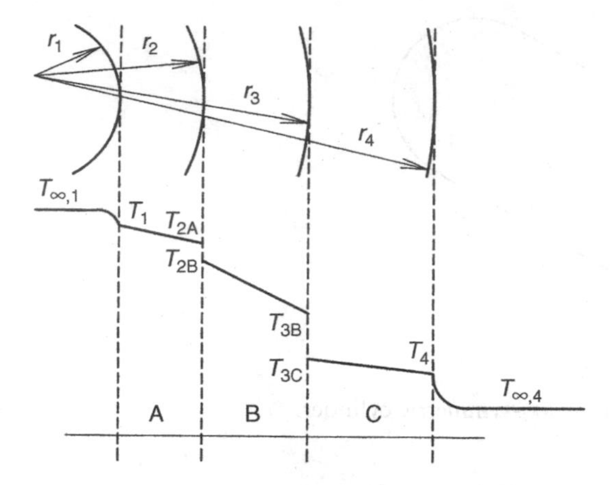

Figure 3.7.2 Long axi-symmetric cylinder with three sections.

resistances on the interfaces. It can be shown that the rate of conduction heat transfer per unit length is then given by

$$\frac{Q}{L} = [T_{\infty,1} - T_{\infty,4}] \Big/ \left[\frac{1}{2\pi r_1 h_1} + \frac{\ln(r_2/r_1)}{2\pi k_A} + \frac{R_{t,cont,AB}}{2\pi r_2} + \frac{\ln(r_3/r_2)}{2\pi k_B} + \frac{R_{t,cont,BC}}{2\pi r_3} + \frac{\ln(r_4/r_3)}{2\pi k_C} + \frac{1}{2\pi r_4 h_4} \right] \tag{3.7.9}$$

where h_1 and h_4 are the respective heat transfer coefficients and $T_{\infty,1}$ and $T_{\infty,4}$ are the free stream fluid temperatures at the inner and the outer surface respectively, and $R_{t,cont,AB}$ and $R_{t,cont,BC}$ are the contact resistances.

The overall heat transfer coefficient or the overall thermal resistance $R_{t,T}$, which are both based on the rate of conduction heat flux must be carefully considered, since they must be defined on the basis of the appropriate surface area. If we define the heat flux on the basis of the inner surface area, $2\pi r_1 L$, we can then show that

$$q_{r_1} = \frac{T_{\infty,1} - T_{\infty,4}}{R_{t,T}} \tag{3.7.10}$$

where

$$R_{t,T} = \frac{1}{h_1} + \frac{r_1}{k_A}\ln(r_2/r_1) + \frac{r_1}{r_2}R_{t,cont,AB} + \frac{r_1}{k_B}\ln(r_3/r_2) + \frac{r_1}{r_3}R_{t,cont,BC} + \frac{r_1}{k_C}\ln(r_4/r_3) + \frac{r_1}{r_4}\frac{1}{h_4} \tag{3.7.11}$$

It should be noted that the overall thermal resistance can be defined on the basis of any radial area.

Example 3.7.1 Long composite axi-symmetric cylinder with three sections.

A long composite axi-symmetric cylinder, shown in Fig. 3.7.2, consists of three sections, each with a constant thermal conductivity. The composite cylinder separates two fluids with known free stream temperatures. The heat transfer coefficients between the fluids and the respective external surfaces of the composite cylinder are also known. Finally, the appropriate contact resistances have also been obtained. There is no heat generation anywhere in the composite cylinder. Assume that steady-state conditions have been reached, and that the problem can be approximated by one-dimensional conduction. Determine the temperature distribution in the composite cylinder.

Analysis

Since the problem is based on the analysis presented in Section 3.7.3, the total thermal resistance $R_{t,T}$ based on the inner radius r_1 can be calculated from Eq. (3.7.11), and the heat flux q_{r_1} then obtained from Eq. (3.7.10). However, as discussed above, the rates of heat transfer per unit length across all elements of the system are identical and equal to $2\pi r_1 q_{r_1}$. Hence

$$\frac{q_{r_1}}{h_1} = T_{\infty,1} - T_1$$

and

$$T_1 = T_{\infty,1} - \frac{q_{r_1}}{h_1}$$

Substituting for Q/L in Eq. (3.7.8) we obtain

$$2\pi r_1 q_{r_1} = -\frac{2\pi k_A}{\ln(r_2/r_1)}(T_{2A} - T_1)$$

and

$$T_{2A} = T_1 - r_1 q_{r_1} \frac{\ln(r_2/r_1)}{k_A}$$

At the first interface, similarly to Eq. (3.5.15),

$$q_{r_2} = \frac{T_{2A} - T_{2B}}{R_{t,cont,AB}}$$

and since

$$q_{r_2} 2\pi r_2 = q_{r_1} 2\pi r_1$$

we obtain

$$T_{2B} = T_{2A} - q_{r_1} \frac{r_1}{r_2} R_{t,cont,AB}$$

Similarly

$$T_{3B} = T_{2B} - r_1 q_{r_1} \frac{\ln(r_3/r_2)}{k_B}$$

$$T_{3C} = T_{3B} - q_{r_1} \frac{r_1}{r_3} R_{t,cont,BC}$$

$$T_4 = T_{3C} - r_1 q_{r_1} \frac{\ln(r_4/r_3)}{k_C}$$

Finally, as shown in Section 3.7.1, the temperature profile in each section is given by Eq. (3.7.5). For example, in section A the temperature profile is given by

$$T = T_1 + \frac{T_{2A} - T_1}{\ln(r_2/r_1)} \ln\left(\frac{r}{r_1}\right)$$

and similarly for sections B and C.

Solution

The solution is given in the workbook Ex03-07-01.xls. Note that there are three worksheets in this workbook: *Input data and results, Temperatures* and *Distribution and graph*.

The first worksheet, *Input data and results*, allows for the input of the data, and presents the numerical results for the surface temperature of the three sections and the overall heat transfer rate per unit length, and the temperature distributions in a graphical form. The input data are entered in **cells (B3:B15)**, with the data already entered used for illustration only. The heat transfer rate per unit length Q/L is given in **cell B18**, and the surface temperatures in **cells (B20:B25)**. The graph shows the temperature profiles in the composite cylinder. The graph and its labels are dynamically linked to the solution, so that the reader can see the influence of the various parameters on the resultant temperature profiles. The presentation of the results is improved by clicking **Full Screen** in **View**.

The second worksheet, *Temperatures*, is used to calculate the total thermal resistance, $R_{t,T}$, the heat flux on the inner surface, q_{r_1}, the heat transfer rate per unit length Q/L, and the surface temperatures of the three sections.

The last worksheet, *Distribution and graph*, is used calculate the temperature profiles in each section, to scale the temperatures and distances for the preparation of the graph, and to provide the appropriate labelling.

Discussion

The solution, shown in the worksheet *Input data and results*, can be used to experiment with the input data and to observe their influence. Additionally, all the investigations discussed in Example 3.5.1 for a composite wall can be

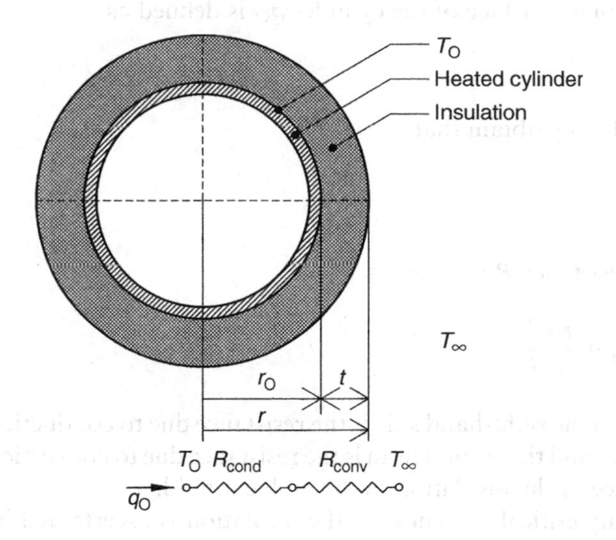

Figure 3.7.3 Long axi-symmetric heated cylinder with insulation.

also undertaken for the composite cylinder. Once again the reader is invited to experiment further.

Example 3.7.2 Critical thickness of thermal insulation.

Figure 3.7.3 shows a heated cylinder, of outer radius r_O, with thermal insulation wrapped round it. The outer surface of the cylinder is maintained at constant temperature T_O. The insulation, which has a constant thickness t and constant thermal conductivity k, has its outer surface exposed to ambient fluid at temperature T_∞, with a uniform convection heat transfer coefficient h. Assuming that steady-state conditions have been reached and that the system can be approximated by a long composite cylinder, investigate the influence of the insulation thickness on the rate of heat transfer from the cylinder.

Analysis

Using the approach developed in Section 3.7.3, Eq. (3.7.9) can be simplified for the present case as

$$\frac{Q}{L} = \frac{T_O - T_\infty}{(\ln(r/r_O)/2\pi k) + (1/2\pi rh)}$$

where r is the radius of the outer surface of the insulation, given as

$$r = r_O + t$$

The heat flux at the outer surface of the cylinder q_O is defined as

$$Q = 2\pi r_O L q_O$$

from which we can finally obtain that

$$q_O = \frac{T_O - T_\infty}{R}$$

where the thermal resistance R is given as

$$R = \frac{r_O}{k}\ln\left(\frac{r}{r_O}\right) + \frac{r_O}{r}\frac{1}{h}$$

where the first term on the right-hand side is the resistance due to conduction through the insulation and the second term is the resistance due to convection from the outer surface of the insulation to the ambient fluid.

The presence of any critical thickness of the insulation is ascertained by the differentiation of the expression for the thermal resistance R with respect to r. The critical thickness of the insulation is then obtained as

$$t_C = \frac{k}{h} - r_O$$

If $t_C > 0$, then we can show that the critical thickness of insulation corresponds to the minimum of the thermal resistance R. Hence the rate of heat transfer Q increases as the thickness of insulation t increases towards t_C, it reaches its maximum for $t = t_C$ and then starts decreasing. If, however, $k/hr_O \leq 1$, the minimum thermal resistance is obtained for $t = 0$. In these cases the rate of heat transfer Q decreases monotonically as the thickness of the insulation increases.

Solution

The solution is given in the workbook Ex03-07-02.xls. Note that there are three worksheets in this workbook: *Input data and results, Calculations*, and *Figure*.

The first worksheet, *Input data and results*, allows for the input of the data, and presents the numerical results for the critical thickness of insulation t_C, the actual rate of heat transfer per unit length Q/L, and the maximum possible rate of heat transfer per unit length Q/L_{MAX}. For $t_C > 0$ the worksheet also presents the variation of the total thermal resistance R, the conductive thermal resistance, and the convective thermal resistance. The input data are entered in **cells (B3:B8)**, with the data already entered used for illustration only. The results for the critical thickness of insulation t_C are given in **cell B11**, the actual rate of heat transfer per unit length Q/L in **cell B12** and the maximum possible rate of heat transfer per unit length Q/L_{MAX} in

cell B13. The presentation of the results is improved by clicking **Full Screen** in **View**.

The second worksheet, *Calculations*, is used to calculate the results and the last worksheet, *Figure*, is used to prepare the graph.

Discussion

The solution, shown in the worksheet *Input data and results*, can be used to experiment with the design parameters and to observe their influence. The following observations can be made:

1 Increasing the thickness of insulation does not necessarily decrease the rate of heat transfer from the cylinder; the problem must be carefully analysed to determine the optimum insulation thickness.

Other situations can be modelled and the reader is invited to experiment further.

3.8 Spherical systems

3.8.1 *Introduction*

In many cases it is possible to study spherical systems with one-dimensional models with temperature gradients in the radial direction only. This will be achieved if the system, including its boundary and initial conditions, possesses spherical symmetry.

We consider steady-state conduction in such a system. The general heat diffusion equation (Eq. (3.2.15)) reduces to

$$\frac{1}{r^2}\frac{\mathrm{d}}{\mathrm{d}r}\left(kr^2\frac{\mathrm{d}T}{\mathrm{d}r}\right) + q_\mathrm{g} = 0 \tag{3.8.1}$$

and this must be again solved, subject to suitable boundary conditions.

Once again, we consider first the case of no internal heat generation. This simplifies Eq. (3.8.1) to

$$k\frac{\mathrm{d}T}{\mathrm{d}r} = \frac{A_\mathrm{c}}{r^2} \tag{3.8.2}$$

where A_c is a constant independent of r. A comparison of Eq. (3.2.12) with the above equation shows that in this case the radial heat flux q_r is indirectly proportional to the square of the radial coordinate r. We again assume that the thermal conductivity k is independent of temperature and obtain the following general solution

$$T = T_1 + A_c \int_{r_1}^{r} \frac{1}{k\,r^2}\,dr \tag{3.8.3}$$

where T_1 is the temperature on the radial coordinate r_1. Assuming next that the thermal conductivity is constant, Eq. (3.8.3) can be solved to obtain

$$T = T_1 + (T_2 - T_1)\frac{1 - r_1/r}{1 - r_1/r_2} \tag{3.8.4}$$

where T_2 is the temperature on the radial coordinate r_2.

3.8.2 *Hollow sphere*

Consider a hollow sphere with full spherical symmetry, shown in Fig. 3.8.1, whose outer surface, of radius r_2, is maintained at temperature T_2 and whose inner surface, of radius r_1, is maintained at temperature T_1.

The temperature variation within the cylinder is then given by Eq. (3.8.4). The rate of heat transfer Q across any cylindrical surface can then be expressed from Eqs (3.1.1) and (3.2.12) as

$$Q = -kA\frac{dT}{dr} \tag{3.8.5}$$

where

$$A = 4\pi r^2 \tag{3.8.6}$$

Combining the above equations it becomes clear that in this case Q is a constant; in other words, the rate of conduction heat transfer in the radial

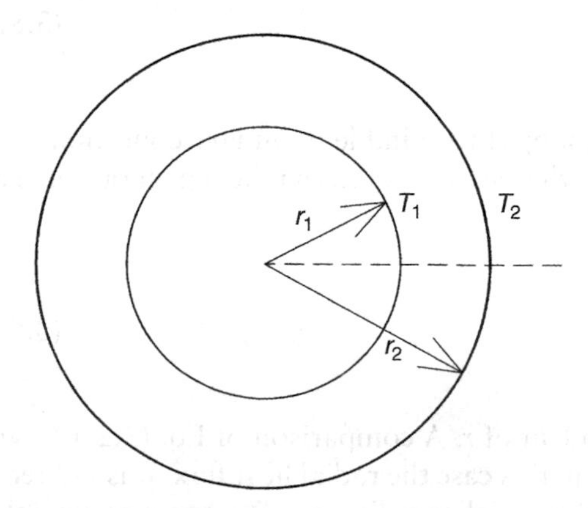

Figure 3.8.1 Hollow sphere with full spherical symmetry.

direction is constant and equal to

$$Q = -\frac{4\pi k}{1/r_1 - 1/r_2}(T_2 - T_1) \tag{3.8.7}$$

3.8.3 *Composite sphere*

Consider a composite sphere with full spherical symmetry, such as the one shown in Fig. 3.8.2. Similarly to Section 3.7.3 it can be shown that the rate of conduction heat transfer is given by

$$\begin{aligned} Q = [T_{\infty,1} - T_{\infty,4}] \Big/ \Bigg[& \frac{1}{4\pi r_1^2 h_1} + \frac{1/r_1 - 1/r_2}{4\pi k_A} + \frac{R_{t,cont,AB}}{4\pi r_2^2} \\ & + \frac{1/r_2 - 1/r_3}{4\pi k_B} + \frac{R_{t,cont,BC}}{4\pi r_3^2} \\ & + \frac{1/r_3 - 1/r_4}{4\pi k_C} + \frac{1}{4\pi r_4^2 h_4} \Bigg] \end{aligned} \tag{3.8.8}$$

Once again, the overall heat transfer coefficient or the overall thermal resistance $R_{t,T}$, which are both based on the rate of conduction heat flux must be carefully considered, since they must be defined on the basis of the appropriate surface area. If we define the rate of heat flux on the basis of the inner surface area, $4\pi r_1^2$, we can then show that

$$q_{r_1} = \frac{T_{\infty,1} - T_{\infty,4}}{R_{t,T}} \tag{3.8.9}$$

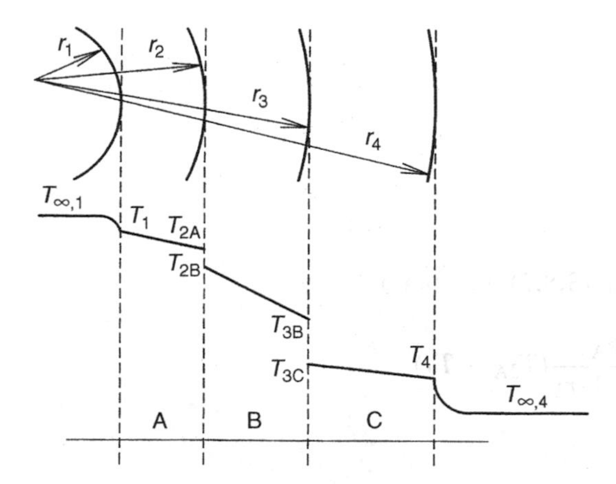

Figure 3.8.2 Composite sphere with three sections and full spherical symmetry.

where

$$R_{t,T} = \frac{1}{h_1} + \frac{r_1^2}{k_A}\left(\frac{1}{r_1} - \frac{1}{r_2}\right) + \left(\frac{r_1}{r_2}\right)^2 R_{t,cont,AB} + \frac{r_1^2}{k_B}\left(\frac{1}{r_2} - \frac{1}{r_3}\right)$$
$$+ \left(\frac{r_1}{r_3}\right)^2 R_{t,cont,BC} + \frac{r_1^2}{k_C}\left(\frac{1}{r_3} - \frac{1}{r_4}\right) + \left(\frac{r_1}{r_4}\right)^2 \frac{1}{h_4} \qquad (3.8.10)$$

Once again, it should be noted that the overall thermal resistance can be defined on the basis of any radial area.

Example 3.8.1 Composite sphere with three sections.

A composite sphere, shown in Fig. 3.8.2, consists of three sections, each with a constant thermal conductivity. The composite sphere separates two fluids with known free stream temperatures. The heat transfer coefficients between the fluids and the respective external surfaces of the composite sphere are also known. Finally, the appropriate contact resistances have also been obtained. There is no heat generation anywhere in the composite sphere. Assume that steady-state conditions have been reached, and that the problem can be approximated by one-dimensional conduction. Determine the temperature distribution in the composite sphere.

Analysis

Since the problem is based on the analysis presented in Section 3.8.3, the total thermal resistance $R_{t,T}$ based on the inner surface area can be calculated from Eq. (3.8.10), and the heat flux q_{r_1} then obtained from Eq. (3.8.9). However, as discussed above, the rate of heat transfer in the radial direction is constant across all elements of the system and equal to $4\pi r_1^2 q_{r_1}$. Hence

$$\frac{q_{r_1}}{h_1} = T_{\infty,1} - T_1$$

and

$$T_1 = T_{\infty,1} - \frac{q_{r_1}}{h_1}$$

Substituting for Q in Eq. (3.8.7) we obtain

$$4\pi r_1^2 q_{r_1} = -\frac{4\pi k_A}{1/r_1 - 1/r_2}(T_{2A} - T_1)$$

and

$$T_{2A} = T_1 - r_1^2 q_{r_1} \frac{(1/r_1 - 1/r_2)}{k_A}$$

At the first interface, similarly to Eq. (3.5.15),

$$q_{r_2} = \frac{T_{2A} - T_{2B}}{R_{t,cont,AB}}$$

and since

$$q_{r_2} 4\pi r_2^2 = q_{r_1} 4\pi r_1^2$$

we obtain

$$T_{2B} = T_{2A} - q_{r_1} \left(\frac{r_1}{r_2}\right)^2 R_{t,cont,AB}$$

Similarly

$$T_{3B} = T_{2B} - r_1^2 q_{r_1} \frac{(1/r_2 - 1/r_3)}{k_B}$$

$$T_{3C} = T_{3B} - q_{r_1} \left(\frac{r_1}{r_3}\right)^2 R_{t,cont,BC}$$

$$T_4 = T_{3C} - r_1^2 q_{r_1} \frac{(1/r_3 - 1/r_4)}{k_C}$$

Finally, as shown in Section 3.8.1, the temperature profile in each section is given by Eq. (3.8.4). For example, in section A the temperature profile is given by

$$T = T_1 + (T_2 - T_1) \frac{1 - r_1/r}{1 - r_1/r_2}$$

and similarly for sections B and C.

Solution

The solution is given in the workbook Ex03-08-01.xls. Note that there are three worksheets in this workbook: *Input data and results, Temperatures*, and *Distribution and graph*.

The first worksheet, *Input data and results*, allows for the input of the data, and presents the numerical results for the surface temperature of the three sections and the overall heat transfer rate, and the temperature distributions in a graphical form. The input data are entered in **cells** **(B3:B15)**, with the data already entered used for illustration only. The heat transfer rate is given in **cell B18**, and the surface temperatures in **cells** **(B20:B25)**. The graph shows the temperature profiles in the composite cylinder. The graph and its labels are dynamically linked to the solution, so that the reader can see the influence of the various parameters on the resultant temperature profiles. The presentation of the results is improved by clicking **Full Screen** in **View**.

The second worksheet, *Temperatures*, is used to calculate the total thermal resistance, $R_{t,T}$, the heat flux on the inner surface, q_{r_1}, the heat transfer rate Q, and the surface temperatures of the three sections.

The last worksheet, *Distribution and graph*, is used calculate the temperature profiles in each section, to scale the temperatures and distances for the preparation of the graph and to provide the appropriate labelling.

Discussion

The solution, shown in the worksheet *Input data and results*, can be used to experiment with the input data and to observe their influence. Additionally, all the investigations discussed in Example 3.5.1 for a composite wall can be also undertaken for the composite sphere. Once again the reader is invited to experiment further.

3.9 Extended surfaces

3.9.1 Introduction

In many engineering applications we employ solid elements that experience energy transfer not only by conduction within its surface boundaries but also by convection from the surfaces and the surrounding fluid. Experiments show that the rate of convective heat transfer generally increases with increasing surface area between the solid element and the surrounding fluid. Hence in order to enhance convective heat transfer rates the surface area of solid elements is frequently increased artificially. The general term for this application is *extended surface*. Since in many cases the extended surface is achieved by attaching thin solid elements to the surface of the primary solid element, these extended surfaces are called *fins*. Such fins are used in many practical applications, such as fins on the tubes of heat exchangers or on the cylinders of air-cooled engines, as shown in Fig. 3.9.1 (Bejan, 1993; Incropera and DeWitt, 1996).

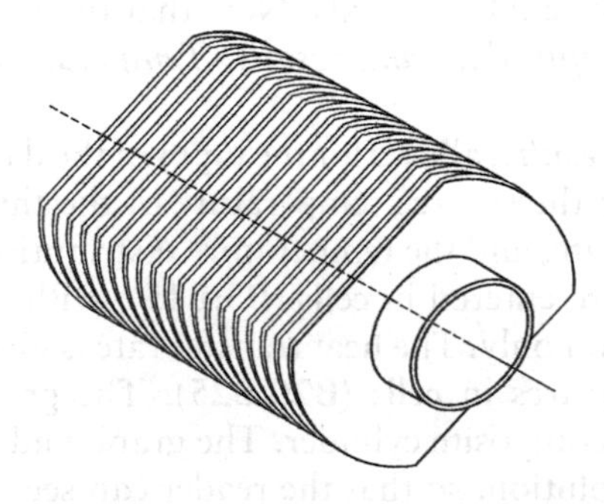

Figure 3.9.1 Fins on the cylinder of an air-cooled engine.

The conductive heat transfer processes within the fin are usually two-dimensional (and can be three-dimensional), so that, strictly, full multi-dimensional analyses should be employed to solve this problem. However, since the fin thickness is very small compared with its length, the temperature changes in the longitudinal direction are much larger than the temperature changes in the transverse direction. Furthermore, the fins generally operate in steady-state conditions. Hence, for many practical applications one-dimensional steady-state conduction can be assumed to describe energy transfer within the fin. Even though we are dealing with one-dimensional steady-state conduction we cannot use directly the equations derived above, and a modified approach must be adopted.

The analysis undertaken in this section is for conducting–convecting fins, since such systems are commonly used for terrestrial applications. However, for space applications conducting–radiating fins are employed and such systems are considered in Chapter 10.

3.9.2 *Analysis*

We will consider a general one-dimensional fin shown in Fig. 3.9.2 and analyse a differential element dx thick (with a cross-sectional area A and surface area dS) and distance x from the base of the fin. Applying conservation of energy on the differential element we obtain

$$Q_x = Q_{x+\mathrm{d}x} + \mathrm{d}Q_\mathrm{C} \tag{3.9.1}$$

where Q_x and $Q_{x+\mathrm{d}x}$ are the rates of conduction heat transfer into and out of the differential element respectively and $\mathrm{d}Q_\mathrm{C}$ is the rate of convection

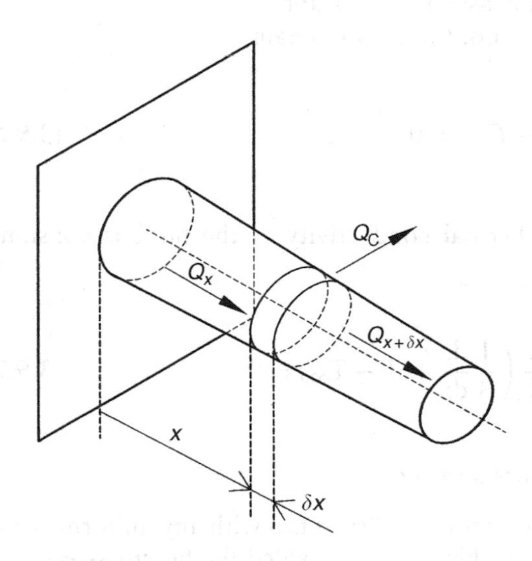

Figure 3.9.2 General one-dimensional fin.

heat transfer to the surrounding medium through the surface area of the differential element.

Using Taylor expansion (similar to Eq. (3.2.1)) we can write

$$Q_{x+\delta x} = Q_x + \frac{dQ_x}{dx}\,dx \tag{3.9.2}$$

Furthermore, using Eqs (3.1.1) and (3.1.3) we can write

$$Q_x = -kA\frac{dT}{dx} \tag{3.9.3}$$

Substituting Eqs (3.9.2) and (3.9.3) into Eq. (3.9.1) we obtain

$$\frac{d}{dx}\left(kA\frac{dT}{dx}\right) - \frac{dQ_C}{dx} = 0 \tag{3.9.4}$$

Next we have to consider the contribution of convection, usually given as

$$dQ_C = h\,dS(T - T_\infty) \tag{3.9.5}$$

where h is the heat transfer coefficient and T_∞ the temperature of the surrounding fluid, both usually assumed constant.

Substituting Eq. (3.9.5) into Eq. (3.9.4) we obtain

$$\frac{d}{dx}\left(kA\frac{dT}{dx}\right) - h\frac{dS}{dx}(T - T_\infty) = 0 \tag{3.9.6}$$

Finally, assuming that the thermal conductivity of the fin k is constant, Eq. (3.9.6) simplifies to

$$\frac{d^2T}{dx^2} + \left(\frac{1}{A}\frac{dA}{dx}\right)\frac{dT}{dx} - \frac{h}{k}\left(\frac{1}{A}\frac{dS}{dx}\right)(T - T_\infty) = 0 \tag{3.9.7}$$

3.9.3 *Fins with uniform cross-section*

The analysis presented in this section applies to fins with any uniform cross-section, such as those shown in Fig. 3.9.3, provided the fin cross-sectional area A and its perimeter P are constant. In such a case $dS = P\,dx$, and

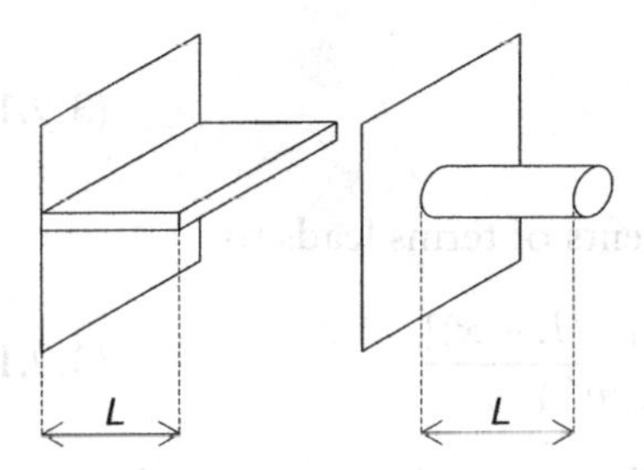

Figure 3.9.3 Fins with uniform cross-section.

Eq. (3.9.7) simplifies to

$$\frac{\mathrm{d}^2T}{\mathrm{d}x^2} - \frac{hP}{kA}(T - T_\infty) = 0 \tag{3.9.8}$$

We can simplify further by defining the *excess* temperature as

$$\theta = T - T_\infty \tag{3.9.9}$$

and

$$m^2 = \frac{hP}{kA} \tag{3.9.10}$$

to obtain

$$\frac{\mathrm{d}^2\theta}{\mathrm{d}x^2} - m^2\theta = 0 \tag{3.9.11}$$

The solution of this differential equation is

$$\theta = C_1\mathrm{e}^{-mx} + C_2\mathrm{e}^{mx} \tag{3.9.12}$$

where the constants C_1 and C_2 depend upon the boundary conditions. The first condition can be determined unambiguously by specifying the temperature T_0 (or $\theta_0 = T_0 - T_\infty$) at the base of the fin ($x = 0$). To specify the second boundary condition at the tip of the fin (at $x = L$, where L is the length of the fin) is more difficult, since several cases should be considered and judgement must be exercised.

We will consider here the third boundary condition of Eq. (3.3.3) by specifying that

$$-kA\left.\frac{\mathrm{d}T}{\mathrm{d}x}\right|_{x=L} = hA(T_\mathrm{L} - T_\infty) \tag{3.9.13}$$

where T_L is the temperature at the tip of the fin. If θ_L is the excess temperature at the tip of the fin, the boundary condition given by Eq. (3.9.13) can be written as

$$-k\frac{d\theta}{dx}\bigg|_{x=L} = h\theta_L \tag{3.9.14}$$

Solving for C_1 and C_2, and re-arrangements of terms leads to

$$\frac{\theta}{\theta_0} = \frac{\cosh[m(L-x)] + (h/mk)\sinh[m(L-x)]}{\cosh(mL) + (h/mk)\sinh(mL)} \tag{3.9.15}$$

It should be noted that in many practical cases the heat convected from the tip of the fin is negligible and Eq. (3.9.15) can be simplified to

$$\frac{\theta}{\theta_0} = \frac{\cosh[m(L-x)]}{\cosh(mL)} \tag{3.9.16}$$

The advantage of Eq. (3.9.16) is its simplicity, so that this equation is used in many industrial applications, even though the assumption of negligible heat convection from the tip of the fin may not necessarily hold. However, this can be compensated by extending the length of the fin such that the surface area of the new extended fin excluding the surface area of the tip equals to the surface area of the original fin with the surface area of the tip. It can be shown that for the rectangular and pin fins shown in Fig. 3.9.3 this extension is respectively $0.5wt/(w+t)$ and $0.5r$, where w and t are the width and the thickness of the rectangular fin, and r is the radius of the pin fin.

3.9.4 *Performance of fins*

Performance of fins is considered in this section. All the aspects will be demonstrated on the simple results of Eq. (3.9.16). Since in steady-state conditions the total heat dissipated by convection from the fin Q_F equals to the heat conducted through the base of the fin, thus

$$Q_F = -kA\frac{dT}{dx}\bigg|_{x=0} = -kA\frac{d\theta}{dx}\bigg|_{x=0} \tag{3.9.17}$$

Substituting Eq. (3.9.16) into Eq. (3.9.17) we obtain

$$Q_F = M\tanh(mL) \tag{3.9.18}$$

where

$$M^2 = \theta_0^2 hPkA \tag{3.9.19}$$

An assessment of the contribution of the fin to increased heat transfer rates can be made by evaluating the fin effectiveness ϕ defined as the ratio of the fin heat transfer rate Q_F to the heat transfer rate which would have existed without the fin through the area equivalent to the cross-sectional area of the base of the fin. Assuming that the heat transfer coefficient is constant and thus unaffected by the presence or the absence of the fin, the fin effectiveness can be calculated as

$$\phi = \frac{Q_F}{hA\theta_0} \tag{3.9.20}$$

Substituting Eqs (3.9.18) and (3.9.19) into Eq. (3.9.20) we obtain

$$\phi = \left(\frac{kP}{hA}\right)^{1/2} \tanh(mL) \tag{3.9.21}$$

For fins to be effective, their effectiveness must be greater than 1, and in order to justify them economically the fin effectiveness must be considerably greater than 1. Two important conclusions can be made immediately from Eq. (3.9.21). First the fin effectiveness increases with the length of the fins. Second the fin effectiveness also increases with the ratio P/A, which implies, for example, that the aspect ratio of rectangular fins should be as large as possible, thus leading to thin fins. Both of these conclusions must be considered simultaneously with economic considerations, since long and thin fins could be expensive. The design which maximises economic return must be adopted.

The limiting effectiveness is obtained for long fins, when Eq. (3.9.21) simplifies to

$$\phi = \left(\frac{kP}{hA}\right)^{1/2} \tag{3.9.22}$$

This equation shows two further important aspects: the fin effectiveness increases with increasing thermal conductivity of the fin material k and with decreasing heat transfer coefficient h. Hence fins are particularly useful when the film heat transfer coefficient h is small, such as in natural convection in gases, which is discussed later. This also explains why fins are placed on the gas side of the heat exchangers, rather than the liquid side, which has a much higher heat transfer coefficient.

The design aim should be to make the whole heat transfer surface as effective as possible, and thus to increase the number of fins. However, care must be taken to ensure that one of the original assumptions of this work, namely that the presence of fins does not affect the heat transfer coefficient h, is still valid. Fins should not be placed too closely so that they interfere with the flow and thus considerably decrease the heat transfer coefficient. As in all other branches of engineering a compromise is required.

Another measure of fin thermal performance is provided by the fin efficiency η, defined as the ratio of the fin heat transfer rate Q_F to the heat transfer rate from the fin if all fin surface were at the base temperature

θ_0. If we, once again, neglect heat transfer from the tip of the fin, the fin efficiency is

$$\eta = \frac{Q_F}{hPL\theta_0} \tag{3.9.23}$$

which, for the present conditions, can be expressed as

$$\eta = \frac{1}{mL}\tanh(mL) \tag{3.9.24}$$

As mL increases from 0 to ∞, the fin efficiency η decreases from 1 to 0; the fin efficiency increases with decreasing mL.

3.9.5 Complex fins

The above section presented several methods for characterising the performance of certain simple fins. However, fins used in practice are more complex, since other factors, such as the ease of manufacture, must be considered (Bejan, 1993; Incropera and DeWitt, 1996).

For fins with non-uniform cross-sectional area we have to use the general description of thermal performance of the fins, given by Eq. (3.9.7). Example 3.9.1 considers such a fin. More complex fins can be analysed numerically, as described in Chapter 4.

Example 3.9.1 Annular fins with constant thickness.

Annular fins with a constant thickness are used in many applications, such as those used in the design of cross-flow heat exchangers to enhance their performance. A typical example is shown in Fig. 3.9.4, which gives the design parameters: the radius of the base of the fin r_1, the radius of the tip of the fin r_2, the fin thickness t, and the fin pitch p. Such a fin is considered by Incropera and DeWitt (1996), who showed that for negligible heat convection from the tip of the fin (*adiabatic tip*), the fin heat transfer rate Q_F can be expressed as

$$Q_F = 2\pi k r_1 t\theta_0 m_1 \frac{K_1(m_1r_1)I_1(m_1r_2) - I_1(m_1r_1)K_1(m_1r_2)}{K_0(m_1r_1)I_1(m_1r_2) + I_0(m_1r_1)K_1(m_1r_2)}$$

where $m_1^2 = 2h/kt$, I_0 and K_0 are modified zero order Bessel functions of the first and second kinds, respectively, and I_1 and K_1 are modified first order Bessel functions of the first and second kinds respectively. Consider the assumption of the adiabatic fin tip, and investigate the influence of the design parameters on the fin efficiency η and the fin effectiveness ϕ.

Analysis

As discussed in Section 3.9.4, the assumption of the adiabatic fin tip does not necessarily hold when the thickness of the fin is comparable to the tip radius.

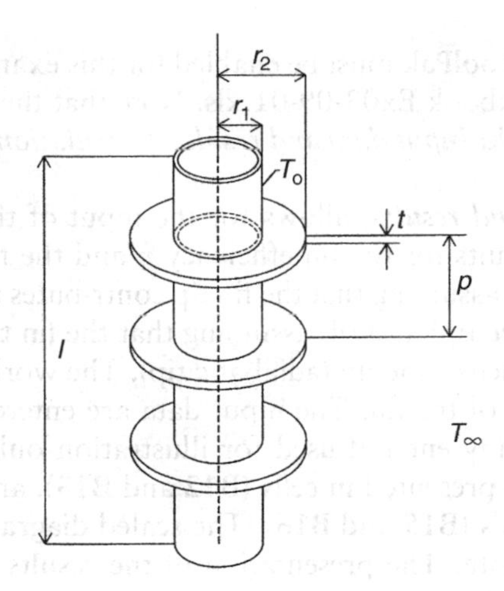

Figure 3.9.4 Cylindrical surface with equally spaced annular fins.

However, as also discussed there a correction can be made by extending the length of the fin, or, more appropriately, the radius of the fin tip. We assume that the surface area of the corrected fins (but excluding the surface area of the fin tip) is equal to the total surface area of the original film, which includes the surface area of the fin tip:

$$A = 2\pi\left(r_2^2 - r_1^2\right) + 2\pi r_2 t$$
$$A_c = 2\pi\left(r_{2C}^2 - r_1^2\right)$$
$$2\pi\left(r_2^2 - r_1^2\right) + 2\pi r_2 t = 2\pi\left(r_{2C}^2 - r_1^2\right)$$

from which the corrected tip radius r_{2C} can be obtained as

$$r_{2C} = r_2(1 + t/r_2)^{0.5}$$

The fin effectiveness ϕ is given, similarly to Eq. (3.9.20), as

$$\phi = \frac{Q_F}{h 2\pi r_1 t \theta_0}$$

The fin efficiency η is given, similarly to Eq. (3.9.23), but based on the total fin area, as

$$\eta = \frac{Q_F}{h 2\pi\left(r_{2C}^2 - r_1^2\right)\theta_0}$$

Solution

It should be noted that the **Analysis ToolPak** must be enabled for this example. The solution is given in the workbook Ex03-09-01.xls. Note that there are three worksheets in this workbook: *Input data and results*, *Calculations*, and *Figure*.

The first worksheet, *Input data and results*, allows for the input of the data, and presents the numerical results for the fin efficiency η and the fin effectiveness ϕ for two conditions: (i) assuming that the fin tip contributes to the performance of the fin (convective tip) and (ii) assuming that the fin tip does not contribute to the performance of the fin (adiabatic tip). The worksheet also presents a scaled diagram of the fin. The input data are entered in **cells (B4:B8)**, with the data already entered used for illustration only. The results for the convective tip are presented in **cells (B12** and **B13)**, and the results for the adiabatic tip in **cells (B15** and **B16)**. The scaled diagram is dynamically linked to the input data. The presentation of the results is improved by clicking **Full Screen** in **View**.

The second worksheet, *Calculations*, is used to calculate the results and the last worksheet, *Figure*, is used to prepare the diagram of the fin.

Discussion

The solution, shown in the worksheet *Input data and results*, can be used to experiment with the design parameters and to observe their influence. The following observations can be made:

1 As the aspect ratio $(r_2 - r_1)/t$ decreases the contribution of heat transfer from the tip of the fin increases. The influence of thermal conductivity k and the heat transfer coefficient h can also be investigated.
2 Increasing the thermal conductivity of the fin material increases both the fin efficiency and the fin effectiveness.
3 Increasing the fin thickness increases the fin efficiency but decreases the fin effectiveness.

Other situations can be modelled and the reader is invited to experiment further.

3.9.6 Performance of finned surfaces

The analysis presented so far has concentrated on the performance of a single isolated fin. Such arrangements are rarely used; it is more common to use an *array* of fins on a base surface, and hence the effectiveness and the efficiency of *finned* surfaces are required. In order to simplify the analysis we will assume that all fins are identical and that the presence of the fins does not affect the value of the heat transfer coefficient h. This assumption is usually satisfied, provided the fins are not too close, so as to interfere with the development of the flow pattern.

We consider the finned surface shown in Fig. 3.9.4, where p denotes the fin pitch and all other symbols have their usual meaning.

Analogously to Eq. (3.9.20), the overall effectiveness of the finned surface ϕ_0 is defined as

$$\phi_0 = \frac{Q_{FS}}{hA_{BS}\theta_0} \tag{3.9.25}$$

where A_{BS}, the total area of the base surface (without any fins present), is given as

$$A_{BS} = A_{SE} + NA_{BF} \tag{3.9.26}$$

where A_{SE} is the area of the exposed part of the base (with the fins present), N is the number of fins, each with base area A_{BF}, and where Q_{FS}, the total rate of heat transfer from the exposed part of the base and all fins, is given as

$$Q_{FS} = hA_{SE}\theta_0 + NQ_F \tag{3.9.27}$$

where Q_F is the rate of heat transfer from each fin. Noting further, that the effectiveness of each fin ϕ_F is given as

$$\phi_F = \frac{Q_F}{hA_{BF}\theta_0} \tag{3.9.28}$$

We next combine Eqs (3.9.25) to (3.9.28) to obtain

$$\phi_0 = 1 + (\phi_F - 1)\frac{NA_{BF}}{A_{BS}} \tag{3.9.29}$$

Similarly we can also define the overall efficiency of the finned surface η_0 as

$$\eta_0 = \frac{Q_{FS}}{hA_{TS}\theta_0} \tag{3.9.30}$$

where A_{TS}, the total surface area, including the area of the exposed part of the base and the surface area of all N fins, each with a surface area A_{SF}, is given as

$$A_{TS} = A_{SE} + NA_{SF} \tag{3.9.31}$$

Noting further that the efficiency of each fin η_F is given as

$$\eta_F = \frac{Q_F}{hA_{SF}\theta_0} \tag{3.9.32}$$

the overall efficiency is then calculated as

$$\eta_0 = 1 - (1 - \eta_F)\frac{NA_{SF}}{A_{TS}} \tag{3.9.33}$$

It can be also shown that the overall effectiveness of the finned surface ϕ_0 is lower than the effectiveness of each fin ϕ_F, and that the overall efficiency of the finned surface η_0 is greater than the efficiency of each fin η_F. The former result is more intuitive, and further insight can be obtained by investigating the behaviour of the overall effectiveness ϕ_0. The use of overall effectiveness and efficiency is discussed further in Example 3.9.2.

Example 3.9.2 Performance of a surface with an array of annular fins.

Consider a cylindrical surface with outer radius r_1 and length l, with N integrally cast and equally spaced annular fins, shown in Fig. 3.9.4. The annular fins, with constant thickness t and tip radius r_2, have constant thermal conductivity k. The outer surface of the cylinder is maintained at constant temperature T_0. The finned cylinder is exposed to ambient fluid at temperature T_∞, with a uniform convection heat transfer coefficient h. Assuming that steady-state conditions have been reached and using the one-dimensional analysis developed above, investigate the influence of the design parameters on heat transfer rate from the cylinder Q_{FS}.

Analysis

The total rate of heat transfer from the finned cylinder Q_{FS} can be obtained from Eq. (3.9.25) as

$$Q_{FS} = \phi_0 h A_{BS}(T_0 - T_\infty)$$

Since A_{BS}, the total area of the base surface, is

$$A_{BS} = 2\pi r_1 l$$

the total rate of heat transfer can be expressed as

$$Q_{FS} = \phi_0 h 2\pi r_1 l(T_0 - T_\infty)$$

The overall effectiveness is given by Eq. (3.9.29) as

$$\phi_0 = 1 + (\phi_F - 1)\frac{N A_{BF}}{A_{BS}}$$

where A_{BF}, the base area of each fin, is

$$A_{BF} = 2\pi r_1 t$$

and where the effectiveness of each fin ϕ_F can be obtained from Example 3.9.1.

Finally, it should be noted that, as discussed above, there are limits on the number of fins. If the clearance between the fins is too small, the basic assumption of the analysis, that is of constant heat transfer coefficient, does not hold. The physical limit N_{MAX} is reached when the clearance between the fins reaches zero; hence

$$N_{MAX} = l/t$$

Solution

It should be noted that the **Analysis ToolPak** must be enabled for this example. The solution is given in the workbook Ex03-09-02.xls. Note that there are three worksheets in this workbook: *Input data and results*, *Calculations*, and *Figure*.

The first worksheet, *Input data and results*, allows for the input of the data, and presents the numerical results for the fin effectiveness ϕ_F, the overall effectiveness of the finned surface ϕ_O, and the overall rate of heat transfer Q_{FS}. The worksheet also gives the results for the maximum number of fins, N_{MAX}, the fin efficiency η_F and the overall efficiency of the finned surface η_O. The worksheet also presents a scaled diagram of three fins on the base surface. The input data are entered in **cells** (**B4:B11** and **B13**) (the number of fins N must be smaller than the maximum number of fins, N_{MAX}), with the data already entered used for illustration only. The results are presented as follows: the maximum number of fins in **cell B12**, the fin pitch in **cell B13**, the fin effectiveness ϕ_F in **cell B17**, the overall effectiveness of the finned surface ϕ_O in **cell B18**, the overall rate of heat transfer Q_{FS} in **cell B19**, the fin efficiency η_F in **cell B20**, and the overall efficiency of the finned surface η_O in **cell B21**. The scaled diagram is dynamically linked to the input data. The presentation of the results is improved by clicking **Full Screen** in **View**.

The second worksheet, *Calculations*, is used to calculate the results and the last worksheet, *Figure*, is used to prepare the diagram of the finned surface.

Discussion

The solution, shown in the worksheet *Input data and results*, can be used to experiment with the design parameters and to observe their influence. The following observations can be made:

1 as the number of fins increases the overall effectiveness of the finned surface ϕ_O increases asymptotically to the fin effectiveness ϕ_F;
2 decreasing the fin thickness t increases the fin effectiveness ϕ_F and allows more fins to be incorporated, thus increasing the overall effectiveness of the finned surface ϕ_O and, of course, the overall rate of heat transfer Q_{FS}.

Other situations can be modelled and the reader is invited to experiment further.

Problems

Problem 3.1

A composite wall, shown in Fig. 3.P.1, consists of a central section, with dimensions 10 cm by 5 cm and thermal conductivity $k_C = 150\,\text{W/m K}$, which is surrounded by 1 cm thick insulation with thermal conductivity $k_I = 1\,\text{W/m K}$. Assume that steady-state conditions have been reached, that the faces $ABCD$ and $A'B'C'D'$ are thermally insulated and that the problem

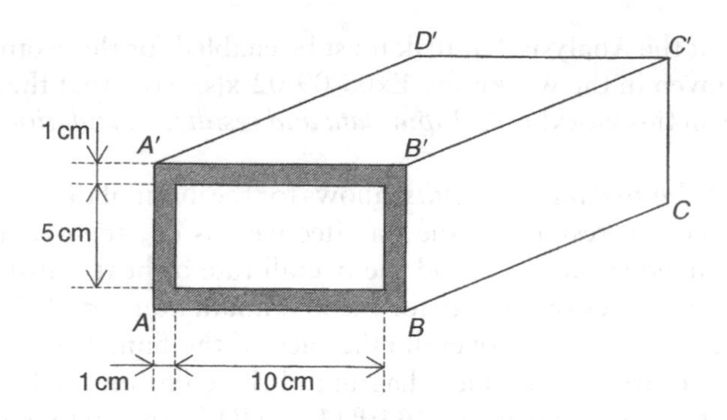

Figure 3.P.1 A composite wall.

can be approximated by one-dimensional conduction. Find the heat transfer rate per unit length of the wall, using both the adiabatic and the isothermal approximations. Discuss the results.

Problem 3.2
Consider the composite wall of Problem 3.1 and discuss the options which are available to decrease the heat transfer rate per unit length of the wall by 50%.

Problem 3.3
Consider a long cylindrical tube with inner radius $r_I = 40$ mm, outer radius $r_O = 70$ mm, and a thermal conductivity $k_T = 50$ W/m K. Thermal insulation, of a constant thickness $t = 50$ mm and a thermal conductivity $k_I = 5$ W/m K, is wrapped round it. The temperature of the inner surface of the tube is 400 K, and the outer surface of the insulation is exposed to ambient fluid at temperature 200 K and uniform convection heat transfer coefficient of 200 W/m^2 K. Assuming that the contact resistance between the cylindrical tube and the thermal insulation is negligible, determine the temperature of the outer surface of the cylindrical tube.

Problem 3.4
Consider the arrangement in Problem 3.3, and determine the temperature of the outer surface of the cylindrical tube, assuming that the contact resistance between the cylindrical tube and the thermal insulation is $R = 10^{-4}$ m^2 K/W.

Problem 3.5
A design requirement for the arrangement in Problem 3.3 is that the temperature of the outer surface of the cylindrical tube is 400 K. A possible solution is to wrap a thin electrical heater around the outer surface of the tube between the tube and the thermal insulation. Assuming that the thickness of the heater is negligible, that the contact resistance between the cylindrical tube and the heater $R_{TH} = 0.5 \times 10^{-4}$ m^2 K/W and that the contact resistance between

the heater and the thermal insulation $R_{\mathrm{HI}} = 1.5 \times 10^{-4}\,\mathrm{m^2\,K/W}$, determine the required heater power per unit length.

Problem 3.6

Consider a heated sphere, of outer radius r_{O}, with thermal insulation wrapped round it. The outer surface of the sphere is maintained at constant temperature T_{O}. The insulation, which has a constant thickness t and constant thermal conductivity k, has its outer surface exposed to ambient fluid at temperature T_{∞}, with a uniform convection heat transfer coefficient h. Assuming that steady-state conditions have been reached and that the system can be assumed to possess spherical symmetry, investigate the influence of the insulation thickness on the rate of heat transfer from the sphere.

Problem 3.7

A rectangular fin, shown in Fig. 3.P.2, is 3 mm thick and 20 mm long and constructed of aluminium with thermal conductivity $k_{\mathrm{A}} = 200\,\mathrm{W/m\,K}$. Under typical operating conditions the temperature of the base of the fin is 500 K (i.e. the fin is attached to a wall with surface temperature 500 K). The fin is exposed to an ambient environment with temperature of 300 K and convection heat transfer coefficient of $100\,\mathrm{W/m^2\,K}$. Calculate the fin effectiveness, the fin efficiency, and the heat transfer rate per unit width of the fin.

Problem 3.8

The fin considered in Problem 3.7 is an integral part of the wall, and thus there is no contact resistance between the surface of the wall and the base of the fin. Such an arrangement is achieved, for example, by machining the wall together with the fin. However, fins may be also manufactured separately and then attached to the wall by a *press fit* process. In such cases thermal resistance is introduced between the surface of the wall and the base of

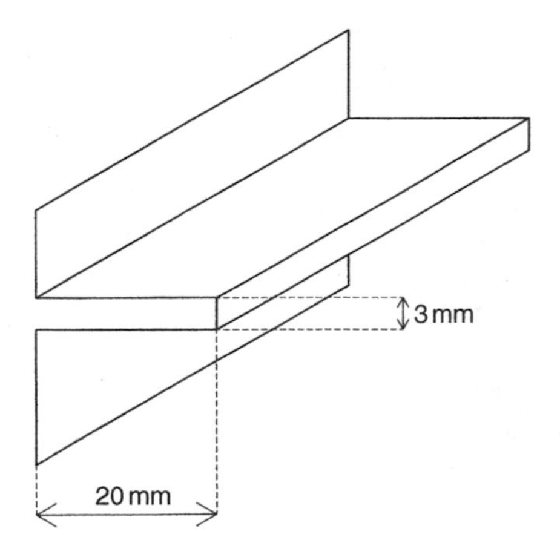

Figure 3.P.2 A rectangular fin.

the fin. Assume that the contact resistance $R = 0.5 \times 10^{-4}\ \mathrm{m^2\,K/W}$, and calculate the fin effectiveness, the fin efficiency, and the heat transfer rate per unit width of the fin considered in Problem 3.7. Discuss the influence of the contact resistance.

Problem 3.9

A wall 100 mm by 100 mm, contains N equally spaced fins, considered in Problems 3.7 and 3.8. Assuming the ambient conditions of Problem 3.7, determine the influence of the number of fins N on the overall effectiveness and the overall efficiency for the two cases: (i) the integral fin of Problem 3.7 and (ii) the press fit fin of Problem 3.8. Discuss the results.

References

Bejan, A. (1993) *Heat Transfer*, John Wiley & Sons, New York.

Brodkey, R. S. and Hershey, H. C. (1988) *Transport Phenomena: A Unified Approach*, McGraw-Hill, New York.

Carslaw, H. S. and Jaeger, J. C. (1976) *Conduction of Heat in Solids*, Oxford University Press, Oxford.

Holman, J. P. (1990) *Heat Transfer*, 7th edn, McGraw-Hill, New York.

Incropera, F. P. and De Witt, D. P. (1996) *Fundamentals of Heat and Mass Transfer*, 4th edn, John Wiley & Sons, New York.

Kreith, F. and Bohn, M. S. (1993) *Principles of Heat Transfer*, 5th edn, West Publishing Company, New York.

Luikov, A. V. (1968) *Analytical Heat Diffusion Theory*, Academic Press, New York.

Özisik, M. N. (1993) *Heat Conduction*, 2nd edn, John Wiley & Sons, New York.

Perry, R. H. (1998) *Perry's Chemical Engineers' Handbook*, 7th edn, McGraw Hill, New York.

Rohsenow, W. M. and Hartnett, J. P. (1973) *Handbook of Heat Transfer*, McGraw-Hill, New York.

4 Multi-dimensional, steady-state conduction

4.1 Introduction

The general forms of the governing equations are discussed in the previous chapter. Current approaches to solving the governing equations use either analytical or numerical techniques. It is often stated that the former techniques are exact, whereas the latter are only approximate. However, this is only the case for very narrow class of theoretical problems, since only some theoretical problems can be solved analytically in a closed form. The solutions are generally not available in closed forms, and usually involve complex mathematical series. In order to obtain numerical results these series must be evaluated, and hence the numerical results only approximate the asymptotic solutions, but the accuracy can be increased by increasing the number of terms of the series which are being evaluated.

Furthermore, engineering problems, which involve real materials and real boundary conditions, can only be solved analytically if a large number of simplifying assumptions is made. Hence analytical solutions thus obtained will only be approximate. Nevertheless, the analytical techniques can provide a real insight into many problems, since we can obtain solutions for a *class* of problems, and then investigate the impact of the various parameters on the behaviour of the whole system.

The numerical techniques are approximate in that they can only provide approximate solutions at discrete points. However, they can be used to solve real engineering problems, involving complex geometries, boundary conditions, and material properties. Furthermore, and as discussed previously, their accuracy can be increased. Finally, properly designed solutions also enable numerical simulation of the influence of various parameters, numerical experimentation, and design optimisation.

The various analytical techniques are not considered in this book, and the reader is referred to the many available textbooks (Carslaw and Jaeger, 1976; Bejan, 1993; Kreith and Bohn, 1993; Özisik, 1993). It should also be noted that there are available many commercial software packages which allow the solutions of even the most complex problems. A search of the web will identify the major developers and suppliers. The major disadvantage of these packages is their complexity; they are not suitable for the undergraduate study of engineering problems. The reader is referred to Chapter 2 for further discussion on this topic.

The numerical technique used in this book is the finite difference method. The major advantage of this approach is that it is quite intuitive, particularly in the solutions of two dimensional problems, and thus well suited for undergraduate study. The finite difference equations are derived in this chapter. The direct and iterative methods of solutions are discussed in Chapter 2; the Excel environment is well suited for such computations to be handled with ease.

4.2 Finite difference approximation in two-dimensional systems: Cartesian system

4.2.1 *Internal nodes*

Energy balances for an infinitesimally small control volume when heat is only transferred by conduction were considered previously. Assuming steady-state and a two-dimensional case, Eq. (3.2.4) simplifies to

$$(q_x - q_{x+dx})\,dy + (q_y - q_{y+dy})\,dx + q_g\,dx\,dy = 0 \tag{4.2.1}$$

The equation can be also used for *any* control volume with dimensions Δx and Δy

$$(q_x - q_{x+\Delta x})\Delta y + (q_y - q_{y+\Delta y})\Delta x + q_g\Delta x\Delta y = 0 \tag{4.2.2}$$

Since the equation is based on the assumption that heat is transferred only by conduction, it can be used to formulate energy balances for a finite control volume associated with the *interior* nodes.

Consider the interior node shown in Fig. 4.2.1. If the control volume about the node is *sufficiently small*, the respective heat fluxes are given as

$$q_x = k_{m-0.5,n}\frac{T_{m-1,n} - T_{m,n}}{\Delta x} \tag{4.2.3}$$

$$q_{x+\Delta x} = k_{m+0.5,n}\frac{T_{m,n} - T_{m+1,n}}{\Delta x} \tag{4.2.4}$$

$$q_y = k_{m,n-0.5}\frac{T_{m,n-1} - T_{m,n}}{\Delta y} \tag{4.2.5}$$

$$q_{y+\Delta y} = k_{m,n+0.5}\frac{T_{m,n} - T_{m,n+1}}{\Delta y} \tag{4.2.6}$$

Equations (4.2.3)–(4.2.6) are general equations which do take into account that the thermal conductivity is not necessarily constant, and hence that its appropriate values must be specified. However, if the thermal conductivity is

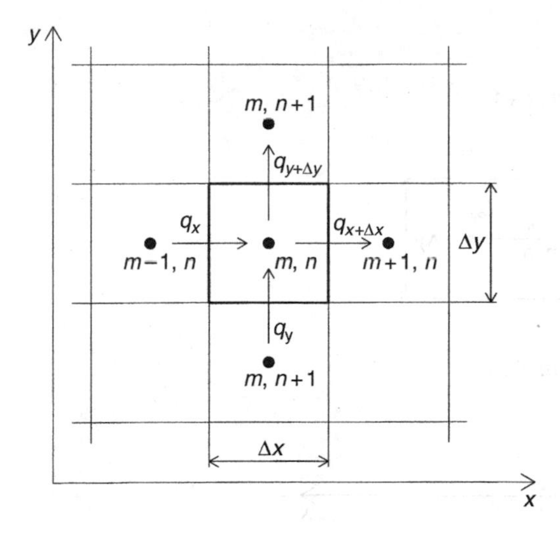

Figure 4.2.1 Internal node in a two-dimensional Cartesian coordinate system.

constant and equal to k and if, furthermore, the nodal system is chosen such that $\Delta x = \Delta y$, substitution of Eqs (4.2.3)–(4.2.6) into Eq. (4.2.2) shows that

$$T_{m+1,n} + T_{m-1,n} + T_{m,n+1} + T_{m,n-1} - 4T_{m,n} + \frac{q_g(\Delta x)^2}{k} = 0 \qquad (4.2.7)$$

If there is no heat generation, Eq. (4.2.7) simplifies to Eq. (2.11.7) derived using an alternative method in Section 2.11.2.

4.2.2 *External nodes with various boundary conditions*

Case 1: Plane surface with a convective boundary condition.

Referring to Fig. 4.2.2 and assuming constant thermal conductivity, the energy balance for the control volume is given by

$$q_x \Delta y - q_c \Delta y + q_y \frac{\Delta x}{2} - q_{y+\Delta y}\frac{\Delta x}{2} + q_g \frac{\Delta x}{2}\Delta y = 0 \qquad (4.2.8)$$

where the convective rate of heat transfer is given by

$$q_c = h(T_{m,n} - T_\infty) \qquad (4.2.9)$$

and all other heat transfer rates are as above. After re-arrangements of terms, and using $\Delta x = \Delta y$ nodal system

$$2T_{m-1,n} + T_{m,n+1} + T_{m,n-1} - 2(\text{Bi} + 2)T_{m,n} + 2\text{Bi}T_\infty + \frac{q_g(\Delta x)^2}{k} = 0 \qquad (4.2.10)$$

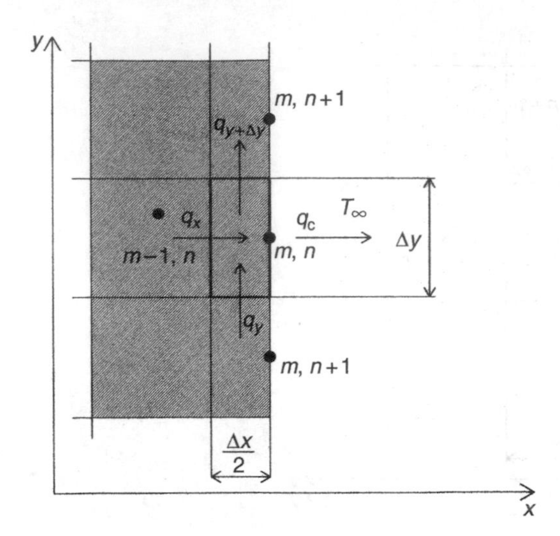

Figure 4.2.2 External plane wall node in a two-dimensional Cartesian coordinate system.

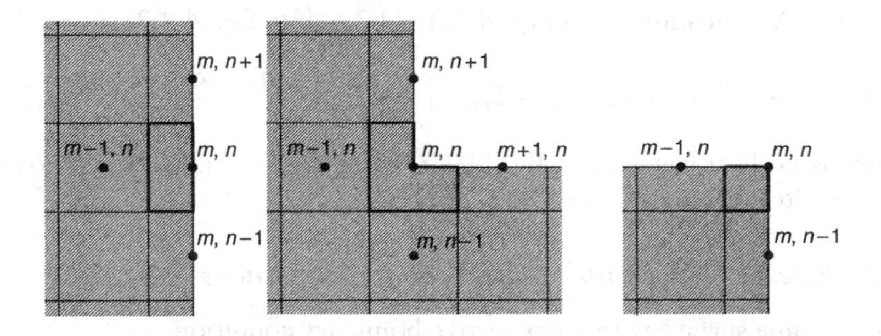

Figure 4.2.3 External nodes in a two-dimensional Cartesian coordinate system.

where Bi is a finite difference form of the Biot number, defined as

$$\mathrm{Bi} = \frac{h\Delta x}{k} \tag{4.2.11}$$

Results for other nodes, shown in Fig. 4.2.3, and other boundary conditions are summarised below.

Case 2: Internal corner with a convective boundary condition.

$$2T_{m+1,n} + 4T_{m-1,n} + 2T_{m,n+1} + 4T_{m,n-1} - 4(\mathrm{Bi} + 3)T_{m,n} + 4\mathrm{Bi}T_\infty + \frac{3q_g(\Delta x)^2}{k} = 0 \tag{4.2.12}$$

Case 3: External corner with a convective boundary condition.

$$2T_{m-1,n}+2T_{m,n-1}-4(\mathrm{Bi}+1)T_{m,n}+4\mathrm{Bi}T_{\infty}+\frac{q_g(\Delta x)^2}{k}=0 \quad (4.2.13)$$

Case 4: Plane surface with uniform heat flux q_E.

$$2T_{m-1,n}+T_{m,n+1}+T_{m,n-1}-4T_{m,n}+\frac{2q_E\Delta x}{k}+\frac{q_g(\Delta x)^2}{k}=0 \quad (4.2.14)$$

Case 5: Internal corner with uniform heat flux q_E.

$$2T_{m+1,n}+4T_{m-1,n}+2T_{m,n+1}+4T_{m,n-1}-12T_{m,n}$$
$$+\frac{4q_E\Delta x}{k}+\frac{3q_g(\Delta x)^2}{k}=0 \quad (4.2.15)$$

Case 6: External corner with uniform heat flux q_E.

$$2T_{m-1,n}+2T_{m,n-1}-4T_{m,n}+\frac{4q_E\Delta x}{k}+\frac{q_g(\Delta x)^2}{k}=0 \quad (4.2.16)$$

Example 4.2.1 Rectangular duct with a coarse grid.

Consider a duct with a rectangular cross-section shown in Fig. 4.2.4. The duct has a constant thermal conductivity k, its outer surface is at constant temperature T_o and its inner surface is at a different constant temperature T_I. Furthermore, there is no heat generation anywhere in the duct. Assume that steady-state conditions have been reached, and that the problem can be approximated by two-dimensional conduction. Determine the temperature distribution in the duct and the rate of conduction heat transfer through the duct.

Analysis

A nodal system with $\Delta x = \Delta y$ and shown in Fig. 4.2.4 is chosen. For all internal nodes, Eq. (4.2.7) can be simplified as

$$4T_{m,n} = T_{m+1,n} + T_{m-1,n} + T_{m,n+1} + T_{m,n-1}$$

Since in this nodal system each surface node is distance $\Delta x/2$ from its associated central node, the thermal resistance between the central node and its associated surface node is half of the thermal resistance between the central node and its associated internal nodes. Following the analysis in Section 4.2.2, we can show that the temperature of each node associated with two surface nodes, such as $T_{2,2}$, is given by

$$6T_{2,2} = T_{2,3} + T_{3,2} + 2T_{2,1} + 2T_{1,2}$$

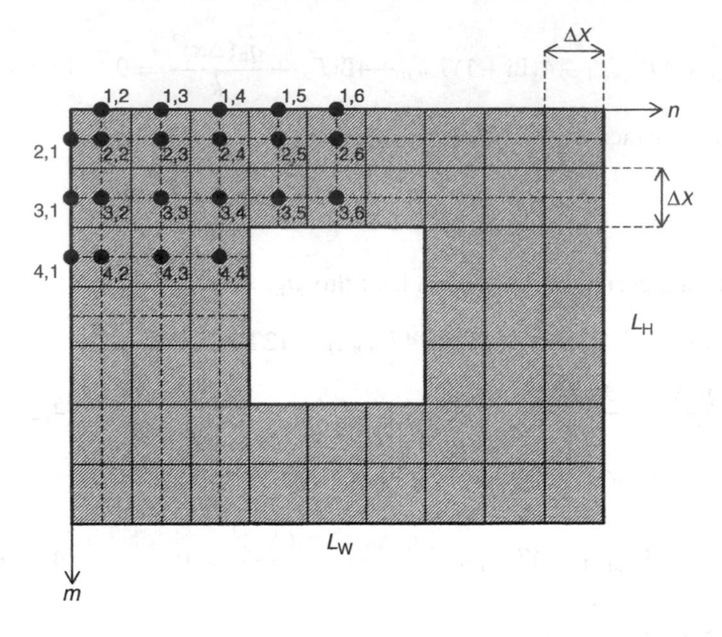

Figure 4.2.4 Rectangular duct.

which for the external corner temperature $T_{2,2}$ of the example simplifies to

$$6T_{2,2} = T_{2,3} + T_{3,2} + 4T_o$$

Similarly, the temperature of each node associated with one surface node, such as $T_{2,3}$ is given by

$$5T_{2,3} = T_{2,4} + T_{3,3} + T_{2,2} + 2T_{1,3}$$

which for the outer wall temperature $T_{2,3}$ of the example simplifies to

$$5T_{2,3} = T_{2,4} + T_{3,3} + T_{2,2} + 2T_o$$

It should be noted that this also applies to the inner wall temperatures, and that because of the nodal system chosen there are no internal corners.

Hence a system of simultaneous equations can be set up and solved, as discussed in Chapter 2. An iterative method is used in this example.

The heat flux from an outer surface node associated with one surface node, such as the node 2,3, is given by

$$q_{2,3} = k\frac{T_{2,3} - T_o}{\Delta x/2}$$

and the rate of heat transfer from the outer surface $\Delta Q_{2,3}$ for such a node is given by

$$\Delta Q_{2,3} = k\frac{T_{2,3} - T_{\text{o}}}{\Delta x/2}\Delta x L_{\text{L}}$$

where L_{L} is the length of the duct. Hence the rate of heat transfer from the outer surface per unit length of the duct associated with node 2,3 is given by

$$\Delta Q_{2,3}/L_{\text{L}} = 2k(T_{2,3} - T_{\text{o}})$$

It can be similarly shown that the rate of heat transfer from an outer surface node associated with two surface nodes, such as the node 2,2, per unit length of the duct is given by

$$\Delta Q_{2,2}/L_{\text{L}} = 4k(T_{2,2} - T_{\text{o}})$$

The rate of conduction heat transfer from the outer surface of the duct per unit length of the duct Q/L_{L} is obtained by summing the contributions of all outer surface nodes. The rate of conduction heat transfer from the inner surface of the duct per unit length of the duct is similarly obtained.

Solution

The solution is given in the workbook Ex04-02-01.xls for one particular geometry of the duct: the outer surface consisting of 9 by 7 cells and the inner surface consisting of 3 by 3 cells. There are four worksheets in this workbook: *Input, Main, Raster*, and *col_code*.

The first worksheet, *Input*, allows for the input of the data, and presents the numerical results for the rate of conduction heat transfer per unit length of the duct. The inner surface temperature T_{I} is entered in **cell B3**, the outer surface temperature T_{o} is entered in **cell B4**, the thermal conductivity k is entered in **cell B5**, and the number of iterations in each run is entered in **cell B9**, with the data already entered used for illustration only. The rate of conduction heat transfer from the outer surface of the duct per unit length of the duct is given in **cell B12** and the rate of conduction heat transfer from the inner surface of the duct per unit length of the duct is given in **cell B13**. It should be noted that the conduction heat transfer rates in **cell B12** and **cell B13** are only valid for the present duct geometry. If the geometry is different, the appropriate heat transfer rates in the worksheet *col_code* must be derived, using the approach discussed above, and the reader is encouraged to do that. Other output parameters are discussed later.

The second worksheet, *Main*, presents the diagram of the cross-section of the duct and gives all the calculated temperatures. It should be noted that the structure and the colours of the diagram are important. The diagram must start in the **cell A1**. The blue field, which must be one cell thick, and must completely surround and be immediately adjacent to the yellow field, gives the outer surface temperature T_{o}. The yellow field presents the calculated nodal temperatures in the duct. The brown field, which must be rectangular and not immediately adjacent to the blue field, gives the inner surface temperature T_I. It should be noted that Excel allows the names of colours to be read from the standard palette provided within the software. This facility is particularly helpful when the colours are difficult to determine from the diagrams.

The third worksheet, *Raster*, presents the raster plot, and the last worksheet, *col_code*, gives the colour code used for the raster plot, and is used for some intermediate calculations, such as the rates of conduction heat transfer.

The workbook also contains three VBA programs: *Initial, Iterate*, and *Rasterplot*. The first program, *Initial*, which must be run first, recognises the colours in the worksheet *Main* and determines the size of the duct matrix in **cell B7** and **cell B8** in worksheet *Input* (it should be noted that the duct matrix must be greater than 3 by 3 and smaller than 98 by 98). It also clears the *Raster* area and sets the initial number of iterations to zero (**cell B10** in *Input*). The second program, *Iterate*, calculates iteratively the nodal temperatures in the duct for the number of iterations in each calculation cycle set in **cell B9** in *Input*. The final program, *Rasterplot*, plots the raster plot in the worksheet *Raster*, using the same structure as the structure used in the worksheet *Main*.

The solution is obtained in the following steps: the program is launched with **Macros** enabled. The *Initial* **Macro** is run first, and the input data then entered in the worksheet *Input*. The *Iterate* **Macro** is run next and the raster plot is then obtained by running the *Rasterplot* **Macro**.

Typical results for $T_O = 0°C$ and $T_I = 200°C$ are shown in the top half of Fig. 4.2.5 and Colour Plate III.

Discussion

The solutions can be used to experiment with the various design and numerical calculation parameters and observe their influence.

1. For steady-state conditions the total rate of conduction heat transfer from the duct must be zero. The total rate of conduction heat transfer from the duct per unit length of the duct is given in **cell B14** of the worksheet *Input*. It is also given as a percentage error in **cell B15** of the worksheet *Input*. The percentage error is based on the average conduction heat transfer to or from the duct per unit length of the duct. As discussed above, the total rates of conduction heat transfer from the duct are only valid for the present duct geometry, and cannot be used when a different duct geometry is considered.
2. The percentage error can be used to determine if the number of iterations is sufficient. If the number of iterations appear insufficient the *Iterate* **Macro** can be run again with the same number of iterations in the run or the number of iterations can be increased.

Other conditions may be observed and the reader is invited to experiment further.

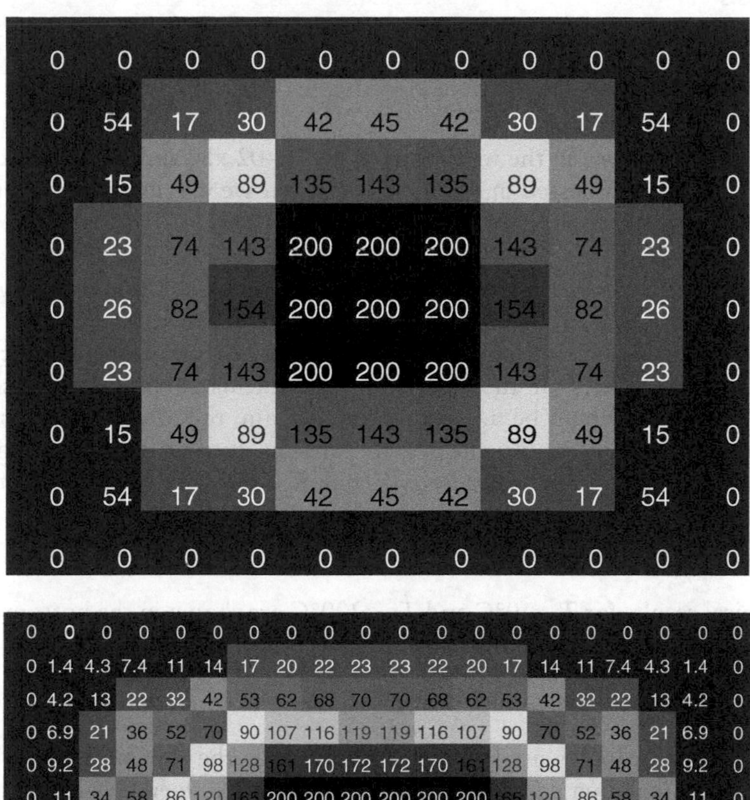

Figure 4.2.5 Steady state, two-dimensional heat conduction analysis for a rectangular duct: Coarse grid (top) and fine grid (bottom). (See Colour Plate III.)

Example 4.2.2 Rectangular duct with a finer grid.

Consider once again the duct discussed in Example 4.2.1 and repeat the calculations with a finer grid.

Analysis

The approach is identical to that discussed in Example 4.2.1.

Solution

The solution is given in the workbook Ex04-02-02.xls, and the approach is identical to that discussed in Example 4.2.1, but we will use twice as many nodes in both directions. The procedure is as follows:

- The diagram of the duct in entered in the worksheet *Main*, ensuring the colour convention. The size of the matrix has increased.
- The rates of conduction heat transfer per unit length of the duct across the two surfaces of the duct are then calculated in the worksheet *col_code*. It should be noted that, once again, the rates of conduction heat transfer are only valid for the present duct geometry, and must be updated in the worksheet *col_code* if a different duct geometry is investigated.
- The solution is then obtained as discussed in Example 4.2.1.

Typical results for $T_O = 0°C$ and $T_I = 200°C$ are shown in the bottom half of Fig. 4.2.5 and Colour Plate III.

Discussion

The solutions can be used to experiment with the various design and numerical calculation parameters and observe their influence.

1 A more detailed and accurate description of the system is obtained with a finer grid.
2 The solutions require much longer computational times when a finer grid is used. The reasons are twofold. First, the number of calculations in each cycle is considerably greater for a finer grid and second, the solution converges more slowly.
3 The size of the grid is determined by compromise between the minimum required accuracy and the maximum practicable computational times.
4 The rate of conduction heat transfer from the outer surface of the duct per unit length of the duct is, for a given shape independent of the size of the duct. Hence, for geometrically similar ducts, the rate of conduction heat transfer from the outer surface of the duct per unit length of the duct depends only on the thermal conductivity of the duct material and the inner and outer surface temperatures.
5 A duct of any size can be modelled with a $\Delta x = \Delta y$ grid, provided that all dimensions are multiples of the grid size Δx.

Other observations can be made and the reader is invited to experiment further.

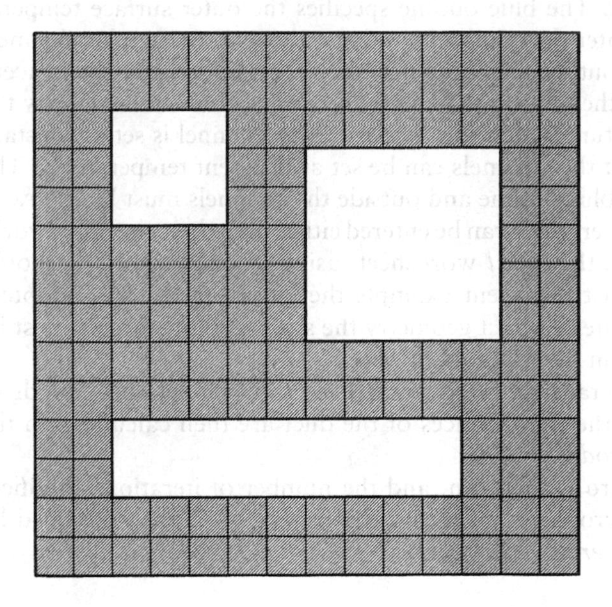

Figure 4.2.6 General multiple channel rectangular duct system.

Example 4.2.3 General multiple channel rectangular duct system.

Consider a general multiple channel rectangular duct system, as shown in Fig. 4.2.6. The system has a constant thermal conductivity k and its outer surface is at constant temperature T_{O}. The temperature of the inner surface of the first duct is T_{I1}, the inner surface of the second duct is T_{I2}, and of the Nth duct is T_{IN}. Furthermore, there is no heat generation anywhere in the duct. Assume that steady-state conditions have been reached, and that the problem can be approximated by two-dimensional conduction. Determine the temperature distribution in the duct and the rate of conduction heat transfer through the duct.

Analysis

The approach is identical to that discussed in Example 4.2.1.

Solution

The solution is given in the workbook Ex04-02-03.xls, and the approach is identical to that discussed in Example 4.2.1. The procedure is as follows:

1 The diagram of the duct in entered in the worksheet *Main*, ensuring the required colour convention. We start with the blue outline of the outer

wall of the duct. The blue outline specifies the outer surface temperature. We then enter the required number of brown rectangular channels within the blue outline, ensuring that they are not immediately adjacent either to each other or to the blue outline. The brown areas specify the surface temperature within the channel. Each channel is set at constant temperature, but the channels can be set at different temperatures. The area within the blue outline and outside the channels must be yellow.

2 The surface temperatures can be entered either directly in the *Main* worksheet or through the *Input* worksheet, using a suitable link. (It should be noted that in the present example the latter approach is adopted, but that for a different duct geometry the surface temperatures must be entered directly in the *Main* worksheet.)

3 If required, the rates of conduction heat transfer per unit length of the duct across the two surfaces of the duct are then calculated in the worksheet *col_code*.

4 The *Initial* **Macro** is then run, and the number of iterations specified. The *Iterate* **Macro** is run next and the raster plot is then obtained by running the *Rasterplot* **Macro**.

Discussion

The solutions can be used to experiment with the various design and numerical calculation parameters and observe their influence.

1 Using the comparison of the rate of conduction heat transfer we can observe that the system converges rapidly with acceptable results being obtained after about 20–30 iterations.

2 The influence of the position of the brown channels can be easily investigated by moving them within the blue envelope, ensuring, as pointed out above, that they are not immediately adjacent either to each other or to the blue outline. Similarly, the influence of the duct geometry can also be investigated. The temperature distributions will be calculated correctly. However, because the duct geometry has changed the rates of conduction heat transfer from the outer and inner surfaces of the duct per unit length of the duct are not calculated correctly. These calculations can either be revised, or, since we know that acceptable convergence is obtained after about 20–30 iterations, we disregard the calculations of the heat transfer rates.

Other observations can be made and the reader is invited to experiment further. It is suggested that an investigation of non-rectangular channels, such as L-shaped channels, be undertaken to determine whether or not the present model can be used also in these applications. Care should be taken when conduction heat fluxes from the inner surfaces are calculated, since some of the inner surface nodes will be associated with two surface nodes (as discussed in Example 4.2.1).

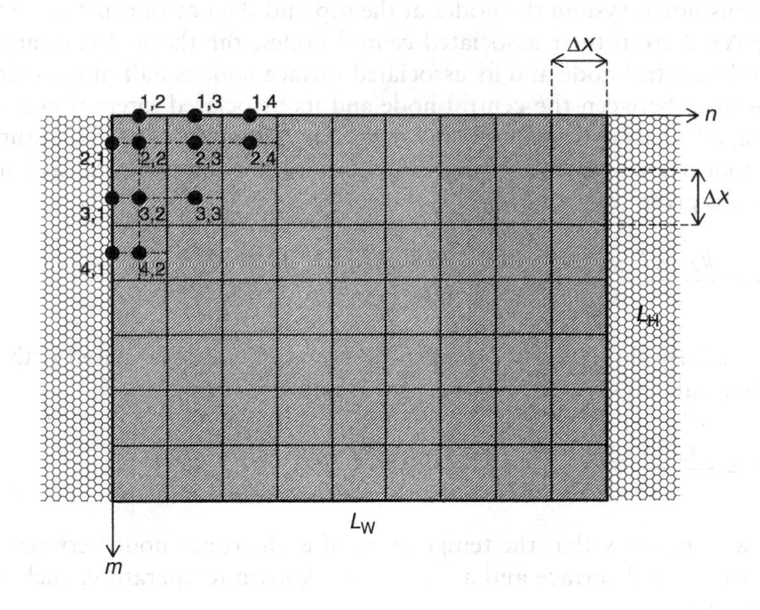

Figure 4.2.7 Rectangular solid element with variable conductivity.

Example 4.2.4 Rectangular solid element with variable conductivity.

Consider a solid element with a rectangular cross-section shown in Fig. 4.2.7. The thermal conductivity within the cross-section is variable, but only a function of the x and y coordinates. The thermal conductivity does not depend on the z coordinate. There is no heat generation anywhere in the element. The temperature of the top surface, T_T, and the temperature of the bottom surfaces, T_B, which are not necessarily constant, are known. The two side surfaces are thermally insulated. Assume that steady-state conditions have been reached, and that the problem can be approximated by two-dimensional conduction. Determine the temperature distribution in the element and the rate of conduction heat transfer across the element.

Analysis

A nodal system with $\Delta x = \Delta y$ and shown in Fig. 4.2.7 is chosen. For all internal nodes, Eqs (4.2.2)–(4.2.6) can be simplified as

$$T_{m,n} = \frac{k_{m-0.5,n}T_{m-1,n} + k_{m+0.5,n}T_{m+1,n} + k_{m,n-0.5}T_{m,n-1} + k_{m,n+0.5}T_{m,n+1}}{k_{m-0.5,n} + k_{m+0.5,n} + k_{m,n-0.5} + k_{m,n+0.5}}$$

where, for example, the thermal conductivity $k_{m-0.5,n}$ is the average of the thermal conductivities of the corresponding adjacent nodes

$$k_{m-0.5,n} = \frac{k_{m-1,n} + k_{m,n}}{2}$$

Since in this nodal system the nodes at the top and the bottom surfaces are distance $\Delta x/2$ from their associated central nodes, the thermal resistance between the central node and its associated surface node is half of the thermal resistance between the central node and its associated internal nodes. Following an analysis similar to above we can show that the temperature of each node associated with a surface node at a given temperature, such as node $T_{2,3}$, is given by

$$T_{2,3} = \frac{k_{2,2.5}T_{2,2} + k_{2.5,3}T_{3,3} + k_{2,3.5}T_{2,4} + 2k_{2,3}T_{1,3}}{k_{2,2.5} + k_{2.5,3} + k_{2,3.5} + 2k_{2,3}}$$

We can similarly show that the temperature of each node associated with a thermally insulated surface node, such as $T_{3,2}$, is given by

$$T_{3,2} = \frac{k_{3,2.5}T_{3,3} + k_{2.5,2}T_{2,2} + k_{3.5,2}T_{4,2}}{k_{3,2.5} + k_{2.5,2} + k_{3.5,2}}$$

Finally, we can show that the temperature of each corner node between a thermally insulated surface and a surface at a known temperature, such as $T_{2,2}$, is given by

$$T_{2,2} = \frac{k_{2,2.5}T_{2,3} + 2k_{2,2}T_{1,2} + k_{2.5,2}T_{3,2}}{k_{2,2.5} + 2k_{2,2} + k_{2.5,2}}$$

Hence a system of simultaneous equations can be set up and solved, as discussed in Chapter 2. An iterative method is used in this example.

Similarly to Example 4.2.1, the rate of conduction heat transfer from the top surface of the element per unit length of the element associated with the node 2,3 is given by

$$\Delta Q_{2,3}/L_{\mathrm{L}} = 2k_{2,3}(T_{2,3} - T_{1,3})$$

where L_{L} is the length of the element. The rate of conduction heat transfer from the top surface of the element per unit length of the element is obtained by summing the contributions of all nodes on the top surface. The rate of conduction heat transfer from the bottom surface of the element per unit length of the element is similarly obtained.

Solution

The solution is given in the workbook Ex04-02-04.xls for one particular geometry of the element: the height of the element consisting of 7 cells and the width of the element consisting of 9 cells. There are five worksheets in this workbook: *Input, k-val, Main, Raster*, and *col_code*.

The first worksheet, *Input*, allows for the input of some of the data, and presents the numerical results for the rate of conduction heat transfer per unit length of the element. The width of the element is entered in **cell B3** and the number of iterations in each run is entered in **cell B9**, with the data already entered used for illustration only. Other input parameters, which must be entered in the worksheet *Main*, are discussed below. The rate of conduction heat transfer from the top surface of the element per unit length of the element is given in **cell B12** and the rate of conduction heat transfer from the bottom surface of the element per unit length of the element is given in **cell B13**. Other output parameters are discussed below.

The second worksheet, *Main*, presents the diagram of the cross section of the solid element and gives all the calculated temperatures. It should be noted that the structure of the diagram is important (the colour is not used in this example). The diagram must start the **cell A2**. The number of columns is proportional to the width of the element and the number of rows is proportional to the height of the element (it should be noted that the matrix must be greater than 3 by 3 and smaller than 50 by 50). A cross-section of any size can be modelled with a $\Delta x = \Delta y$ grid, provided that all dimensions are multiples of the grid size Δx. Hence we determine the appropriate aspect ratio of the cross-section and the size of the matrix and enter it in the *Main* worksheet (shown yellow and outlined). Since we also specified the width of the element in the *Input* worksheet, the height of the element is automatically calculated and is given in **cell B4** of the *Input* worksheet. The required temperatures of the top surface are then entered in the cells immediately above the diagram of the section (shown in green, and for the present example in **cells (A1:I1)** in the *Main* worksheet). The required temperatures of the bottom surface are then entered in the cells immediately below the diagram of the section (shown in green, and for the present example in **cells (A9:I9)** in the *Main* worksheet). All temperatures for the cross-section of the material are initially set to zero.

The third worksheet, *k-val*, is used to enter the required value of the thermal conductivity of the material. The diagram of the cross-section of the solid element from the *Main* worksheet is copied to the *k-val* worksheet, starting, once again, in **cell A2**. Hence for the present example the diagram occupies **cells (A2:I8)**. The position of the diagram is important because the values of the thermal conductivity appropriate to the corresponding cells in the *Main* worksheet are read from this diagram, and because the diagram is also used to determine the total number of cells in the solid element. The number of rows and the number of columns are then given in **cell B7** and **cell B8** of the *Input* worksheet. As discussed above, this information is then used to calculate the height of the element in the *Main* worksheet.

The fourth worksheet, *Raster*, presents the raster plot, and the last worksheet, *col_code*, gives the colour code used for the raster plot, and is used for some intermediate calculations, such as the rates of conduction heat transfer from the top and the bottom surfaces.

The workbook also contains three VBA programs: *Initial*, *Iterate*, and *Rasterplot*. The first program, *Initial*, clears the *Raster* area and sets the initial number of iterations to zero (**cell B10** in *Input*). The second programme, *Iterate*, calculates iteratively the nodal temperatures in the solid element for the number of iterations in each calculation cycle set in **cell B9** in *Input*. The final program, *Rasterplot*, plots the raster plot in the worksheet *Raster*, using the same structure as the structure used in the worksheet *Main*.

The solution is obtained in the following steps: the program is launched with **Macros** enabled. The *Initial* **Macro** is run first, and the input data then entered in the worksheet *Input*. The *Iterate* **Macro** is run next and the raster plot is then obtained by running the *Rasterplot* **Macro**.

Discussion

The solutions can be used to experiment with the various design and numerical calculation parameters and observe their influence.

1 For steady-state conditions the total rate of conduction heat transfer from the solid element must be zero. The total rate of conduction heat transfer from the solid element per unit length of the element is given in **cell B14** of the worksheet *Input*. It is also given as a percentage error in **cell B15** of the worksheet *Input*. The percentage error is based on the average conduction heat transfer to and from the duct per unit length of the duct.
2 The percentage error can be used to determine if the number of iterations is sufficient. If the number of iterations appear insufficient the *Iterate* **Macro** can be run again with the same number of iterations in the run or the number of iterations can be increased.
3 As further experience is obtained, a reasonable estimate of the temperatures in the solid cross-section can be made and entered (rather than setting these temperatures at zero). This will allow a more rapid convergence and thus require fewer iterations.
4 The adiabatic condition on the side surfaces can be used to model elements with planes of symmetry, since the heat fluxes are zero over the planes of symmetry.
5 Complex composite walls, such as those in Example 3.5.2, can be investigated by appropriately sub-dividing the cross-section into rectangular regions with different but constant thermal conductivities.

Other conditions may be observed and the reader is invited to experiment further.

Example 4.2.5 Rectangular solid element with heat generation.

Consider a solid element with a rectangular cross-section shown in Fig. 4.2.8. The rate of heat generation within the cross-section is variable, but only a function of the x and y coordinates. The rate of heat generation does not

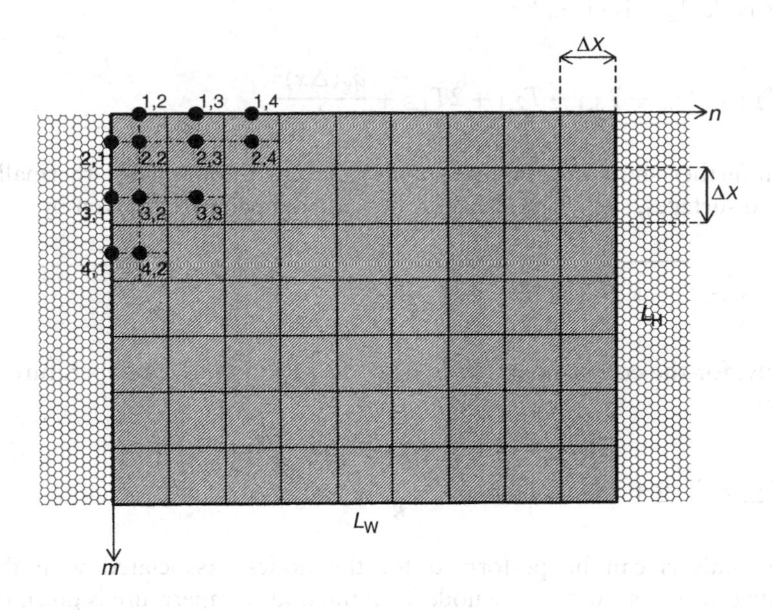

Figure 4.2.8 Rectangular solid element with heat generation.

depend on the z coordinate. The thermal conductivity is constant. The top and bottom surfaces are exposed to ambient fluids at respective temperature T_{T0} and T_{B0} with a uniform and constant convection heat transfer coefficient h. The two side surfaces are thermally insulated. Assume that steady-state conditions have been reached, and that the problem can be approximated by two-dimensional conduction. Determine the temperature distribution in the element and the rate of convection heat transfer from the element.

Analysis

A nodal system with $\Delta x = \Delta y$ and shown in Fig. 4.2.8 is chosen. For all internal nodes, Eqs (4.2.2)–(4.2.6) can be simplified as

$$4T_{m,n} = T_{m+1,n} + T_{m-1,n} + T_{m,n+1} + T_{m,n-1} + \frac{q_g(\Delta x)^2}{k}$$

Since in this nodal system the nodes at the top and the bottom surfaces are distance $\Delta x/2$ from their associated central nodes, the thermal resistance between the central node and its associated surface node is half of the thermal resistance between the central node and its associated internal nodes. Following an analysis similar to above we can show that the temperature of each node associated with a surface node at a given temperature, such as node $T_{2,3}$, is given by

$$5T_{2,3} = T_{2,2} + T_{3,3} + T_{2,4} + 2T_{1,3} + \frac{q_g(\Delta x)^2}{k}$$

We can further show that for each node adjacent to one of the thermally insulated surfaces, such as node $T_{3,2}$, the node temperature is given by

$$3T_{3,2} = T_{4,2} + T_{3,3} + T_{2,2} + \frac{q_g(\Delta x)^2}{k}$$

Similarly, for the corner nodes, such as node $T_{2,2}$, the node temperature is given by

$$4T_{2,2} = T_{3,2} + T_{2,3} + 2T_{1,2} + \frac{q_g(\Delta x)^2}{k}$$

Similar analysis can be performed for the nodes, associated with the convective surfaces, such as the node $T_{1,3}$; the node temperature is given by

$$(2 + \mathrm{Bi})T_{1,3} = 2T_{2,3} + \mathrm{Bi}T_{\mathrm{TO}}$$

where Bi is the Biot number, given by Eq. (4.2.11).

Hence a system of simultaneous equations can be set up and solved, as discussed in Chapter 2. An iterative method is used in this example.

The convective heat flux from the top surface node associated with one surface node, such as the node 2,3, is given by

$$q_{2,3} = h(T_{1,3} - T_{\mathrm{TO}})$$

and the rate of convective heat transfer $\Delta Q_{2,3}$ for such a node is given by

$$\Delta Q_{2,3} = h(T_{1,3} - T_{\mathrm{TO}})\Delta x L_{\mathrm{L}}$$

where L_{L} is the length of the element. Hence the rate of convective heat transfer from the top surface per unit length of the element associated with node 2,3 is given by

$$\Delta Q_{2,3}/L_{\mathrm{L}} = h(T_{1,3} - T_{\mathrm{TO}})\Delta x$$

The rate of convection heat transfer from the top surface of the element per unit length of the element Q/L_{L} is obtained by summing the contributions of all top surface nodes. The rate of convection heat transfer from the bottom surface of the element per unit length of the element is similarly obtained.

Finally, the rate of heat generated in each node ΔQ_g, assuming that each node is L_L long, is given by

$$\Delta Q_g = q_g \Delta x \Delta y L_L$$

and the rate of heat generated per unit length of the element is then obtained as

$$\Delta Q_g / L_L = q_g \Delta x \Delta y$$

The total rate of heat generated is then obtained by summing the contributions of all nodes.

Solution

The solution is given in the workbook Ex04-02-05.xls for one particular geometry of the element: the height of the element consisting of 7 cells and the width of the element consisting of 11 cells. There are five worksheets in this workbook: *Input, Main, Q-gen, Raster*, and *col_code*.

The first worksheet, *Input*, allows for the input of some of the data, and presents the numerical results for the heat balance per unit length of the element. The width of the element is entered in **cell B3**, the thermal conductivity of the element in **cell B7**, the coefficient of convection heat transfer in **cell B8** and the number of iterations in each run is entered in **cell B12**, with the data already entered used for illustration only. Other input parameters, which must be entered in the worksheets *Main* and *Q-gen*, are discussed below. The rate of convection heat transfer from the top surface of the element per unit length of the element is given in **cell B15** and the rate of convection heat transfer from the bottom surface of the element per unit length of the element is given in **cell B16**. The rate of convection heat transfer from both surfaces of the element per unit length of the element is given in **cell B17**. The rate of heat generated in the element per unit length of the element is given in **cell B18**. Other output parameters are discussed below.

The second worksheet, *Main*, presents the diagram of the cross-section of the solid element and gives all the calculated temperatures. It should be noted that the structure and the colours of the diagram are important, since this information is used, as discussed below, to determine the number of cells in the element. The diagram of the cross-section must start the **cell B3**. The number of columns is proportional to the width of the element and the number of rows is proportional to the height of the element (it should be noted that the matrix of the cross-section must be greater than 3 by 3 and smaller than 50 by 50). A cross-section of any size can be modelled with a $\Delta x = \Delta y$ grid, provided that all dimensions are multiples of the grid size Δx. Hence we determine the appropriate aspect ratio of the cross-section and the size of the matrix and enter it in yellow within the *Main* worksheet. Since we also specified the width of the element in the *Input* worksheet, the grid size and height of the element are automatically calculated and are given respectively in **cell B5** and **cell B4** of the *Input* worksheet.

The required temperatures of the ambient fluid above the top surface are then entered in the green cells above the diagram of the section (shown for the present example in **cells (B1:L1)** in the *Main* worksheet). The required temperatures of the ambient fluid below the bottom surface are entered in the green cells below the diagram of the section (shown for the present example in **cells (B11:L11)** in the *Main* worksheet). The calculated temperatures of the top surface will be given in the blue cells above the diagram of the cross-section (shown for the present example in **cells (B2:L2)** in the *Main* worksheet). Similarly, the calculated temperatures of the bottom surface will be given in the blue cells below the diagram of the cross-section (shown for the present example in **cells (B10:L10)** in the *Main* worksheet). Finally, the thermal insulation of the two side surfaces is indicated in black (for the present example, **cells (A1:A11)** and **cells (M1:M11)** in the *Main* worksheet respectively). All temperatures for the cross-section of the material and the top and the bottom surfaces are initially set to zero.

The third worksheet, *Q-gen*, is used to enter the required value of the rate of heat generation per unit volume of the material in MW/m^3. The diagram of the solid cross-section from the *Main* worksheet is copied to the *Q-gen* worksheet, starting, once again, in **cell B3** (however, if the whole diagram is copied it will start in **cell A1**). The values of the rate of heat generation per unit volume of the material are then entered as required in the yellow cells (**cells (B3:L9)**). The position of the diagram is important because the values of the rate of heat generation per unit volume of the material appropriate to the corresponding cells in the *Main* worksheet are read from this diagram.

The fourth worksheet, *Raster*, presents the raster plot, and the last worksheet, *col_code*, gives the colour code used for the raster plot, and is used for some intermediate calculations, such as the Biot number.

The workbook also contains three VBA programs: *Initial, Iterate*, and *Rasterplot*. The first program, *Initial*, determines the size of the matrix from the worksheet *Main* by calculating the number of rows and the number of columns in the cross-section of the element (**cell B10** and **cell B11** of the *Input* worksheet respectively), clears the *Raster* area and sets the initial number of iterations to zero (**cell B13** in the *Input* worksheet). The second program, *Iterate*, calculates iteratively the nodal temperatures in the solid element for the number of iterations in each calculation cycle set in **cell B12** in the *Input* worksheet. The program also calculates the rate of convection heat transfer from the top surface of the element per unit length of the element, the rate of convection heat transfer from the bottom surface of the element per unit length of the element and the rate of heat generated in the element per unit length of the element. The final program, *Rasterplot*, plots the raster plot in the worksheet *Raster*, using the same structure as the structure used in the worksheet *Main*.

The solution is obtained in the following steps: the program is launched with **Macros** enabled. The *Initial* **Macro** is run first, and the input data entered in the worksheet *Input*. The *Iterate* **Macro** is run next and the raster plot is then obtained by running the *Rasterplot* **Macro**.

Discussion

The solutions can be used to experiment with the various design and numerical calculation parameters and observe their influence.

1 For steady-state conditions the total rate of convection heat transfer from the solid element per unit length of the element is equal to the rate of heat generated in the element per unit length of the element. This information is presented in the *Input* worksheet in **cell B17** and **cell B18** respectively. It also gives a percentage error in **cell B19** of the *Input* worksheet. The percentage error is based on the average of the total rate of convection heat transfer from the solid element per unit length of the element and the rate of heat generated in the element per unit length of the element.
2 The percentage error can be used to determine if the number of iterations is sufficient. If the number of iterations appear insufficient the *Iterate* **Macro** can be run again with the same number of iterations in the run or the number of iterations can be increased.
3 A large number of iterations may be required for the solution to converge. As experience is obtained, a reasonable estimate of the temperatures in the solid cross-section can be made and entered (rather than setting these temperatures at zero). This will allow a more rapid convergence and thus require fewer iterations.
4 The adiabatic condition on the side surfaces can be used to model elements with planes of symmetry, since the heat fluxes are zero over the planes of symmetry.

Other conditions may be observed and the reader is invited to experiment further.

4.3 Finite difference approximation in two-dimensional systems: cylindrical system

4.3.1 Internal nodes

Consider a two-dimensional cylindrical system, as shown in Fig. 4.3.1. Similarly to Eq. (4.2.2) the energy balance for any control volume with dimensions $r\Delta\theta$ and Δr can be written as

$$q_r(r - 0.5\Delta r)\Delta\theta - q_{r+\Delta r}(r + 0.5\Delta r)\Delta\theta + q_\theta \Delta r - q_{\theta+\Delta\theta}\Delta r + q_g r\Delta\theta\Delta r = 0 \tag{4.3.1}$$

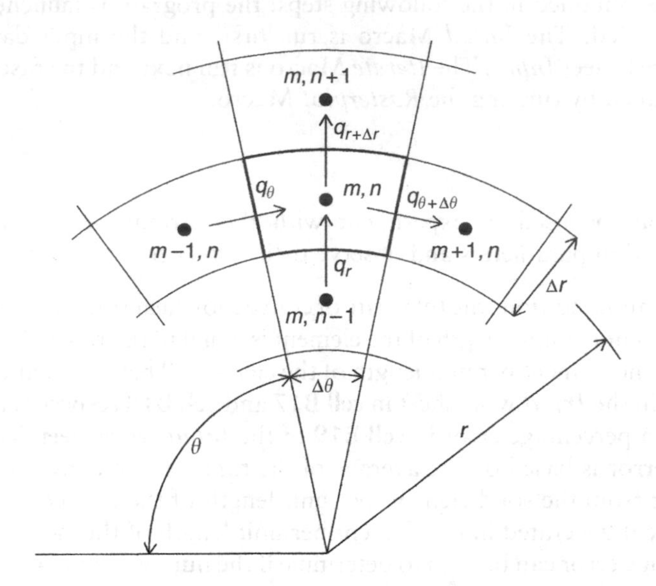

Figure 4.3.1 Internal node in a two-dimensional cylindrical coordinate system.

Similarly to the treatment for Cartesian coordinates, for any sufficiently small interior node the heat fluxes are given as

$$q_r = k_{m,n-0.5}\frac{T_{m,n-1} - T_{m,n}}{\Delta r} \tag{4.3.2}$$

$$q_{r+\Delta r} = k_{m,n+0.5}\frac{T_{m,n} - T_{m,n+1}}{\Delta r} \tag{4.3.3}$$

$$q_\theta = k_{m-0.5,n}\frac{T_{m-1,n} - T_{m,n}}{r\Delta\theta} \tag{4.3.4}$$

$$q_{\theta+\Delta\theta} = k_{m+0.5,n}\frac{T_{m,n} - T_{m+1,n}}{r\Delta\theta} \tag{4.3.5}$$

Finally, if the thermal conductivity is constant and equal to k, substituting Eqs (4.3.2)–(4.3.5) into Eq. (4.3.1) yields

$$T_{m+1,n}\frac{\Delta r}{r\Delta\theta} + T_{m-1,n}\frac{\Delta r}{r\Delta\theta} + T_{m,n+1}\frac{(r+0.5\Delta r)\Delta\theta}{\Delta r} + T_{m,n-1}\frac{(r-0.5\Delta r)\Delta\theta}{\Delta r} - 2T_{m,n}\left(\frac{r\Delta\theta}{\Delta r} + \frac{\Delta r}{r\Delta\theta}\right) + \frac{q_g r\Delta\theta\,\Delta r}{k} = 0 \tag{4.3.6}$$

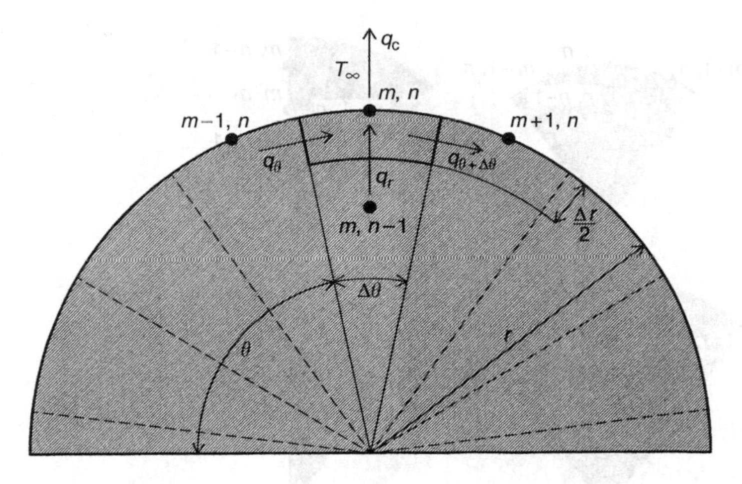

Figure 4.3.2 External cylindrical wall node in a two-dimensional cylindrical coordinate system.

4.3.2 *External nodes with various boundary conditions*

Case 1: Cylindrical wall with a convective boundary condition.

Referring to Fig. 4.3.2 and assuming constant thermal conductivity, the energy balance for the control volume shown is given by

$$q_r\left(r-\frac{\Delta r}{2}\right)\Delta\theta - q_c r\Delta\theta + q_\theta\frac{\Delta r}{2} - q_{\theta+\Delta\theta}\frac{\Delta r}{2} + q_g\left(r-\frac{\Delta r}{4}\right)\Delta\theta\frac{\Delta r}{2} = 0 \tag{4.3.7}$$

Similarly to the approach above, the finite difference equation can be written as

$$\begin{aligned}
&\frac{\Delta r}{r\Delta\theta}T_{m+1,n} + \frac{\Delta r}{r\Delta\theta}T_{m-1,n} + \frac{2\,(r-0.5\Delta r)\,\Delta\theta}{\Delta r}T_{m,n-1} \\
&\quad - 2\left(\frac{\Delta r}{r\Delta\theta} + \frac{(r-0.5\Delta r)\,\Delta\theta}{\Delta r} + \frac{hr\Delta\theta}{k}\right)T_{m,n} + \frac{2hr\Delta\theta T_\infty}{k} \\
&\quad + \frac{q_g}{k}\left(r-\frac{\Delta r}{4}\right)\Delta\theta\Delta r = 0
\end{aligned} \tag{4.3.8}$$

Similar equations can be set up for a cylindrical wall with a prescribed heat flux. As shown in Fig. 4.3.3 other external nodes can be identified, and appropriate finite difference equations then developed.

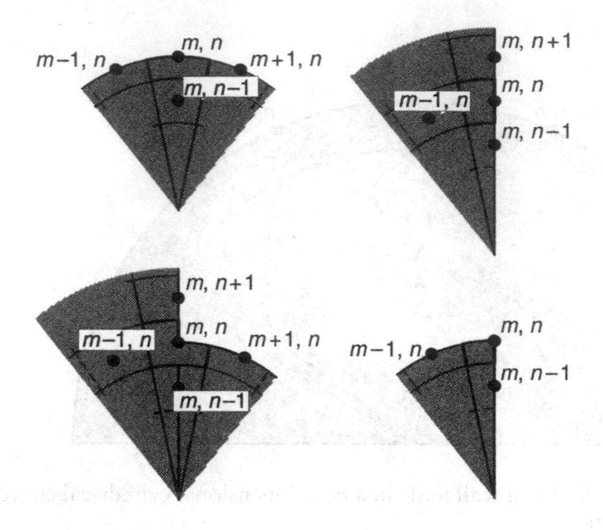

Figure 4.3.3 External nodes in a two-dimensional cylindrical coordinate system.

4.3.3 Central node

The central node of a cylindrical system is a source of singularity and cannot be described as any other internal node.

Consider the central node in two-dimensional cylindrical system, as shown in Fig. 4.3.4. The energy balance for a control volume of radius $\Delta r/2$ and unit thickness can be written as

$$-(q_{r1,2} + q_{r2,2} + \cdots + q_{rn,2})\frac{\Delta r}{2}\Delta\theta + q_g\pi\left(\frac{\Delta r}{2}\right)^2 = 0 \quad (4.3.9)$$

where n is the number of identical segments adjacent to the control volume, q_g is the rate of heat generated per unit volume in the control volume and, for example, $q_{r1,2}$ is the radial heat flux across the boundary of the control volume to the volume of node 1, 2 given as

$$q_{r1,2} = k\frac{T_{1,1} - T_{1,2}}{\Delta r} \quad (4.3.10)$$

Assuming constant thermophysical properties and constant rate of heat generated per unit volume and equal to q_g, Eq. (4.3.9) with Eq. (4.3.10) simplify to

$$-\frac{k}{\Delta r}[(T_{1,1}-T_{1,2}) + (T_{1,1}-T_{2,2}) + \cdots + (T_{1,1}-T_{n,2})] + nq_g\frac{1}{2}\left(\frac{\Delta r}{2}\right) = 0 \quad (4.3.11)$$

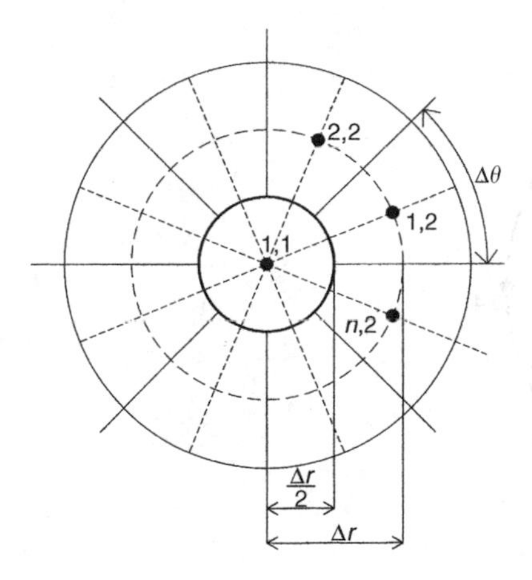

Figure 4.3.4 Central node in a two-dimensional cylindrical coordinate system.

Simplifying the above we finally obtain

$$\frac{1}{n}(T_{1,2} + T_{2,2} + T_{3,2} + \cdots + T_{n,2}) - T_{1,1} + \frac{q_g(\Delta r)^2}{4k} = 0 \qquad (4.3.12)$$

Example 4.3.1 Solid cylinder with heat generation.

Consider a solid element with a circular cross-section of diameter D shown in Fig. 4.3.5. The rate of heat generation per unit volume q_g within the cross-section and the thermal conductivity of the material k are constant. The outside surface of the cylinder is exposed to an ambient fluid at temperature T_∞. The convection heat transfer coefficient h is a function of the longitude θ, but independent of the axial coordinate z. Assume that steady-state conditions have been reached, and that the problem can be approximated by two-dimensional conduction. Determine the temperature distribution in the solid element, using a suitable grid system.

Analysis

A nodal system shown in Fig. 4.3.5 is chosen. In this example

$$\Delta r = \frac{D}{6}$$

$$\Delta\theta = \frac{\pi}{4}$$

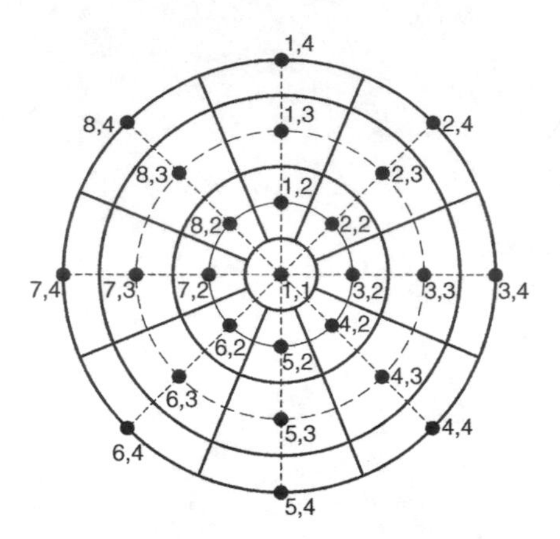

Figure 4.3.5 Solid element with cylindrical cross-section.

Re-arranging Eq. (4.3.12), the temperature of the central node $T_{1,1}$ can be written as

$$T_{1,1} - \frac{1}{8}T_{1,2} - \frac{1}{8}T_{2,2} - \frac{1}{8}T_{3,2} - \frac{1}{8}T_{4,2} - \frac{1}{8}T_{5,2}$$
$$- \frac{1}{8}T_{6,2} - \frac{1}{8}T_{7,2} - \frac{1}{8}T_{8,2} = \frac{q_g(\Delta r)^2}{4k}$$

Similarly, re-arranging Eq. (4.3.6) and noting that for the first layer of internal nodes ($n = 2$) $r = \Delta r$, the temperatures $T_{m,2}$ of the eight nodes can be written as

$$2\left(1 + \frac{1}{(\Delta\theta)^2}\right)T_{m,2} - \frac{1}{(\Delta\theta)^2}T_{m+1,2}$$
$$- \frac{1}{(\Delta\theta)^2}T_{m-1,2} - 1.5T_{m,3} - 0.5T_{1,1} = \frac{q_g(\Delta r)^2}{k}$$

Furthermore, noting that for the second layer of internal nodes ($n = 3$) $r = 2\Delta r$, the temperatures $T_{m,3}$ of the eight nodes can be written as

$$2\left(1 + \frac{1}{4(\Delta\theta)^2}\right)T_{m,3} - \frac{1}{4(\Delta\theta)^2}T_{m+1,3}$$
$$- \frac{1}{4(\Delta\theta)^2}T_{m-1,3} - 1.25T_{m,4} - 0.75T_{m,2} = \frac{q_g(\Delta r)^2}{k}$$

Finally, re-arranging Eq. (4.3.8) and noting that for the surface nodes ($n = 4$) $r = 3\Delta r$, the temperatures $T_{m,4}$ of the eight nodes can be written as

$$\left(\frac{2}{8.25(\Delta\theta)^2} + \frac{5}{2.75} + \frac{6}{2.75}\frac{h_m\Delta r}{k}\right)T_{m,4}$$
$$-\frac{1}{8.25}\frac{1}{(\Delta\theta)^2}T_{m+1,4} - \frac{1}{8.25}\frac{1}{(\Delta\theta)^2}T_{m-1,4} - \frac{5}{2.75}T_{m,3}$$
$$= \frac{6}{2.75}\frac{h_m\Delta r T_\infty}{k} + \frac{q_g}{k}\Delta r^2$$

where h_m is the convective heat transfer coefficient associated with the node $(m, 4)$.

Hence a system of twenty-five simultaneous liner equations can be set up and solved, as discussed in Chapter 2. The matrix inversion method is used in this example.

Solution

The solution is given in the workbook Ex04-03-01.xls. There are three worksheets in this workbook: *Input data and results, Parameters*, and *Calculations*.

The first workbook, *Input data and results*, allows for the input of the data, and presents the numerical results for the temperatures of the 25 nodes. The cylinder diameter D is entered in **cell B2**, the rate of heat generation per unit volume q_g is entered in **cell B3**, the thermal conductivity of the material k is entered in **cell B4** and the temperature of the ambient fluid T_∞ is entered in **cell B5**. The local values of the convective heat transfer coefficient h_m are entered in the blue cells, starting with h_1 at the top and continuing clockwise with $h_2, h_3, \ldots, h_8$. (It should be noted that the data already entered are for illustrations only.) The calculated nodal temperatures $T_{m,n}$ are given in the yellow cells, as indicated in Fig. 4.3.5, with, for example, $T_{1,4}$ at the top and continuing clockwise with $T_{2,4}, T_{3,4}, \ldots, T_{8,4}$. The presentation of the results is improved by clicking **Full Screen** in **View**.

The second worksheet, *Parameters*, is used to calculate the intermediate parameters, and the final worksheet, *Calculations*, presents the various matrices, calculates the inversion matrix and gives the numerical values of the nodal temperatures $T_{m,n}$ in the yellow cells (**cells (AF3:AF27)**).

Entering new values of the input data automatically re-calculates the nodal temperatures.

Discussion

The solutions can be used to experiment with the various design and numerical calculation parameters and observe their influence.

- One-dimensional steady-state conduction in an axi-symmetric cylinder can be modelled by setting all values of the convective heat transfer coefficient h_m to the same value h.
- Simple heat balance then indicates that

$$\frac{Dq_g}{4h(T_S - T_\infty)} = 1$$

  where T_S is the surface temperature. Hence the constant value of the convective heat transfer coefficient h can be chosen so as to achieve the required values of the surface temperature T_S.
- It can be shown that for one-dimensional steady-state conduction in a long axi-symmetric cylinder with constant thermal conductivity and constant rate of heat generation per unit volume, the internal temperatures T are given by

$$T - T_S = \frac{q_g}{4k}\left[\left(\frac{D}{2}\right)^2 - r^2\right]$$

  where r is the radial distance from the centre. The theoretical results can be compared with the numerical results and the adequacy of the chosen grid considered.
- Heat balances with the convection heat transfer coefficient h as a function of the longitude θ, can also be calculated and discussed.

Other observations can be made and the reader is invited to experiment further.

4.4 Finite difference approximation in three-dimensional systems

The finite difference equations are only developed for the Cartesian system. However, as pointed out above, this is shown for illustrative purposes. As mentioned in Section 4.1, commercial software is generally available to solve these systems.

Similarly to Eqs (4.2.1) and (4.2.2), energy balance can be written for any control volume with dimensions Δx, Δy, and Δz as

$$(q_x - q_{x+\Delta x})\Delta y \Delta z + (q_y - q_{y+\Delta y})\Delta x \Delta z + (q_z - q_{z+\Delta z})\Delta x \Delta y + q_g \Delta x \Delta y \Delta z = 0 \tag{4.4.1}$$

Using the symbol o for the discretisation along the z-axis, similarly to Eq. (4.2.3), for any sufficiently small interior node the heat flux can be given as

$$q_x = k_{m-0.5,n,o} \frac{T_{m-1,n,o} - T_{m,n,o}}{\Delta x} \tag{4.4.2}$$

and all other heat fluxes can be written using this convention. If the thermal conductivity is constant and if, furthermore, the nodal system is chosen such that $\Delta x = \Delta y = \Delta z$, the final results can be written as

$$\begin{aligned} &T_{m+1,n,o} + T_{m-1,n,o} + T_{m,n+1,o} + T_{m,n-1,o} + T_{m,n,o+1} \\ &\quad + T_{m,n,o-1} - 6T_{m,n,o} + \frac{q_g(\Delta x)^2}{k} = 0 \end{aligned} \tag{4.4.3}$$

Additional equations can be developed as indicated in Section 4.2 for the two-dimensional system.

Problems

Problem 4.1
Derive a finite difference approximation for an internal node in a solid with variable thermal conductivity using a Cartesian coordinate system, and assuming two-dimensional steady-state conduction.

Problem 4.2
Derive a finite difference approximation for an external corner in a solid with variable thermal conductivity using a cylindrical coordinates system, and assuming two-dimensional steady-state conduction.

Problem 4.3
Show that for geometrically similar ducts, the rate of conduction heat transfer from the outer surface of the duct per unit length of the duct depends only on the thermal conductivity of the duct material and the inner and outer surface temperatures.

Problem 4.4
Termodeck is the name of a new technology that uses hollow core concrete floor planks and the thermal mass provided by the concrete structure for passive heating and cooling of buildings (Fig. 4.P.1). The technology, originating in Sweden, gained rapid popularity in Europe and is now being exploited worldwide. The system uses the holes made within concrete flooring planks as ducts through which air may be circulated. Through modification of air temperature and flow rate internal air temperatures may be altered accordingly. Depending on environmental factors, air to ventilate the building is drawn from the most acceptable source, that is, from the outside (for cooling) or from a solar heated atrium (for heating). The concrete floor and ceiling slabs act as energy exchange devices and either add or remove heat from the ventilated air depending on the ambient temperatures and building

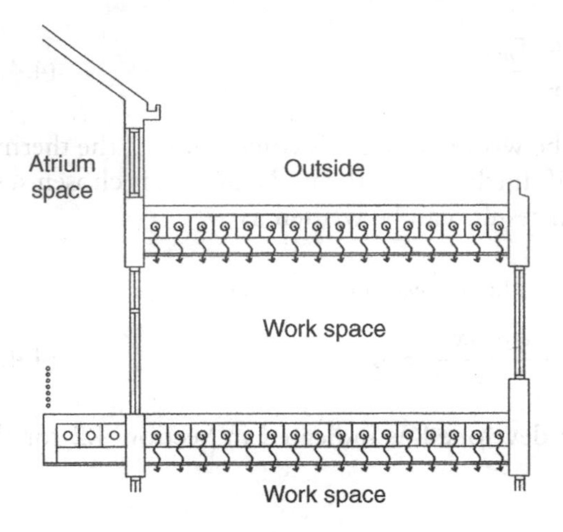

Figure 4.P.1 Cross-section of a building using Termodeck technology.

environmental conditions. A control system operates the HVAC systems. The cross-section of a building that is to be heated by means of Termodeck technology and the detail design of the concrete plank that would form the ceiling and floor are shown in Fig. 4.P.2. A typical section of the plank contains square holes within the concrete planks (each 60×60 mm with a 220 mm pitch). Air emerging from the solar heated atrium at 60°C is fed into the square holes at a flow rate such that, for steady state conditions the internal surface of the Termodeck block is at a uniform temperature of 50°C. The top and the bottom surfaces of the concrete planks are observed to be at 30°C. Assuming that the two side surfaces of the plank are thermally insulated and that steady-state two-dimensional conduction can be used, determine the temperature distribution within the concrete material and the rate of conduction heat transfer per unit length of the concrete plank, using an appropriate grid. Use the following thermal conductivity of concrete: $k = 0.8\,\mathrm{W/m^2\,K}$.

Problem 4.5

A composite wall, shown in Fig. 4.P.3, is made up of 16 mm thick wood panels, 48 mm × 120 mm wooden studs, that are placed at 400 mm pitch, with in an in-fill of wall insulation and 16 mm gypsum board. The outside walls are at 25°C and 5°C, respectively. Assuming steady-state conditions, compute the temperature distribution within the wall and the conduction heat transfer through the wall based on: (i) one-dimensional adiabatic heat flow analysis, (ii) one-dimensional isothermal heat flow analysis, and (iii) two-dimensional heat flow analysis. Use the following thermal conductivities: $k_{\mathrm{wood}} = 0.2\,\mathrm{W/m^2\,K}$, $k_{\mathrm{insulation}} = 0.05\,\mathrm{W/m^2\,K}$, and $k_{\mathrm{gypsum}} = 0.17\,\mathrm{W/m^2\,K}$.

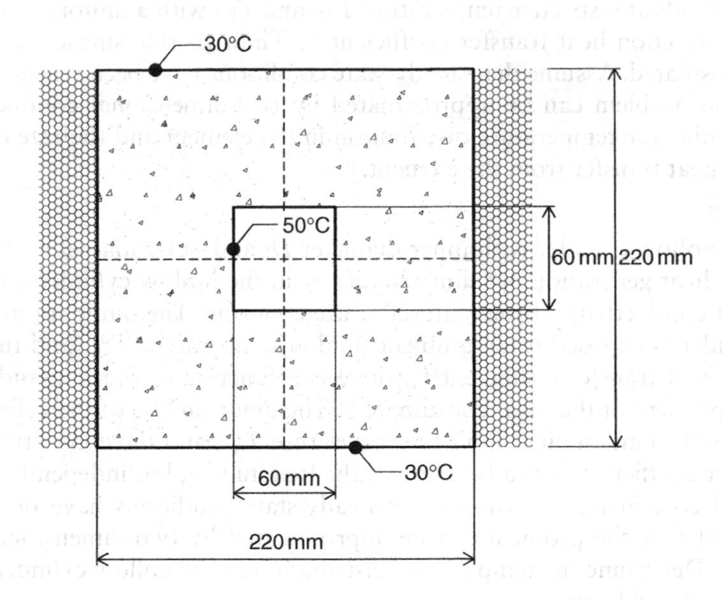

Figure 4.P.2 Design of the concrete plank for Termodeck technology.

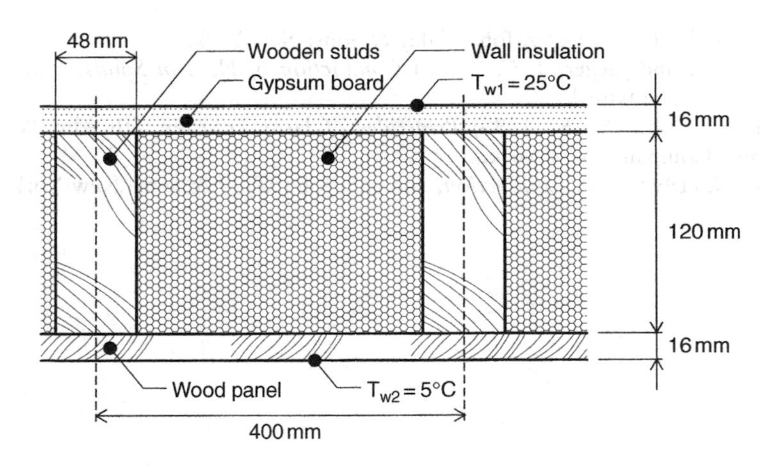

Figure 4.P.3 A composite wall.

Problem 4.6

Consider a solid element with a rectangular cross-section shown in Fig. 4.2.8. The rate of heat generation within the cross-section and the thermal conductivity of the material are variable, but only a function of the x and y coordinates. The rate of heat generation and the thermal conductivity do not depend on the z coordinate. The top and bottom surfaces are exposed to ambient fluids at respective temperature T_{T0} and T_{B0} with a uniform and constant convection heat transfer coefficient h. The two side surfaces are thermally insulated. Assume that steady-state conditions have been reached, and that the problem can be approximated by two-dimensional conduction. Determine the temperature distribution in the element and the rate of convection heat transfer from the element.

Problem 4.7
Consider a hollow cylinder with inner diameter D_{I} and outer diameter D_{o}. The rate of heat generation per unit volume q_{g} in the hollow cylinder and the thermal conductivity of the material k are constant. The outer surface of the cylinder is exposed to an ambient fluid at temperature T_{∞}^{o}, and the convection heat transfer coefficient h_{o} there is a function of the longitude θ, but independent of the axial coordinate z. The inner surface of the cylinder is exposed to an ambient fluid at temperature T_{∞}^{I}, and the convection heat transfer coefficient h_I is a function of the longitude θ, but independent of the axial coordinate z. Assume that steady-state conditions have been reached, and that the problem can be approximated by two-dimensional conduction. Determine the temperature distribution in the hollow cylinder, using a suitable grid system.

References

Bejan, A. (1993) *Heat Transfer*, John Wiley & Sons, New York.
Carslaw, H. S. and Jaeger, J. C. (1976) *Conduction of Heat in Solids*, Oxford University Press, Oxford.
Kreith, F and Bohn, M. S. (1993) *Principles of Heat Transfer*, 5th edn, West Publishing Company, New York.
Özisik, M. N. (1993) *Heat Conduction*, 2nd edn, John Wiley & Sons, New York.

5 Transient conduction

5.1 Introduction

So far we have considered only steady-state conduction. It should be pointed out that even though steady-state conduction does give many useful engineering results, the assumption of steady-state conditions is *always* only an approximation. The reasons are, for example, that the physical boundaries of the thermal system are always subjected to some thermal perturbation and that some heat may be randomly generated within the boundaries of the thermal system. This results in *unsteady* or *transient* processes. Nevertheless, in many cases the thermal perturbations are small and their effects have negligible influence on the thermal behaviour of the system. The system can be approximated by assuming that the system is in thermally steady-state, and hence the influence of time can be neglected. Steady-state temperature distributions and steady-state heat fluxes can then be calculated.

However, in many engineering situations, involving heating or cooling processes, such an approximation cannot be made and the time dependence of the temperature and flux distributions has to be determined. This is particularly important when the transient processes are rapid, resulting in high temperature gradients, which can induce considerable transient thermal stresses in engineering components. Examples of such situations are the quenching of machine components to acquire desired material properties, rapid heating of a heat shield on space vehicles during their re-entry to Earth atmosphere and changes in solar radiation on solid masonry.

The general forms of the governing equations are discussed in Chapter 3. Current approaches to solving the governing equations use either analytical or numerical techniques. The advantages and limitations of the full analytical solutions are discussed above. Once again the full analytical techniques are not considered in this book, and the reader is referred to the many available textbooks (Carslaw and Jaeger, 1976; Bejan, 1993; Kreith and Bohn, 1993; Özisik, 1993; Taine and Petit, 1993).

However, one approximate analytical technique is briefly outlined. This technique, termed the *lumped capacitance method*, can be used for conditions in which the temperature gradients within the solid are small. The thermal process is, of course, still transient, in that the temperature within the solid does vary with time, but an assumption is made that the temperature within the solid is spatially uniform during the transient process. This is a simple, but a powerful method for solving a large class of transient conduction processes. Nevertheless, considerable experience with these systems is required in order to estimate the errors introduced in adopting this method. The method is not considered in this book, and, once again, the reader is referred to the many available textbooks (Holman, 1990; Incropera and DeWitt, 1996).

Furthermore, and as discussed in Section 4.1, it should also be noted that there are available several excellent commercial software packages which allow the solutions of even the most complex problems. The major disadvantage of these packages is their complexity; they are not suitable for the undergraduate study of engineering problems.

As pointed out previously, this book uses the numerical techniques to solve the transient conduction problem, and as before only the finite difference method is discussed. Finally, the bulk of this chapter concentrates on the explicit method, whose major advantage is that it is more intuitive than the implicit method, and thus more suited for undergraduate study.

5.2 Finite difference approximation in one-dimensional systems: explicit method in Cartesian system

5.2.1 Internal nodes

Once again consider Eqs (3.2.4) and (3.2.5) for a one-dimensional system, shown in Fig. 5.2.1.

$$(q_x - q_{x+\mathrm{d}x}) + q_\mathrm{g}\,\mathrm{d}x = \rho c \frac{\partial T}{\partial t}\,\mathrm{d}x \tag{5.2.1}$$

Equation (5.2.1) can be written for any sufficiently small control volume Δx and sufficiently small time increment Δt as

$$(q_x - q_{x+\Delta x}) + q_\mathrm{g}\Delta x = \rho c \frac{T_{t+\Delta t} - T_t}{\Delta t}\Delta x \tag{5.2.2}$$

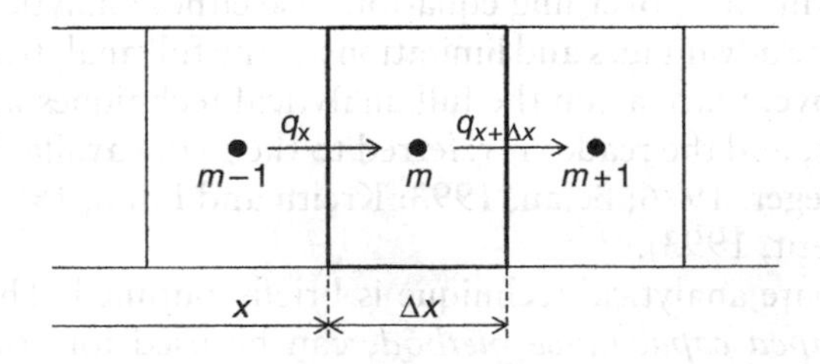

Figure 5.2.1 Internal node in a one-dimensional Cartesian coordinate system.

Furthermore, the time can be written by introducing superscript p as

$$t = p\Delta t \tag{5.2.3}$$

Equation (5.2.2) can then be re-written as

$$\rho_m c_m \frac{T_m^{p+1} - T_m^p}{\Delta t} = k_{m-0.5} \frac{T_{m-1}^p - T_m^p}{(\Delta x)^2} + k_{m+0.5} \frac{T_{m+1}^p - T_m^p}{(\Delta x)^2} + q_g \tag{5.2.4}$$

If the thermophysical properties are constant, Eq. (5.2.4) can be expressed as

$$T_m^{p+1} = \text{Fo}(T_{m+1}^p + T_{m-1}^p) + (1 - 2\text{Fo})T_m^p + \frac{q_g \Delta t}{\rho c} \tag{5.2.5}$$

where Fo is a finite difference form of the Fourier number, defined as

$$\text{Fo} = \frac{\alpha \Delta t}{(\Delta x)^2} \tag{5.2.6}$$

It should be noted that in the formulation of Eq. (5.2.4) the spatial behaviour (the first two terms on the right-hand side of the equation) is evaluated at the *previous*, p time. This approach, based on the forward-difference approximation to the time behaviour, defines the *explicit* method of solution. The method is explicit because the unknown nodal temperature at the new time, $p + 1$, depends only on the known nodal temperatures at the previous time, p. Hence, starting with the known initial temperature of all nodes at time zero, $t = 0\Delta t$ ($p = 0$), all nodal temperatures can be calculated for time, $t = 1\Delta t$ ($p = 1$), and from these nodal temperatures all nodal temperatures can be calculated for time, $t = 2\Delta t$ ($p = 2$). The transient temperature distribution can thus be obtained by marching out in time at intervals Δt.

As always in these calculations the size of the spatial and temporal intervals is chosen by compromise between acceptable accuracy and computational capacity. Generally, the accuracy and the computational requirements increase with decreasing values of Δx and Δt. Acceptable accuracy can be checked by observing the effects of decreased values of Δx and Δt.

The explicit method is intuitively satisfying since it clearly indicates the way of obtaining all unknown new temperatures from all known old temperatures. Nevertheless, this method has one major limitation – the method can become numerically unstable for values of the time interval Δt above certain limiting values. In other words, the time interval Δt must be chosen such that the stability criterion is satisfied.

The determination of the stability criterion can be quite complex, but in this particular case the stability criterion requires that the coefficient associated with T_m^p is greater or equal to zero. Hence, using Eqs (5.2.5) and (5.2.6), it can be shown that the stability is satisfied if

$$\Delta t \leq \frac{0.5(\Delta x)^2}{\alpha} \tag{5.2.7}$$

However, in many other cases the stability criterion is quite complex. Stability can then be determined by trial and error. Since the instability can be clearly observed from the nature of the solution, the temporal interval can be gradually decreased until the stability is achieved.

5.2.2 *External nodes with various boundary conditions*

Case 1: Plane surface with a convective boundary condition.

Referring to Fig. 5.2.2 and assuming constant thermophysical properties, the energy balance for the control volume shown, assuming unit width, is given by

$$q_x - q_c + q_g \frac{\Delta x}{2} = \rho c \frac{T_{t+\Delta t} - T_t}{\Delta t} \frac{\Delta x}{2} \tag{5.2.8}$$

where the convective rate of heat transfer is given by

$$q_c = h(T_m - T_\infty) \tag{5.2.9}$$

Using the approach discussed above, Eq. (5.2.8) can be written as

$$T_m^{p+1} = 2\text{Fo}(T_{m-1}^p + \text{Bi}T_\infty) + (1 - 2\text{Fo} - 2\text{FoBi})T_m^p + \frac{q_g \Delta t}{\rho c} \tag{5.2.10}$$

where Bi is a finite difference form of the Biot number, defined as

$$\text{Bi} = \frac{h\Delta x}{k} \tag{5.2.11}$$

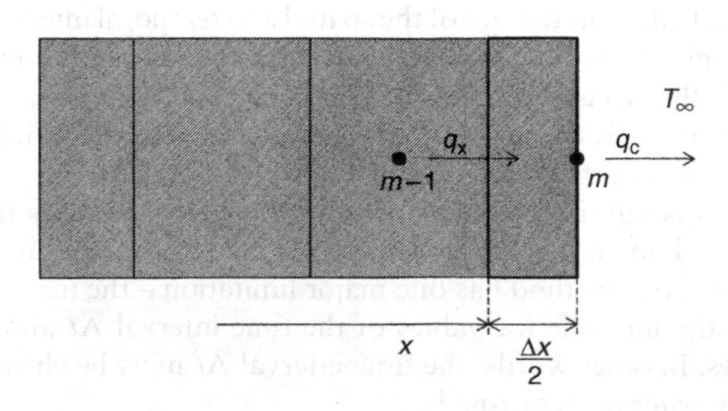

Figure 5.2.2 External plane wall node in a one-dimensional Cartesian coordinate system.

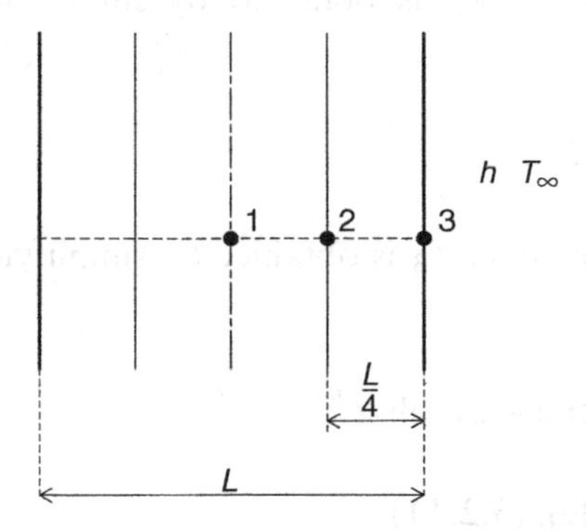

Figure 5.2.3 Symmetrical plane wall with a coarse grid.

Case 2: Plane surface with uniform heat flux q_E.

$$T_m^{p+1} = 2\text{Fo}T_{m-1}^p + (1 - 2\text{Fo})T_m^p + 2\frac{q_E \Delta t}{\rho c \Delta x} + \frac{q_g \Delta t}{\rho c} \tag{5.2.12}$$

Example 5.2.1 Transient behaviour of a plane wall with a coarse grid.

Consider a symmetrical plane wall of thickness L, shown in Fig. 5.2.3. The wall has constant thermophysical properties. Furthermore, there is no heat generation in the wall. The wall is initially at a uniform temperature T_0. The wall is suddenly immersed in an ambient fluid of temperature $T_\infty (T_0 \neq T_\infty)$, with constant convection heat transfer coefficient h. Assume that the system can be approximated by one-dimensional transient conduction, and determine the temperature variation in the plane wall as a function of time, using a suitable grid and time increment.

Analysis

Since the system is symmetrical, we will only consider one half of the plane, with three nodes, and hence

$$\Delta x = \frac{L}{4}$$

Since node 1 is on the plane of symmetry, which equates to zero heat flux there, we can re-arrange Eq. (5.2.12) with $q_E = 0$ to obtain the temperature T_1 as

$$T_1^{p+1} = 2\text{Fo}T_2^p + (1 - 2\text{Fo})T_1^p$$

where the Fourier number Fo is given by Eq. (5.2.6).

The temperature of the internal node T_2 is obtained by simplifying Eq. (5.2.5) as

$$T_2^{p+1} = \text{Fo}(T_1^p + T_3^p) + (1 - 2\text{Fo})T_2^p$$

Finally, the temperature of the surface node, T_3 is obtained by simplifying Eq. (5.2.10) as

$$T_3^{p+1} = 2\text{Fo}(T_2^p + \text{Bi}T_\infty) + (1 - 2\text{Fo} - 2\text{FoBi})T_3^p$$

where the Biot number Bi is given by Eq. (5.2.11).

As discussed in Section 5.2, the solution is obtained by forward time marching method.

Solution

The solution is given in the workbook Ex05-02-01.xls. There are two worksheets in this workbook: *Input data and results* and *Calculations*.

The first workbook, *Input data and results*, allows for the input of the data, and presents the numerical results for the three temperatures T_1, T_2, and T_3 as function of time t.

The thermal conductivity of the material k is entered in **cell B3**, the density of the material is entered in **cell B4**, the specific heat of the material is entered in **cell B5**, the wall thickness is entered in **cell B6**, the temperature of the ambient fluid is entered in **cell B7**, the initial wall temperature is entered in **cell B8**, the convective heat transfer coefficient h is entered in **cell B9**, and the time increment Δt in **cell B11**. (It should be noted that the data already entered are for illustrations only.) The grid size, the Fourier number, and the Biot number, which are automatically calculated, are also given. The calculated nodal temperatures are given in the two graphs, with T_1 in black, T_2 in red, and T_3 in blue, as functions of time t. The two graphs also show the ambient temperature T_∞ in green. The first graph shows the temperatures for the whole time interval chosen, whereas the second graph only shows the temperatures for the first 50 s. The presentation of the results is improved by clicking **Full Screen** in **View**.

The second worksheet, *Calculations*, is used to calculate the temperatures. The initial time is in **cell B5**, the initial temperatures in **cells (C5:E5)**, and the ambient temperature in **cell F5**. The temperatures for the first time increment are calculated, as indicated above in **cells (C6:E6)**. Temperatures for the next time increment are obtained by copying **cells (B6:F6)** to **cells (B7:F7)**. Thus, for example, the temperatures for $t = 1$ s are given in blue **cells (C15:E15)**. The total time interval can be extended by copying the cells further.

Entering new values of the input data automatically re-calculates the nodal temperatures.

Discussion

The solutions can be used to experiment with the various design and numerical calculation parameters and observe their influence.

1 The influence of the thermophysical properties on the response of the wall to transient conditions can be investigated.
2 The stability criterion, discussed in Section 5.2, can be investigated.
3 Increasing the time increment speeds up the solution, but the numerical results are less accurate for very short time intervals.

Other observations can be made and the reader is invited to experiment further.

Example 5.2.2 Transient behaviour of a plane wall with a finer grid.

Consider once again the plane discussed in Example 5.2.1 and repeat the calculations with a finer grid.

Analysis

We use the grid show in Fig. 5.2.4 with $\Delta x = L/8$. The approach is similar to that discussed in Example 5.2.1. The temperature on the plane of symmetry T_1 and the surface temperature T_5 are calculated as the corresponding temperatures above. The temperatures of the internal nodes, T_2, T_3, and T_4, are calculated as the temperature of the internal node above.

Solution

The solution is given in the workbook Ex05-02-02.xls, and the approach is identical to that discussed in Example 5.2.1. However, it should be noted

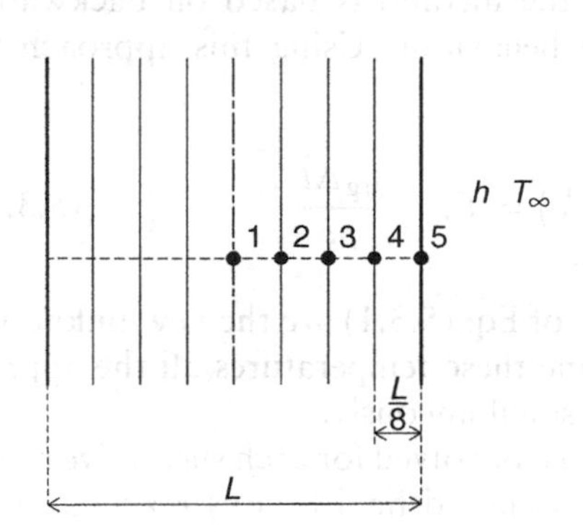

Figure 5.2.4 Symmetrical plane wall with a finer grid.

that the temperatures shown on the two graphs are temperature T_1 (in black), temperature T_3 (in red), and temperature T_5 (in blue).

Discussion

The solutions can be used to experiment with the various design and numerical calculation parameters and observe their influence.

1 A more detailed and accurate description of the system is obtained with a finer grid.
2 The solutions require longer computational times when a finer grid is used. The reasons are twofold. First, the number of calculations in each cycle is considerably greater for a finer grid and second, the solution converges more slowly.
3 The size of the grid is obtained by compromise between the minimum required accuracy and the maximum practicable computational times.

Other observations can be made and the reader is invited to experiment further.

5.3 Finite difference approximation in one-dimensional systems: implicit method in Cartesian system

5.3.1 Internal nodes

As discussed above, the explicit method is intuitively satisfying and, additionally, computationally convenient. However, as also pointed out above, for many geometries the time interval Δt must be extremely small in order to satisfy the stability criterion. Larger time intervals can be used if an *implicit* rather than explicit method is used. In this method the spatial behaviour of Eq. (5.2.4) (the first two terms on the right-hand side of the equation) is evaluated at the *new* time, $p+1$, rather than the previous time p. An alternative explanation of this technique is that the method is based on backward-difference approximation to the time behaviour. Using this approach in Eq. (5.2.4), and after re-arrangements

$$(1+2\text{Fo})T_m^{p+1} - \text{Fo}\left(T_{m+1}^{p+1} + T_{m-1}^{p+1}\right) = T_m^p + \frac{q_g \Delta t}{\rho c} \tag{5.3.1}$$

The temperatures on the left-hand side of Eq. (5.3.1) are the new unknown temperatures, and in order to determine these temperatures all the appropriate nodal equations must be solved simultaneously.

Hence the simultaneous equations must be solved for each successive time, $t = 1\Delta t, 2\Delta t, 3\Delta t, \ldots, p\Delta t$, until the required final time is reached. The implicit technique is more laborious to code, but its major advantage is that it is stable with respect to the choice of the time interval Δt.

5.3.2 *External nodes with various boundary conditions*

Case 1: Plane surface with a convective boundary condition.

Referring to Fig. 5.2.2 and assuming constant thermophysical properties, the energy balance given by Eq. (5.2.8) can be re-written as

$$(1+2\text{Fo}+2\text{FoBi})T_m^{p+1} - 2\text{Fo}T_{m-1}^{p+1} = T_m^p + 2\text{FoBi}T_\infty + \frac{q_g \Delta t}{\rho c} \tag{5.3.2}$$

Case 2: Plane surface with uniform heat flux q_E.

$$(1+2\text{Fo})T_m^{p+1} - 2\text{Fo}T_{m-1}^{p+1} = T_m^p + 2\frac{q_E \Delta t}{\rho c \Delta x} + \frac{q_g \Delta t}{\rho c} \tag{5.3.3}$$

5.4 Finite difference approximation in one-dimensional systems: explicit method in cylindrical system

5.4.1 *Internal nodes*

Consider a one-dimensional cylindrical system (in which the temperature depends only on the radial co-ordinate r and time t) with constant thermophysical properties, as shown in Fig. 5.4.1. The energy balance for a sufficiently small control volume can be written as

$$q_r\left(r - \frac{\Delta r}{2}\right) - q_{r+\Delta r}\left(r + \frac{\Delta r}{2}\right) + q_g r \Delta r = \rho c r \Delta r \frac{T_{t+\Delta t} - T_t}{\Delta t} \tag{5.4.1}$$

Using simplified forms of Eqs (4.3.2) and (4.3.3) and the approach discussed above, Eq. (5.4.1) can be written as

$$T_m^{p+1} = \text{Fo}\left[\left(1 + \frac{\Delta r}{2r}\right)T_{m+1}^p + \left(1 - \frac{\Delta r}{2r}\right)T_{m-1}^p\right] + (1-2\text{Fo})T_m^p + \frac{q_g \Delta t}{\rho c} \tag{5.4.2}$$

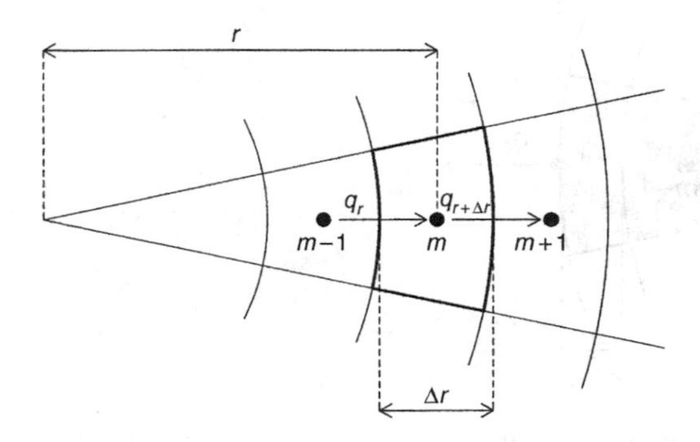

Figure 5.4.1 Internal node in a one-dimensional cylindrical coordinate system.

where Fo is a finite difference form of the Fourier number, defined as

$$\mathrm{Fo} = \frac{\alpha \Delta t}{(\Delta r)^2} \tag{5.4.3}$$

5.4.2 External nodes with various boundary conditions

Case 1: Cylindrical wall with a convective boundary condition.

Referring to Fig. 5.4.2 and assuming constant thermophysical properties, the energy balance for the control volume shown, assuming unit width, is given by

$$q_r\left(r - \frac{\Delta r}{2}\right) - q_c r + q_g\left(r - \frac{\Delta r}{4}\right)\frac{\Delta r}{2} = \rho c\left(r - \frac{\Delta r}{4}\right)\frac{\Delta r}{2}\frac{T_{t+\Delta t} - T_t}{\Delta t} \tag{5.4.4}$$

where the convective rate of heat transfer is given by Eq. (5.2.9). Similarly to above this can be written as

$$\begin{aligned} T_m^{p+1} &= 2\mathrm{Fo}\frac{1-(\Delta r/2r)}{1-(\Delta r/4r)}T_{m-1}^p \\ &+ \left(1 - 2\mathrm{Fo}\frac{1-(\Delta r/2r)}{1-(\Delta r/4r)} - 2\mathrm{FoBi}\frac{1}{1-(\Delta r/4r)}\right)T_m^p \\ &+ 2\mathrm{FoBi}\frac{1}{1-(\Delta r/4r)}T_\infty + \frac{q_g \Delta t}{\rho c} \end{aligned} \tag{5.4.5}$$

with the Biot number Bi defined as

$$\mathrm{Bi} = \frac{h\Delta r}{k} \tag{5.4.6}$$

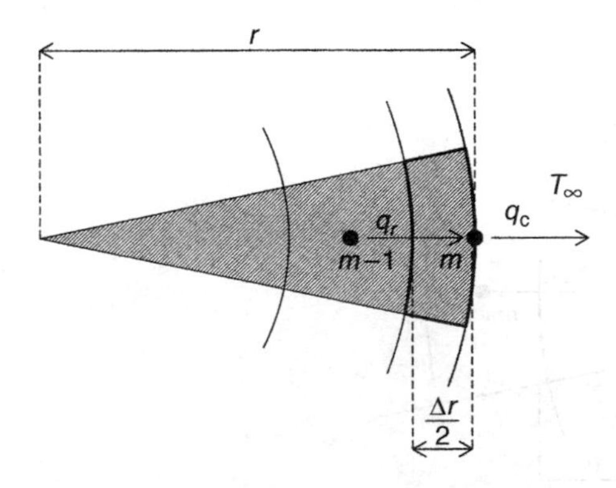

Figure 5.4.2 External cylindrical wall node in a one-dimensional cylindrical coordinate system.

Case 2: Cylindrical wall with uniform heat flux q_E.

$$T_m^{p+1} = 2\text{Fo}\frac{1-(\Delta r/2r)}{1-(\Delta r/4r)}T_{m-1}^p + \left(1 - 2\text{Fo}\frac{1-(\Delta r/2r)}{1-(\Delta r/4r)}\right)T_m^p$$
$$+ 2\frac{q_E \Delta t}{\rho c \Delta r}\frac{1}{1-(\Delta r/4r)} + \frac{q_g \Delta t}{\rho c} \tag{5.4.7}$$

5.4.3 *Central node*

The transient behaviour of the central node is analysed for two-dimensional systems in Section 5.6.3. The result for the two-dimensional system, given by Eq. (5.6.6), can be used to obtain the transient behaviour for the central node of a one-dimensional system as

$$T_1^{p+1} = T_1^p(1 - 4\text{Fo}) + 4\text{Fo}T_2^p + \frac{q_g \Delta t}{\rho c} \tag{5.4.8}$$

Example 5.4.1 Transient behaviour of a solid cylinder with heat generation.

Consider a long solid element with a circular cross-section of diameter D, and constant material thermophysical properties, shown in Fig. 5.4.3. The cylinder is initially at a uniform temperature T_0. The cylinder is suddenly immersed in an ambient fluid of temperature T_∞ ($T_0 \neq T_\infty$), with constant convection heat transfer coefficient h. At the same time heat starts to be generated in the cylinder, with constant rate of heat generation per unit volume q_g. Assume that the system can be approximated by one-dimensional

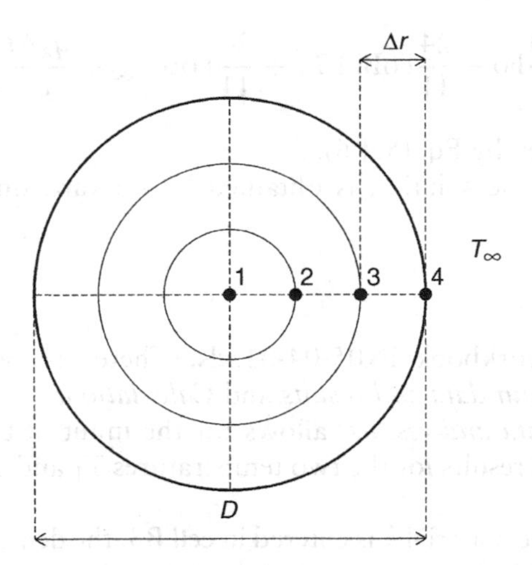

Figure 5.4.3 Solid element with cylindrical cross-section.

transient conduction, and determine the temperature variation in the cylinder as a function of time, using a suitable grid and time increment.

Analysis

Since the system can be approximated by one-dimensional analysis in cylindrical coordinates, we will use the approach developed in Section 5.4 with four nodes; thus

$$\Delta r = \frac{D}{6}$$

The temperature of the central node T_1 is given by Eq. (5.4.8) as

$$T_1^{p+1} = T_1^p(1 - 4\text{Fo}) + 4\text{Fo}T_2^p + \frac{q_g \Delta t}{\rho c}$$

where the Fourier number Fo is given by Eq. (5.4.3).

For the two internal nodes, node 2 and node 3, r equals Δr and $2\Delta r$, respectively and the nodal temperatures T_2 and T_3 are obtained from Eq. (5.4.2) as

$$T_2^{p+1} = \text{Fo}(1.50T_3^p + 0.50T_1^p) + (1 - 2\text{Fo})T_2^p + \frac{q_g \Delta t}{\rho c}$$

$$T_3^{p+1} = \text{Fo}(1.25T_4^p + 0.75T_2^p) + (1 - 2\text{Fo})T_3^p + \frac{q_g \Delta t}{\rho c}$$

Finally, noting that for the surface node, node 4, r equals $3\Delta r$ the surface temperature T_4 is obtained from Eq. (5.4.5) as

$$T_4^{p+1} = \frac{20}{11}\text{Fo}T_3^p + \left(1 - \frac{20}{11}\text{Fo} - \frac{24}{11}\text{FoBi}\right) T_4^p + \frac{24}{11}\text{FoBi}T_\infty + \frac{q_g \Delta t}{\rho c}$$

where the Biot number Bi is given by Eq. (5.4.6).

As discussed in Section 5.2, the solution is obtained by forward time marching method.

Solution

The solution is given in the workbook Ex05-04-01.xls. There are two worksheet in this workbook: *Input data and results* and *Calculations*.

The first workbook, *Input data and results*, allows for the input of the data, and presents the numerical results for the two temperatures T_1 and T_4 as function of time t.

The thermal conductivity of the material k is entered in **cell B3**, the density of the material is entered in **cell B4**, the specific heat of the material is entered in **cell B5**, the cylinder diameter is entered in **cell B6**, the temperature of the ambient fluid is entered in **cell B7**, the initial temperature is entered in **cell B8**, the convective heat transfer coefficient h is entered in **cell B9**, the rate of heat generation per unit volume in **cell B10**, and the time increment Δt in **cell B12**. (It should be noted that the data already entered are for illustrations only.) The grid size, the Fourier number, the Biot number, and the group $q_g \Delta t / \rho c$, which are automatically calculated, are also given. The calculated temperatures are given in the two graphs, with T_1 in black and T_4 in red, as functions of time t. The two graphs also show the ambient temperature T_∞ in green. The first graph shows the temperatures for the whole time interval chosen, whereas the second graph only shows the temperatures for the first 50 s. The presentation of the results is improved by clicking **Full Screen** in **View**.

The second worksheet, *Calculations*, is used to calculate the temperatures. The initial time is in **cell B5**, the initial temperatures in **cells (C5:F5)**, and the ambient temperature in **cell G5**. The temperatures for the first time increment are calculated, as indicated above in **cells (C6:F6)**. Temperatures for the next time increment are obtained by copying **cells (B6:G6)** to **cells (B7:G7)**. Thus, for example, the temperatures for $t = 1$ s are given in blue **cells (C10:F10)**. The total time interval can be extended by copying the cells further.

Entering new values of the input data automatically re-calculates the nodal temperatures.

Discussion

The solutions can be used to experiment with the various design and numerical calculation parameters and observe their influence.

1. The influence of the thermophysical properties on the response of the solid cylinder to transient conditions can be investigated.
2. The stability criterion, discussed in Section 5.2, can be investigated.
3. The steady-state solution can be compared with an appropriate solution from Example 4.3.1.

Other observations can be made and the reader is invited to experiment further.

5.5 Finite difference approximation in two-dimensional systems: explicit method in Cartesian system

5.5.1 Internal nodes

Similarly to the analysis presented in Chapter 4 and Section 5.2, for a system with constant thermophysical properties, the energy balance for a sufficiently small control volume can be written as

$$(q_x - q_{x+\Delta x})\Delta y + (q_y - q_{y+\Delta y})\Delta x + q_g \Delta x \Delta y = \rho c \frac{T_{t+\Delta t} - T_t}{\Delta t} \Delta x \Delta y \tag{5.5.1}$$

If, furthermore, the nodal system is chosen, once again, such that $\Delta x = \Delta y$, Eq. (5.5.1) can be simplified to

$$T_{m,n}^{p+1} = \text{Fo}(T_{m+1,n}^p + T_{m-1,n}^p + T_{m,n+1}^p + T_{m,n-1}^p) + (1 - 4\text{Fo})T_{m,n}^p + \frac{q_g \Delta t}{\rho c} \tag{5.5.2}$$

with Fo defined by Eq. (5.2.6).

5.5.2 External nodes with various boundary conditions

Case 1: Plane surface with a convective boundary condition.

Similarly to above it can be shown that, assuming constant thermophysical properties and using $\Delta x = \Delta y$ nodal system

$$\begin{aligned} T_{m,n}^{p+1} &= \text{Fo}(2T_{m-1,n}^p + T_{m,n+1}^p + T_{m,n-1}^p + 2\text{Bi}T_\infty) \\ &\quad + (1 - 4\text{Fo} - 2\text{FoBi})T_{m,n}^p + \frac{q_g \Delta t}{\rho c} \end{aligned} \tag{5.5.3}$$

Case 2: Internal corner with a convective boundary condition.

$$\begin{aligned} T_{m,n}^{p+1} &= \frac{2}{3}\text{Fo}(T_{m+1,n}^p + 2T_{m-1,n}^p + T_{m,n+1}^p + 2T_{m,n-1}^p + 2\text{Bi}T_\infty) \\ &\quad + \left(1 - 4\text{Fo} - \frac{4}{3}\text{FoBi}\right) T_{m,n}^p + \frac{q_g \Delta t}{\rho c} \end{aligned} \tag{5.5.4}$$

Case 3: External corner with a convective boundary condition.

$$T_{m,n}^{p+1} = 2\text{Fo}(T_{m-1,n}^p + T_{m,n-1}^p + 2\text{Bi}T_\infty) + (1 - 4\text{Fo} - 4\text{FoBi})T_{m,n}^p + \frac{q_g \Delta t}{\rho c} \tag{5.5.5}$$

Case 4: Plane surface with uniform heat flux q_E.

$$T_{m,n}^{p+1} = \text{Fo}(2T_{m-1,n}^p + T_{m,n+1}^p + T_{m,n-1}^p) + (1 - 4\text{Fo})T_{m,n}^p + 2\frac{q_E \Delta t}{\rho c \Delta x} + \frac{q_g \Delta t}{\rho c} \tag{5.5.6}$$

Case 5: Internal corner with uniform heat flux q_E.

$$T^{p+1}_{m,n} = \frac{2}{3}\text{Fo}(T^p_{m+1,n} + 2T^p_{m-1,n} + T^p_{m,n+1} + 2T^p_{m,n-1}) + (1 - 4\text{Fo})T^p_{m,n} + \frac{4}{3}\frac{q_E \Delta t}{\rho c \Delta x} + \frac{q_g \Delta t}{\rho c} \tag{5.5.7}$$

Case 6: External corner with uniform heat flux q_E.

$$T^{p+1}_{m,n} = 2\text{Fo}(T^p_{m-1,n} + T^p_{m,n-1}) + (1 - 4\text{Fo})T^p_{m,n} + 4\frac{q_E \Delta t}{\rho c \Delta x} + \frac{q_g \Delta t}{\rho c} \tag{5.5.8}$$

Example 5.5.1 Transient behaviour of a rectangular duct.

Consider a duct with a rectangular cross-section shown in Fig. 5.5.1. The duct has constant thermophysical properties and there is no heat generation anywhere in the duct. The initial temperature in the cross-section is a function of the x and y coordinates; the temperature does not depend on the z coordinate. The temperature of the outer surface and the temperature of the inner surface are instantaneously and simultaneously set at T_o and T_I, respectively. Assume that the system can be approximated by two-dimensional transient conduction, and determine the temperature variation in the duct as a function of time, using a suitable grid and time increment.

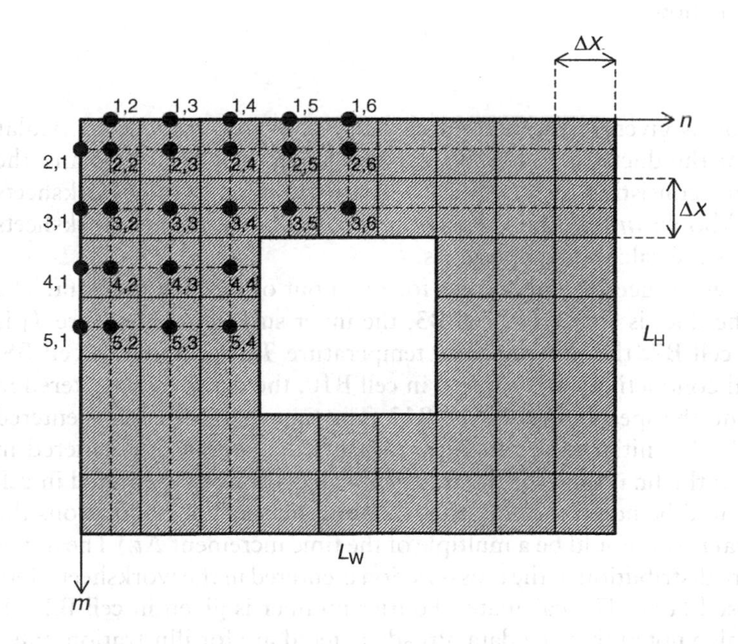

Figure 5.5.1 Rectangular duct.

Analysis

A nodal system with $\Delta x = \Delta y$ and shown in Fig. 5.5.1 is chosen. For all internal nodes, Eq. (5.5.2) can be simplified as

$$T_{m,n}^{p+1} = \text{Fo}(T_{m+1,n}^{p} + T_{m-1,n}^{p} + T_{m,n+1}^{p} + T_{m,n-1}^{p}) + (1 - 4\text{Fo})T_{m,n}^{p}$$

where the Fourier number Fo is given by Eq. (5.2.6).

Since in this nodal system each surface node is distance $\Delta x/2$ from its associated central node, the thermal resistance between the central node and its associated surface node is half of the thermal resistance between the central node and its associated internal nodes. Following the analysis in Section 5.5.1, we can show that the temperature of each node associated with two surface nodes, such as $T_{2,2}$, is given by

$$T_{2,2}^{p+1} = \text{Fo}(4T_\text{o} + T_{2,3}^{p} + T_{3,2}^{p}) + (1 - 6\text{Fo})T_{2,2}^{p}$$

Similarly, the temperature of each node associated with one surface node, such as $T_{2,3}$, is given by

$$T_{2,3}^{p+1} = \text{Fo}(2T_\text{o} + T_{2,4}^{p} + T_{3,3}^{p} + T_{2,2}^{p}) + (1 - 5\text{Fo})T_{2,3}^{p}$$

It should be noted that this also applies to the inner wall temperatures, and that because of the nodal system chosen there are no internal corners.

As discussed in Section 5.2, the solution is obtained by forward time marching method.

Solution

The solution is given in the workbook Ex05-05-01.xls for one particular geometry of the duct: the outer surface consisting of 9 by 7 cells and the inner surface consisting of 3 by 3 cells. There are initially four worksheets in this workbook: *Input*, *Main*, *Raster*, and *col_code*, but other worksheets are added as the calculations progress.

The first worksheet, *Input*, allows for the input of most of the data. The width of the duct is entered in cell **B3**, the inner surface temperature T_I is entered in cell **B7**, the outer surface temperature T_o is entered in cell **B8**, the thermal conductivity k is entered in cell **B10**, the density ρ is entered in cell **B11**, and the specific heat in cell **B12**. The time increment Δt is entered in cell **B17**, the initial time for the transient calculations t_0 is entered in cell **B18** and the final time for the transient calculations t_F is entered in cell **B19**. (It should be noted that in order to allow subsequent calculations the time interval t_F–t_0 should be a multiple of the time increment Δt.) The initial temperature distribution in the cross-section is entered in the worksheet *Main* and discussed later. The calculated Fourier number is given in cell **B21**. It should be also noted that the data already entered are for illustration only.

The second worksheet, *Main*, presents the diagram of the cross-section of the duct and gives all the calculated temperatures. It should be noted that the structure and the colours of the diagram are important, since, as discussed below, this information is used to determine the number of cells in the cross-section of the duct. The diagram must start the **cell A1**. The blue field, which must be one cell thick, and must completely surround and be immediately adjacent to the yellow field, gives the outer surface temperature T_O. The yellow field presents the calculated nodal temperatures in the duct. The brown field, which must be rectangular and not immediately adjacent to the blue field, gives the inner surface temperature T_I. The number of columns in the yellow field is proportional to the width of the duct and the number of rows in the yellow field is proportional to the height of the duct (it should be noted that the matrix of the cross-section must be greater than 4 by 4 and smaller than 50 by 50). A duct with any cross-section can be modelled with a $\Delta x = \Delta y$ grid, provided that all dimensions are multiples of the grid size Δx. Hence we determine the appropriate aspect ratio of the cross-section and the size of the matrix and enter it in yellow in the *Main* worksheet. Since we also specified the width of the element in the *Input* worksheet, the grid size and height of the duct are automatically calculated and are given respectively in **cell B5** and **cell B4** of the *Input* worksheet. The initial temperature distribution in the cross-section is then entered in the yellow cells.

The third worksheet, *Raster*, presents the raster plot, and the worksheet, *col_code*, gives the colour code used for the raster plot.

The workbook also contains three VBA programs: *Initial*, *Iterate*, and *Rasterplot*. The first program, *Initial*, which must be run first, recognises the colours in the worksheet *Main* and determines the size of the duct matrix, which is entered in **cell B14** and **cell B15** in the worksheet *Input*, and also clears the *Raster* area. The second program, *Iterate*, calculates the nodal temperatures from the initial time t_0 to the final time t_F, using the time increment Δt. The final program, *Rasterplot*, plots the raster plot in the worksheet *Raster*, using the same structure as the structure used in the worksheet *Main*.

The solution is obtained in the following steps:

1 The program is launched with **Macros** enabled.
2 The workbook must only include the following worksheets: *Input*, *Main*, *Raster*, and *col_code*; all other worksheets must be deleted.
3 The size of the duct is specified and entered in the worksheet *Main*, ensuring that the structure and colour conventions are followed.
4 The *Initial* **Macro** is run.
5 All design parameters are specified and entered in the worksheets *Input* and *Main*.
6 The *Rasterplot* **Macro** is run.

7 The worksheet *Raster* is copied at the end after the four active worksheets and renamed 0 s, thus giving the temperature profile at the start of the transient calculations.
8 The initial time t_0 is specified and entered.
9 The final time t_F is specified and entered.
10 The time increment Δt is specified and entered, ensuring that the time interval t_F–t_0 is a multiple of the time increment Δt.
11 The *Iterate* Macro is run.
12 When the iterations are completed the *Rasterplot* Macro is run.
13 The worksheet *Raster* is copied at the end after all existing active worksheets and renamed t_F s (e.g. for $t_F = 10$ s, the new name will be 10 s), thus giving the temperature profile at the end of the time interval.
14 The final time of this time interval t_F will become the initial time of the next interval t_0. (It should be noted that the *Iterate* Macro will automatically do this by entering t_F into cell **B18**.)
15 We now return to step 8 and specify and enter the new final time t_F.
16 If required a new time increment Δt is specified and entered.
17 The whole process is continued until steady-state conditions are reached. In the present calculations 10 subsequent time intervals were investigated. The final times t_F and the respective time increments Δt (in seconds) were as follows: (10, 1), (20, 1), (40, 2), (100, 4), (200, 10), (400, 20), (1000, 40), (2000, 40), (4000, 40), and (10000, 40).
18 A blank worksheet *all* was then created between the worksheet *col_code* and the worksheet *0* s. Temperatures $T_{5,2}$, $T_{5,3}$, $T_{5,4}$ (as shown in Fig. 5.5.1) were, for the appropriate time intervals t_F, copied from the corresponding worksheets into worksheet *all*. The results are also presented in a graphical form. The worksheet also shows the steady-state solution obtained from the solution of Example 4.2.1. This shows that virtually steady-state conditions are obtained after about 4000 s.

Typical results are shown in Fig. 5.5.2 and Colour Plate IV.

Discussion

The solutions can be used to experiment with the various design and numerical calculation parameters and observe their influence.

- Steps 1–17 must be followed for all new numerical experiments.
- The influence of the thermophysical properties on the response of the duct to transient conditions can be investigated.
- The influence of a non-uniform distribution of initial temperature in the cross-section of the duct can be investigated by entering those temperatures in the worksheet *Main*.
- The initial variation of temperatures can be extremely rapid, and very short time increments Δt should be used to investigate the initial behaviour.

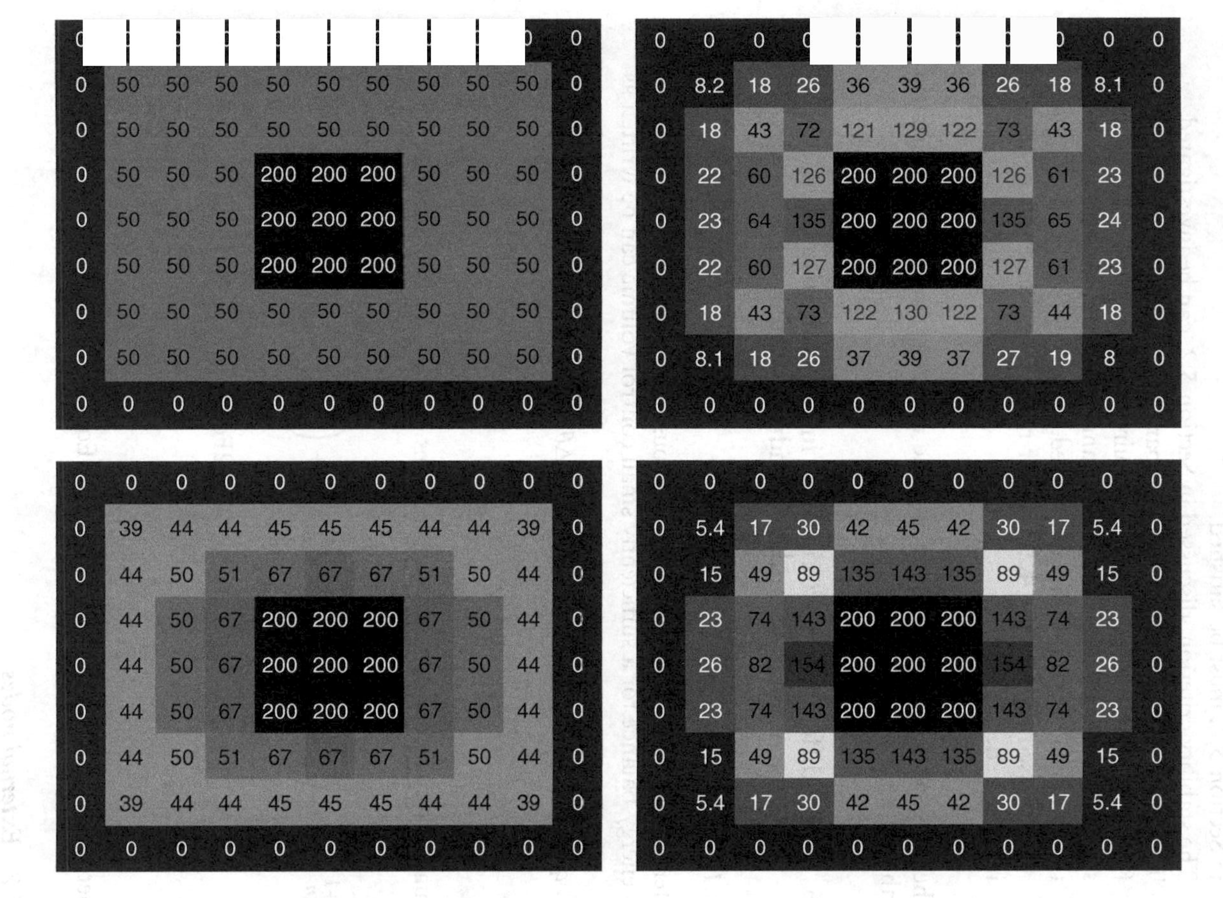

Figure 5.5.2 Transient, two-dimensional heat conduction analysis for a rectangular duct: Initial condition $t = 0$ s (top-left); $t = 40$ s (bottom-left); $t = 400$ s (top-right); and $t = 4000$ s (bottom-right). Note that at $t = 4000$ s steady-state conditions have been achieved as this plot now compares favourably with that shown in Plate III (top-half). (See Colour Plate IV.)

- As the time increases the temperature changes are small and much larger time increments Δt should be used, but the stability criterion, discussed in Section 5.2, must be satisfied.
- The stability criterion, discussed in Section 5.2, can be investigated.
- The steady-state solution can be compared with an appropriate solution from Example 4.2.1. The time required to achieve virtually steady-state conditions can be observed, and the influence of the various thermophysical parameters determined.
- As discussed in Example 4.2.3, other rectangular duct systems can be investigated.

Other observations can be made and the reader is invited to experiment further.

5.6 Finite difference approximation in two-dimensional systems: explicit method in cylindrical system

5.6.1 *Internal nodes*

Referring to Fig. 4.3.1, for a system with constant thermophysical properties, the energy balance for a sufficiently small control volume can be written as

$$\begin{aligned} & q_r(r-0.5\Delta r)\Delta\theta - q_{r+\Delta r}(r+0.5\Delta r)\Delta\theta + q_\theta\Delta r - q_{\theta+\Delta\theta}\Delta r + q_g r\Delta\theta\Delta r \\ & = \rho c\frac{T_{t+\Delta t}-T_t}{\Delta t} r\Delta\theta\Delta r \end{aligned} \tag{5.6.1}$$

Equation (5.6.1) can be re-written in terms of nodal temperatures as

$$\begin{aligned} T_{m,n}^{p+1} = \text{Fo}\Bigg[& \left(\frac{\Delta r}{r\Delta\theta}\right)^2 (T_{m+1,n}^p + T_{m-1,n}^p) + \left(1+\frac{\Delta r}{2r}\right)T_{m,n+1}^p \\ & + \left(1-\frac{\Delta r}{2r}\right)T_{m,n-1}^p\Bigg] + \left[1-2\text{Fo}\left(1+\left(\frac{\Delta r}{r\Delta\theta}\right)^2\right)\right]T_{m,n}^p + \frac{q_g\Delta t}{\rho c} \end{aligned} \tag{5.6.2}$$

where the Fourier number Fo is given by Eq. (5.4.3).

5.6.2 *External nodes*

Case 1: Cylindrical wall with a convective boundary condition.

Referring to Fig. 4.3.2, for a system with constant thermophysical properties, the energy balance for a sufficiently small control volume can be written as

$$q_r\left(r-\frac{\Delta r}{2}\right)\Delta\theta - q_c r\Delta\theta + q_\theta\frac{\Delta r}{2} - q_{\theta+\Delta\theta}\frac{\Delta r}{2} + q_g\left(r-\frac{\Delta r}{4}\right)\Delta\theta\frac{\Delta r}{2}$$
$$= \rho c\frac{T_{t+\Delta t}-T_t}{\Delta t}\left(r-\frac{\Delta r}{4}\right)\Delta\theta\frac{\Delta r}{2} \tag{5.6.3}$$

$$T_{m,n}^{p+1} = \frac{\text{Fo}}{1-(\Delta r/4r)}$$
$$\times\left[\left(\frac{\Delta r}{r\Delta\theta}\right)^2\left(T_{m+1,n}^p + T_{m-1,n}^p\right) + 2\left(1-\frac{\Delta r}{2r}\right)T_{m,n-1}^p + 2\text{Bi}T_\infty\right]$$
$$+\left[1-2\frac{\text{Fo}}{1-(\Delta r/4r)}\left[\left(1-\frac{\Delta r}{2r}\right)+\left(\frac{\Delta r}{r\Delta\theta}\right)^2+\text{Bi}\right]\right]T_{m,n}^p + \frac{q_g\Delta t}{\rho c} \tag{5.6.4}$$

where the Biot number Bi is given by Eq. (5.4.6).

Similar equations can be set up for a cylindrical wall with a prescribed heat flux. As shown in Fig. 4.3.3 other external nodes can be identified, and appropriate finite difference equations then developed.

5.6.3 *Central node*

Consider again the central node in a two-dimensional cylindrical system shown in Fig. 4.3.4. As discussed in Section 4.3.3, the energy balance for a sufficiently small control volume of radius $\Delta r/2$ and unit thickness can be written as

$$-(q_{r1,2}+q_{r2,2}+\cdots+q_{rm,2})\frac{\Delta r}{2}\Delta\theta + q_g\pi\left(\frac{\Delta r}{2}\right)^2 = \rho c\frac{\partial T_{1,1}}{\partial t}\pi\left(\frac{\Delta r}{2}\right)^2 \tag{5.6.5}$$

where q_g is the rate of heat generated per unit volume in the control volume and the radial heat flux $q_{r1,2}$ is defined previously by Eq. (4.3.10).

Equation (5.6.5) can be re-written in terms of the central node temperature $T_{1,1}$ as

$$T_{1,1}^{p+1} = T_{1,1}^p(1-4\text{Fo}) + 4\frac{\text{Fo}}{n}(T_{1,2}^p + T_{2,2}^p + T_{3,2}^p + \cdots + T_{n,2}^p) + \frac{q_g\Delta t}{\rho c} \tag{5.6.6}$$

5.7 Final remarks

Similar equations can be developed for three-dimensional systems, but since no further insight is necessarily obtained, these systems are not considered in this work. The various numerical methods are discussed in detail in Chapter 2.

Problems

Problem 5.1

Consider the wall in Example 5.2.1 and determine the convection heat transfer rate from the wall per unit area of the wall as a function of time.

Problem 5.2

The current sheet glass tempering systems used by industry to toughen glass typically use the technique of blowing air over the glass as it leaves the furnace to cool and temper the glass. For toughening thin glass this involves very high capital and running cost as the fans for such systems are very large and expensive. For a 6 mm glass, the temperature from the furnace exit is typically around 620°C. The surface temperature of this heated glass has to drop to below 520°C within 4 s to reach the desired 'strain point'. Owing to the fact that air is not a very effective medium for heat transfer large volume rates are required to cool the glass surface quickly enough to temper the glass. This requires a large blower, and blowers for these systems are typically of the order of 200–400 kW or more in capacity. A more effective method of cooling the glass surface would be to use the heat of vaporisation of water to extract heat at sufficient rates to temper the glass. This would involve relatively small quantities of water being sprayed in a fine mist-like fashion onto the glass surfaces to quench the glass. The main problem associated with water-cooling is the possibility of large sized droplets contacting the glass which may cause the glass to break. However, with a good design this problem may be addressed, thus enabling large energy savings. In the design of such a facility a transient conduction analysis of the quenching process is to be undertaken. Consider a 1.5 m^2 sheet of glass that is 6 mm thick, which is to be quenched by water at 90°C under film boiling regime (assume convection heat transfer coefficient h of 375 $W/m^2 K$). Determine the transient temperature profiles and the heat flux rates. Compare the above results with conventional cooling undertaken with air blowers, with a convection heat transfer coefficient of 180 $W/m^2 K$. Use the following thermophysical properties of glass: thermal conductivity: 1 W/m K, density: 2800 kg/m^3, and specific heat: 840 J/kg K.

Problem 5.3

Consider the problem introduced in Example 5.5.1, but assume that the temperature of the outer surface and the temperature of the inner surface increase as functions of time. Determine the temperature variation in the duct as a function of time, using a suitable grid and time increment.

Problem 5.4

Derive a finite difference approximation for an external corner in a solid with variable thermal conductivity using a Cartesian coordinate system, and assuming two-dimensional transient conduction.

Problem 5.5

Derive a finite difference approximation for transient conduction for the system considered in Example 4.2.4.

Problem 5.6

Refer to the steady-state two-dimensional conduction problem that was considered in Problem 4.4. During a period of clear sky with high receipt of solar radiation it has been observed that the energy gain within the atrium increases significantly. As a result the temperature of the internal surface of the Termodeck block increases instantaneously to 60°C, with the top and bottom surfaces remaining at 30°C. Determine the temperature distribution as a function of time. Use the following thermophysical properties of concrete: thermal conductivity $k = 0.8\ \mathrm{W/m^2\,K}$, density: $2000\ \mathrm{kg/m^3}$, and specific heat: 840 J/kg K.

Problem 5.7

Derive a finite difference approximation for transient conduction for the system considered in Example 4.3.1.

References

Bejan, A. (1993) *Heat Transfer*, John Wiley & Sons, New York.

Carslaw, H. S. and Jaeger, J. C. (1976) *Conduction of Heat in Solids*, Oxford University Press, Oxford.

Holman, J. P. (1990) *Heat Transfer*, 7th edn, McGraw-Hill, New York.

Incropera, F. P. and De Witt, D. P. (1996) *Fundamentals of Heat and Mass Transfer*, 4th edn, John Wiley & Sons, New York.

Kreith, F and Bohn, M. S. (1993) *Principles of Heat Transfer*, 5th edn, West Publishing Company, New York.

Özisik, M. N. (1993) *Heat Conduction*, 2nd edn, John Wiley & Sons, New York.

Taine, J. and Petit, J. P. (1993) *Heat Transfer*, Prentice Hall, New York.

6 Introduction to convection

6.1 Introduction and classification

Convection heat transfer may be classified as belonging to either of two subgroups, namely, *free* or *forced*. Motions, or fluid flows, which are caused, and sustained, solely by the density gradients created by temperature differences are termed *natural*, or free. In contrast, motions that are induced by the action of an external cause are termed forced.

Free convection, may be observed, for example, around a hot plate or tube that is situated within a quiescent fluid, or within the fluid tubes of a thermosyphon solar water heater. Fluid motion is caused solely by the difference between weight and buoyancy in the gravitational field, with buoyancy differences created through a difference between the surface and fluid temperatures.

The passage of warm air driven through ventilation ducts, hot fluid pumped through pipes, and air blown across the heated elements in an electric fan heater are examples of where heat transfer between the fluid and the surface in question is driven by *forced convection*. The transfer of heat promoted in many thermal engineering systems often comprises both forced and free convection components.

For example, consider the case of the above mentioned ventilation duct. Provided there is a temperature gradient, and, presuming the air within the duct to be above ambient temperature, heat transfer from the flowing internal air to the duct walls will be driven by forced convection, whereas, on the external surface, provided there is no forced cooling or heating, heat will be transferred to the surrounding air by free convection. Heat will, of course, also be transferred to the surroundings by radiation from the duct outer surface, and the total heat transferred will be the sum of the free convective and radiative components. The mechanisms and methods of quantifying radiative heat transfer are presented in Chapter 9.

Convection may be further classified in relation to both the geometry of the system or surface under consideration, and the *type* of flow, that is, either *laminar* or *turbulent*. In the former case, in considering *geometry*, the flow may be considered to be either *external*, for example, flow of a semi-infinite

Table 6.1.1 Typical range of values for the surface convection heat transfer coefficient

Process	*Fluid type*	h (W/m^2 K)
Free convection	Gas	2–25
	Liquid	50–1000
Forced convection	Gas	25–250
	Liquid	50–20000

unbounded fluid over a plate or around a cylinder, or, as is the case for flow inside tubes and ducts, *internal*. An additional classification considers the number of phases of the fluid present, that is, liquid or gas. However, in the present text we limit ourselves solely to the study of heat transfer within single-phase fluids. Both Suryanarayana (1995) and Incropera and DeWitt (2002) present studies of boiling and condensation heat transfer should the reader wish to examine this particular field of study in depth.

Irrespective of the mechanism, either free or forced, which provides the motive force for heat transfer, the geometry, or flow condition, the fundamental convection law, often referred to as Newton's Law of Cooling is

$$q = hA\Delta T \tag{6.1.1}$$

where q is the heat flux (W/m^2), h the *convection heat transfer coefficient* (W/m^2 K), and ΔT the difference between the surface and the fluid temperatures (K). Often, the temperature of both the surface and the bulk fluid is known, thus requiring knowledge of h alone to determine the rate of heat transfer. Where the heat flux is constant, as is, for example, the case for the heating element in an electric kettle, derivation of the convection heat transfer coefficient provides for the estimation of the difference between the element surface and fluid temperatures (Holman, 1990).

As fluid velocities are generally greater for forced convection systems, the surface heat transfer coefficient is also generally larger. Table 6.1.1 presents a comparison of the range of values for h for forced and free convection, for both liquids and gases.

6.2 The convection heat transfer coefficient

As can be seen from Table 6.1.1, while h is generally greater for forced convection, there is a degree of overlap in its values for the two convection processes, and between each of the fluid types. For large temperature differences, the buoyancy forces induced in free convection can create velocities that approach those realised in certain forced convection systems. Also, for viscous liquids, the forced convection heat transfer coefficient may be lower than that established for a free convection system comprising a less viscous solution.

The value of h depends on many factors, including the fluid thermophysical properties, surface roughness and geometry, and resulting boundary layer conditions. The convective heat transfer coefficient is therefore not a property of the fluid, but is determined by the velocity of the flow and fluid properties. In considering the effect of geometry on h, for the same fluid properties, temperatures, and velocity, for free convection, the heat flux from a horizontal flat plate is different to that from, say, a vertical cylinder of the same surface area. As the free convection *boundary layer* develops as a result of fluid density differences in the gravitational field, the orientation of a surface within the gravitational field will affect the boundary layer, and hence also the heat flux.

The study of both free and forced convective heat transfer is therefore primarily concerned with the derivation of the heat transfer coefficient in order to define temperatures, and heat fluxes, within thermal systems operating with fluids. The heat transfer coefficient is highly dependent on the condition of the boundary layer that forms when a fluid possesses motion relative to a surface.

The term boundary layer is conventionally used for the thin region of steep temperature and velocity gradients that exist, bounded by a surface on the one side, and, in the case of free convection, an extensive quiescent fluid on the other. Schlichting (1960) provides a detailed account of factors affecting boundary layer development.

For forced convection the boundary layer is again bounded by a surface on one side, but, more commonly, by a fluid stream on the other. Such flows are termed '*external flows*'. If the surface and fluid temperatures are known, the fluid thermophysical properties may be obtained from available tables for the fluid *film temperature*. The film temperature is generally taken to be the average of the fluid and surface temperatures, as given by Eq. (6.2.1).

$$T_f = (T_S + T_\infty)/2 \tag{6.2.1}$$

where the value of T_f lies between two tabulated temperature values, the required thermophysical properties may be obtained using the interpolation techniques presented in Sections 2.5 and 2.6. For the case of a fluid passing through a closed conduit, '*internal flow*', where, due to the transfer of heat between the fluid and the conduit walls, the fluid free stream temperature is not constant, the fluid properties are evaluated at the *mean bulk temperature* of the fluid. This will be considered in more detail in Chapter 7.

For the case of constant heat flux, where the surface and fluid temperatures are not necessarily known, an initial estimate is made of the film temperature to obtain the values for the fluid properties. The heat transfer coefficient is thence defined, and then used to redefine the film temperature. An iterative process, such as that presented in Section 2.12.2, is then employed to more accurately define the heat transfer coefficient and required temperature values.

While it is possible to derive an analytical solution for h for some of the simpler geometries, it is not possible to obtain an accurate solution for more complex surfaces. The estimation of h therefore relies heavily on the application of empirical relationships that have been obtained through the analysis of experimentally generated data. These relationships, or regressions, defining the convection heat transfer coefficient in relation to the boundary layer condition and fluid thermophysical properties, have been derived through statistical analysis of the aforementioned experimental data.

As, for any case, there will be a certain degree of spread of experimental data around the regression equation, this necessarily implies a margin of error in the value of h obtained via these equations. Suryanarayana (1995) states that, for single-phase fluids, the uncertainty in the value of h calculated using the above mentioned correlations is of the order of $\pm 10\%$. The individual regressions that are most commonly used in the derivation of h for each of the above mentioned sub groups (forced or free, laminar or turbulent, and external or internal) are presented for a range of surface geometries in Chapters 7 and 8 for the individual cases in question. Examples of their application are also given in the usual manner.

As stated above, the heat transfer coefficient is highly dependent on the boundary layer. The study of both forced and free convection is therefore primarily concerned with defining the boundary layer conditions in order to define h. To further develop our understanding of the physical mechanisms that underpin convection energy transfer, and the factors that influence the rate of heat transfer, an overview of those mechanisms and fluid properties that effect the development of the boundary layer is now presented.

6.3 Boundary layers

6.3.1 Introduction

Development of the boundary layer is now presented in relation to forced flow. For certain geometries, where flow is induced through free convection, the boundary layer develops in a similar manner to that now described, and the following principles therefore also apply. In certain cases, however, such as inclined heated and cooled plates, the boundary layer may start to form in the manner described below, but, at some distance x from the leading edge, departs from the surface. It then begins to reform, then departs, and so on. This phenomenon is discussed in more detail in the chapter dedicated to free convection, Chapter 8. As stated earlier, Schlichting (1960) provides a thorough account of boundary layer theory.

If a fluid of known temperature T_∞ passes over a surface at a different temperature, T_S, we know that heat will be transferred between the fluid and the surface. The direction of heat transfer will be dependent on the direction

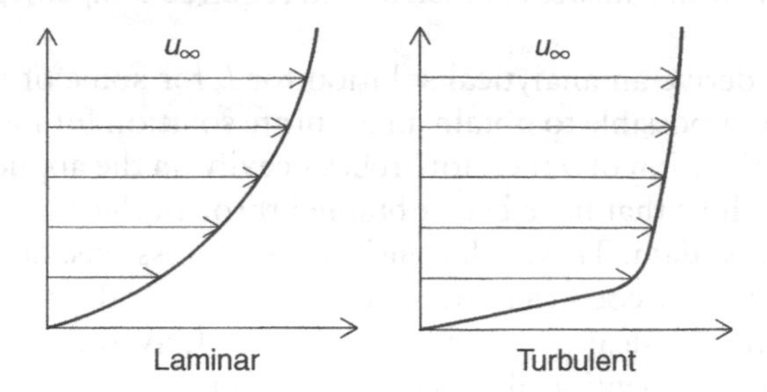

Figure 6.3.1 Typical velocity boundary layer profiles for laminar and turbulent flow.

of the temperature gradient, with heat given to the fluid from the surface if $T_S > T_\infty$. As the fluid possesses motion in relation to the surface, due to the effects of friction between the fluid and the surface, and the viscous forces within the fluid, the fluid particles adjacent to the surface slow down, and, in turn act as to reduce the velocity of those particles in the next fluid layer. At some distance y normal to the surface, as shown in Fig. 6.3.1, this slowing effect becomes negligible and the fluid velocity becomes equal to that of the free stream. It is this region of velocity gradient normal to the surface that is generally termed the boundary layer. In simpler terms, the boundary layer may be defined as the zone of influence across which the transitional momentum and/or thermal effects are experienced by the fluid.

Strictly speaking, the above region is actually the *hydrodynamic* or *velocity boundary layer* and may exist, for forced convection systems, in the absence of temperature differences independently of any *thermal boundary layer*. However, when a temperature gradient does exist, as we shall soon see, while the hydrodynamic and thermal boundary layers are closely linked, they do not necessarily possess the same profile and thickness. The thickness of the velocity boundary layer, hereafter referred to as simply the boundary layer, is often designated by the symbol δ, which is typically defined as the value of y, that is, the distance perpendicular to the surface, for which the fluid velocity $u = 0.99\,u_\infty$ (where the subscript ∞ denotes the free stream conditions outwith the boundary layer). In external free convection, the maximum fluid velocity is generally observed adjacent to the surface and free stream velocity is zero. As can be seen from Fig. 6.3.1, for forced convection over a flat plate, the thickness of the boundary layer changes with the horizontal distance from the leading edge of the surface.

Initially, the boundary layer, developing from the leading edge, is laminar in nature. This term refers to the fact that, within this region, the fluid has a tendency to move as though comprised of distinct layers. The viscous forces within the fluid are equivalent in nature to a *shear stress*, τ, between the fluid layers. If the shear stress is considered to be proportional to the velocity gradient normal to the surface, $\partial u/\partial y$, with the constant of proportionality the *dynamic viscosity*, μ, the shear stress is given by

$$\tau = \mu \frac{\partial u}{\partial y} \tag{6.3.1}$$

The *surface shear stress* may be determined through considering the velocity gradient immediately adjacent to the surface, thus,

$$\tau_S = \mu \left. \frac{\partial u}{\partial y} \right|_{y=0} \tag{6.3.2}$$

Consequently, for external flows, the surface shear stress is a key parameter in defining the local *friction coefficient*,

$$C_f = \frac{\tau_S}{\rho u_\infty^2/2} \tag{6.3.3}$$

Depending on the free stream velocity, u_∞, and fluid properties, at some distance x_{CR} from the leading edge, the flow profile within the boundary layer becomes turbulent. This region of flow may be considered as comprising a narrow laminar sub layer adjacent to the surface, and a larger region of eddies where the motion of fluid particles is random. As such, as can be seen in Fig. 6.3.1, while the boundary layer is thicker, the velocity profile within the turbulent region is much flatter than the parabolic profile observed within the laminar region. It has been shown that the boundary layer condition, that is, laminar or turbulent, may be determined, by considering the magnitude of a dimensionless group of fluid properties called the Reynolds number, Re,

$$\mathrm{Re}_x = \frac{\rho u_\infty x}{\mu} \tag{6.3.4}$$

where ρ is the fluid density (kg/m^3), μ the viscosity (kg/m s), u_∞ the free stream velocity (m/s), and x the distance from the leading edge. The Reynolds number is also sometimes presented in the form given in Eq. (6.3.5),

$$\mathrm{Re}_x = \frac{u_\infty x}{\nu} \tag{6.3.5}$$

where ν is the kinematic viscosity (m^2/s) and is equal to μ/ρ.

The subscript x indicates that the specific value calculated for the Reynolds number corresponds to that distance x considered from the surface leading edge. All fluid properties are taken at the film temperature, Eq. (6.2.1).

For a flat plate, the *critical Reynolds number*, Re_{CR}, being that for which boundary layer flow becomes unstable and moves into the turbulent zone, is known to lie between 10^5 and 3×10^6, with the actual value depending on surface roughness and free stream conditions. A representative value of 5×10^5 is often assumed and will be used for all further calculations in this text. By applying this value, a value may be obtained for the *critical distance*, x_{CR}, from the surface leading edge, at which point boundary layer flow may be assumed to become turbulent.

6.3.2 *Velocity boundary layer thickness*

Again considering forced flow, when fluid particles come into contact with the surface they become trapped and stop moving. As outlined above, these stationary particles then act, through the influence of viscous forces, as to retard the motion of those particles passing adjacent to this surface layer, and so on until a point is reached where the retardation effect is negligible and the particles assume the free stream velocity. The extent of this retarding effect, that is, the boundary layer thickness, is dependent on the density ρ, the dynamic viscosity μ, the free stream velocity u_∞, and the distance from the surface leading edge, x. These parameters, as outlined above and presented in Eq. (6.3.4), combine to produce the dimensionless grouping termed the Reynolds number. If the boundary layer thickness is defined as the distance perpendicular to a surface at which the fluid velocity is 99% of that of the free stream, for laminar flow ($\mathrm{Re}_x < \mathrm{Re}_{CR}$), this distance, or thickness, may be obtained through Eq. (6.3.6) (Incropera and DeWitt, 2002). Similarly, for turbulent flow ($\mathrm{Re}_x > \mathrm{Re}_{CR}$), the thickness may be obtained through Eq. (6.3.7).

$$\delta = \frac{5x}{\mathrm{Re}_x^{0.5}} \tag{6.3.6}$$

$$\delta = \frac{0.37x}{\mathrm{Re}_x^{0.2}} \tag{6.3.7}$$

A plot of boundary layer thickness versus distance from the leading edge was presented in Example 2.5.2. The effect on the thickness δ of varying other fluid properties, such as dynamic viscosity, may be investigated by manipulating the data in the workbook Ex02-05-02.xls, associated with Example 2.5.2.

For flow between surfaces, or, for example, within pipes, at some distance x from the leading edge, the boundary layer may extend outwards (or radially inwards for pipes) from the surface to such an extent that it meets the boundary layer developing from the opposite surface. This is shown in Fig. 6.3.2 for the case of flow within a cylinder. At the point where the boundary layers converge, the flow is deemed to be fully developed, and may further become turbulent if the critical value of Re is exceeded ($\mathrm{Re}_{CR} = 2300$ for pipes).

For free convection, where the free stream/quiescent fluid velocity is zero, the velocity profile will assume a form quite different from that observed for

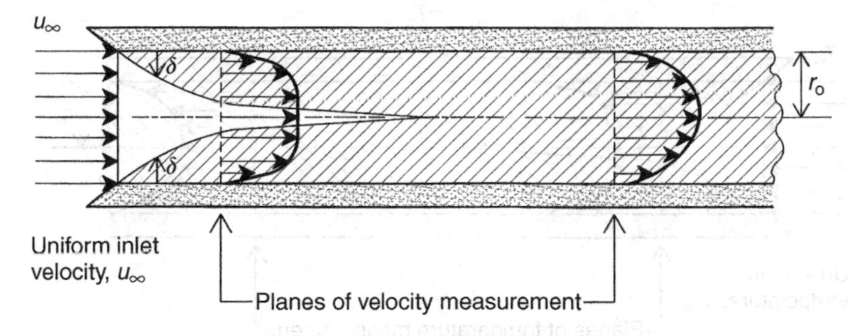

Figure 6.3.2 Velocity boundary layer development in a circular tube of radius r_o.

forced convection. Due to the respective friction and viscosity effects of the surface and the quiescent fluid, the higher fluid velocities will be realised at a point intermediate between the surface wall and boundary layer edge.

Thus far, boundary layer thickness has been considered solely in relation to the velocity profile with fluid properties, namely the density and dynamic viscosity, taken for the film temperature. As stated previously, whenever there is flow over a surface, a velocity boundary layer will form. The velocity boundary layer will therefore develop irrespective of fluid and surface temperatures. However, in the presence of temperature differences, a temperature profile will develop normal to the surface. A discussion on the nature of the profile and the associated method employed in the calculation of the thickness of the thermal boundary layer is now presented.

6.3.3 *Thermal boundary layer*

While, for forced convection heat transfer, the velocity profiles will assume the form of those shown in Fig. 6.3.1, the thermal profile will be quite different, Bejan (1993). The thermal boundary layer may be defined as the region where temperature gradients are present in the flow, with its thickness designated as δ_T. For the case of a heated flat plate, where $T_S > T_\infty$, as shown in Fig. 6.3.3, the fluid temperature will be greatest adjacent to the surface and decease with distance from the wall. The temperature profile within the thermal boundary layer is therefore seen to be the converse of that for the velocity. While the form of the respective profiles may be different, both result from the properties and nature of the fluid motion adjacent to the surface.

The ratio of their respective thicknesses has been shown to depend on the Prandtl number, Pr, formulated by Nusselt and named after Ludwig Prandtl (1875–1953), a German scientist who was instrumental in introducing the theory of boundary layer formation. This dimensionless parameter, as given in Eq. (6.3.8), provides a measure of the ratio of the rate at which momentum

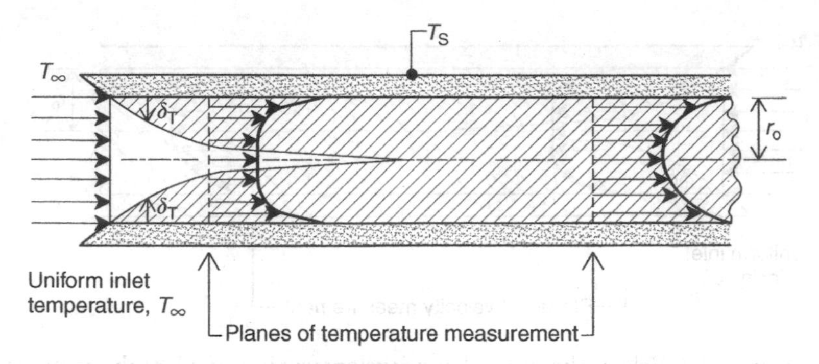

Figure 6.3.3 Thermal boundary layer development in a heated circular tube.

may diffuse through the fluid to that of its thermal diffusivity.

$$\Pr = \frac{\nu}{\alpha} = \frac{C_P \mu}{k} \tag{6.3.8}$$

The Prandtl number, obtained from fluid property tables for the fluid film temperature, therefore provides the link between the velocity and thermal boundary layers. For forced convection and laminar flow, as will be discussed in Section 7.2.1, through analysis and experimentation, the relationship between the thickness of the thermal and velocity boundary layers is as given by Eq. (6.3.9).

$$\delta_T = \frac{\delta}{\Pr^{1/3}} \tag{6.3.9}$$

For a Prandtl number of less than 1 (Pr for air ~ 0.7) for a given distance $x < x_{CR}$, the thermal boundary layer will therefore be thicker than the velocity boundary layer. For other gases, such as Ammonia (Pr $= 0.887$ at 300 K), or steam (Pr $= 1.06$ at 380 K) the Prandtl number approaches unity. If Pr $= 1$, that is, $\nu = \alpha$, developing from the surface leading edge, the velocity and thermal boundary layers will grow at the same rate, and hence $\delta = \delta_T$. For liquid metals, the Prandtl number is much smaller (Pr for mercury at 300 K $= 0.0248$, McAdams, 1954), and the thermal energy diffusion rate is therefore much greater than that of momentum.

For fully developed turbulent flow, the transfer of momentum and energy by eddies in the boundary layer outweighs that by molecular and thermal diffusion. The thermal and hydrodynamic boundary layer thicknesses are therefore the same, as given by

$$\delta_T = \delta \tag{6.3.10}$$

For a more detailed study of boundary layer theory and development, the reader is again directed towards one of the most frequently referred texts on this subject, namely, Schlichting (1960).

6.4 Governing equations and relationships

6.4.1 *Introduction*

As highlighted earlier, analytical determination of the convective heat transfer coefficient is restricted to a few of the simpler surface geometries. It is therefore more practical to use empirically defined relationships that have been derived from, and subsequently validated through, the analysis of experimentally obtained data. These empirical relationships or equations are often presented in the form of a group of dimensionless parameters that are themselves linked through their common use of fluid property variables.

In most convective heat transfer regressions there are typically three interrelated dimensionless parameters, derived from fluid properties and flow conditions, that are required to obtain an estimate of the convection heat transfer coefficient. In forced convection, as previously described, the Reynolds number defines the flow condition, and the Prandtl number, introduced via Eq. (6.3.6), provides a measure of the ratio of momentum and thermal diffusivities. The Nusselt number, as defined in Eq. (6.4.1), contains the convective heat transfer coefficient as one of its' parameters, and is a measure of the dimensionless temperature gradient at the surface/fluid interface. It provides a measure of the ratio of the convective to conductive heat transfer coefficients.

$$\mathrm{Nu} = \frac{hd}{k} \tag{6.4.1}$$

Where the Reynolds number defines the flow condition for forced convection flows, the Grashof number plays a similar role for *buoyancy-induced flows*. This parameter will be discussed in more detail in Section 6.4.3. A summary of the dimensionless groups commonly used in convective heat transfer analyses is given in Box 6.4.1.

6.4.2 *Forced convection regressions*

In forced convection, considering the variables that comprise the three above mentioned parameters, namely, the Reynolds, Prandtl, and Nusselt numbers, we expect the convective heat transfer coefficient to be dependent on the velocity of the fluid, its viscosity, thermal conductivity, density, specific heat, and the *characteristic length* of the surface over which the fluid is flowing. Again, through analysis of experimental data, it has been found that the Nusselt number may be expressed in the general form given in Eq. (6.4.2)

$$\mathrm{Nu} = C\,\mathrm{Re}^m\,\mathrm{Pr}^n \tag{6.4.2}$$

Box 6.4.1 Summary of dimensionless groups commonly used in convection heat transfer analyses

Group	*Interpretation*	*Definition/equation*
Grashof number (Gr)	Ratio of buoyancy to viscous forces	$\dfrac{g\beta(T_S - T_\infty)L^3}{\nu^2}$
Nusselt number (Nu_L)	Dimensionless temperature gradient at a surface. Also ratio of convective to conductive heat transfer	$\dfrac{hL}{k_f}$
Peclet number (Pe)	Dimensionless independent heat transfer parameter	RePr
Prandtl number (Pr)	Ratio of momentum and thermal diffusivities	$\dfrac{c_p\mu}{k} = \dfrac{\nu}{\alpha}$
Rayleigh number (Ra)	Product of Grashof and Prandtl numbers	$\dfrac{g\beta(T_S - T_\infty)L^3}{\alpha\nu}$
Reynolds number (Re)	Ratio of inertial and viscous forces	$\dfrac{VL}{\nu}$
Stanton number (St)	Modified Nusselt number	$\dfrac{h}{\rho V c_p} = \dfrac{Nu_L}{Re_L Pr}$

The constant '*C*', and exponents '*m*' and '*n*', whose respective values will be presented in detail in Chapter 7, have been shown to be dependent on the Reynolds number and the surface geometry. Application of the above equation, with the appropriate values for the aforementioned constant and exponent values, has been shown to afford a reasonable regression with experimental data. However, alternative regressions have been derived for specific scenarios. These will also be discussed in detail in Chapter 7.

6.4.3 *Natural convection regressions*

As for forced convection, the regressions relating the heat transfer coefficient to fluid properties are again expressed in dimensionless form and depend on the interrelationship between three dimensionless parameters. Whereas the Reynolds number, which includes a measure of the fluid free stream velocity and the ratio of inertial to viscous forces, was used to define the flow condition within the boundary layer in forced convection, an alternative dimensionless group, termed the Grashof number, is applied in free convection systems. The Grashof number, as given in Eq. (6.4.3), provides a measure of the ratio of buoyancy forces, resulting from temperature induced density differences, to fluid viscous forces. It is these buoyancy forces that induce the motion of fluid in natural convection systems. As such, it therefore plays a similar role in free convection to that of the Reynolds number in forced convection.

$$\mathrm{Gr} = \frac{g\beta\rho^2(T_S - T_\infty)L^3}{\mu^2} \tag{6.4.3}$$

where β is the volumetric thermal expansion coefficient (K^{-1}), and g is the gravitational constant (m/s^2). As $\nu = \mu/\rho$, the above expression is often simplified to that given in Eq. (6.4.4)

$$\mathrm{Gr} = \frac{g\beta(T_S - T_\infty)L^3}{\nu^2} \tag{6.4.4}$$

While Eq. (6.4.4) will be used hereafter for calculating the Grashof number, Eq. (6.4.3), through its inclusion of the density term, is included to highlight those factors, presented below, that affect the convection heat transfer coefficient.

If one considers the variables included in the above Grashof number in addition to those in the Nusselt and Prandtl numbers, it can be seen that the convective heat transfer coefficient is a function of:

- the characteristic length of the surface, L,
- the fluid thermal conductivity, k,
- fluid viscosity, μ, which acts as to retard flow,
- specific heat at constant pressure, c_p,
- gravitational acceleration, g,
- temperature difference, $T_S - T_\infty$,
- rate of change of density with respect to temperature, $(\partial\rho/\partial T)$.

This latter variable is contained in the Grashof number (Eq. (6.4.3)) in the form of the product of the coefficient of thermal expansion, β, and the square of the density, ρ^2.

As discussed in Section 6.2, all fluid properties are generally evaluated at the film temperature. However, as will be seen in both Chapters 7 and 8, some of the regressions presented require that properties be obtained for both the film and surface temperatures. Unless otherwise stated, all fluid properties are film temperature based.

The volumetric thermal expansion coefficient, β, for an ideal gas, can be shown to be equal to the reciprocal of the film temperature in Kelvin (see Box 6.4.2).

$$\beta = \frac{1}{T} \tag{6.4.5}$$

For non ideal gases and liquids the value of β must be obtained from the appropriate property tables.

Box 6.4.2 Derivation for the coefficient of thermal expansion for ideal gases

The manner in which β is obtained depends on the gas. For an ideal gas, $\rho = p/RT$, and hence,

$$\beta = -1/\rho(\partial\rho/\partial T)_p = (1/\rho)(p/RT^2) = 1/T$$

6.4.4 *Characteristic length*

The characteristic length, L, is a representative measure of the length of the surface over which the fluid flows. For example, for a flat plat with external parallel flow, as will be discussed in Section 7.2, the characteristic length is simply the length of the plate. However, for other surface geometries, such as those presented by angular or spherical objects, the characteristic length is taken as the length of the subject cross-section presented perpendicular to the direction of fluid flow. For external forced convection, a summary of the characteristic length, employed in the analysis of heat transfer from cylinders and spheres is given in Table 7.2.2. In this context, the characteristic length is usually referred to as 'D'.

References

Bejan, A. (1993) *Heat Transfer*, John Wiley and Sons, New York.
Holman, J. P. (1990) *Heat Transfer*, 7th edn, McGraw-Hill.
Incropera, F. P. and DeWitt, P. (2002) *Fundamentals of Heat and Mass Transfer*, 5th edn, John Wiley and Sons, New York.
McAdams, W. H. (1954) *Heat Transmission*, 3rd edn, McGraw-Hill, New York.
Schlichting, H. (1960) *Boundary Layer Theory*, 4th edn, McGraw-Hill, New York.
Suryanarayana, N. V. (1995) *Engineering Heat Transfer*, West Publishing Company, New York.

7 Forced convection

7.1 Introduction

The main factors affecting the rate of *convective heat transfer* were introduced in Chapter 6. Convection heat transfer was classified according to the mode, that is, either *free* or *forced*, the type of flow, either *laminar* or *turbulent*, and the flow geometry, either *internal* or *external*. In *free convection*, the temperature difference between a body and the quiescent surroundings induces a density differential in the fluid surrounding the body. Gravitational effects acting on this density differential create a buoyancy force. This buoyancy force has the impetus to cause displacement in the surrounding fluid thus establishing fluid motion around the body in the form of free convection currents. A full analysis of the factors that affect the *free convection heat transfer coefficient* for different flow regimes and geometries is presented in Chapter 8.

In this chapter, however, we shall concern ourselves with convection heat transfer in situations where fluid flow is induced by some external means such as a pump or a fan. The heat transfer promoted by this means is referred to as *forced convection* and may be studied, as outlined in Chapter 6, in terms of two sub-topics, namely, external and internal flow. In the case of external flow, *boundary layers* are generally able to develop freely,whereas, for internal flow, as will be discussed in Section 7.3, the constraints imposed by adjacent surfaces affect their development. Hence, a separate analysis is required for each of these flow regimes. The importance of the interrelationship of the three dimensionless parameters, namely, the Reynolds, Prandtl, and Nusselt numbers, in defining *the convective heat transfer coefficient* for *forced convection* systems was introduced in Chapter 6. The general form of the equation relating these three parameters, inferred from dimensional analysis and experimental measurements, is given as in Eq. (7.1.1)

$$\overline{\mathrm{Nu}_L} = C\,\mathrm{Re}_L^m\,\mathrm{Pr}^n \tag{7.1.1}$$

where the subscript L (or as is the case for cylinders and tubes D), and the bar over the Nusselt number, indicates that the Nusselt and Reynolds numbers used are defined for the surface characteristic length (or diameter) and are therefore average values. In this form Eq. (7.1.1) will yield a value for the average convective heat transfer coefficient for the whole surface. Where a value of h is required for a specific location on a given surface, Eq. (7.1.1) is presented in the form given in Eq. (7.1.2), where the subscript x denotes that the Nusselt and Reynolds numbers are calculated for that specific distance 'x' from the leading edge. For heated or cooled horizontal plates

$$\mathrm{Nu}_x = C\mathrm{Re}_x^m \mathrm{Pr}^n \tag{7.1.2}$$

Generally, in engineering heat transfer problems, we are concerned with obtaining a value for the total heat flux from or to the surface in question, and Eq. (7.1.1) is more commonly used. However, as we shall see in Example 7.2.2 knowledge of the local heat transfer coefficient is also often required. As the manner in which the fluid boundary layer develops depends on surface geometry, for the same free stream velocity and fluid and surface temperatures, the convective heat transfer coefficient, and hence Nusselt number, will also vary. The coefficient 'C', and exponents 'm', and 'n' in Eqs (7.1.1) and (7.1.2) will also therefore depend on the shape of the surface and condition of the flow. While an analytical approach may be applied for some of the simpler geometries, as discussed in Chapter 6, for more complex surfaces, the above mentioned coefficient and exponents can be obtained through regressing the data obtained from experimental measurements in terms of the appropriate dimensionless parameters. For the majority of cases, for fluids with a Prandtl number between 0.6 and 50, the value of 'n', being the exponent for the Prandtl number, has been found to have a value of 1/3. If $\mathrm{Pr} \ll 1$, as for liquid metals, $n = 1/2$ (Suryanarayana, 1995).

The main aims of the present chapter are therefore to determine and present the convection coefficient 'C' and exponent 'm' for different flow and surface geometries. This will allow an estimation of the heat transfer coefficient to be obtained for a given situation, and, hence, provide the tools for solving many common forced convection heat transfer problems.

While, for the majority of flow conditions and surface geometries the respective values of the above mentioned coefficient 'C', and exponents 'm' and 'n' are derived through regression of convection heat transfer experimental data, an analytical solution may be applied to the simplest forced convection case of *external laminar flow over a flat plate*. The case of external flow and the analytical solution for *laminar flow across a flat plate* is therefore considered prior to that of internal flow.

7.2 External flow

7.2.1 Analytical solution for laminar parallel flow across a flat plate

As an introduction to the solutions for forced convection external flow heat transfer, the case of laminar parallel flow along a flat plate is now presented.

This situation represents perhaps the most basic of the analytical boundary layer heat transfer solutions. It is assumed that the free stream velocity is constant and that the surface is semi-infinite, and at a constant temperature. As the plate temperature is uniform, both the thermal and hydrodynamic boundary layers will develop from the surface leading edge.

The motion of a fluid in which there are both velocity and temperature gradients must obey certain natural laws. Mass and energy must be conserved at each point in the fluid, and *Newton's second law of motion* must also be satisfied. While it is not within the scope of the present text to detail the derivation of the equations that satisfy these laws, the reader is directed towards Schlichting (1960), and Incropera and DeWitt (2002) for a fuller derivation. Assuming steady, two-dimensional flow of an incompressible fluid with constant properties, the aforementioned equations may be reduced to a simpler form. These simplified equations are now presented to provide a foundation for the foregoing analysis:

$$\frac{\partial u}{\partial x} + \frac{\partial v}{\partial y} = 0 \qquad \text{(continuity/balance of mass)} \tag{7.2.1}$$

$$u\frac{\partial u}{\partial x} + v\frac{\partial u}{\partial y} = \nu\frac{\partial^2 u}{\partial y^2} \qquad \text{(momentum)} \tag{7.2.2}$$

$$u\frac{\partial T}{\partial x} + v\frac{\partial T}{\partial y} = \alpha\frac{\partial^2 T}{\partial y^2} \qquad \text{(energy)} \tag{7.2.3}$$

Again assuming constant fluid properties, conditions in the *hydrodynamic boundary layer* will be independent of temperature. It is therefore possible to solve Eqs (7.2.1) and (7.2.2) for the hydrodynamic boundary layer, independently of Eq. (7.2.3), the energy equation. The solution of these equations will then afford a solution to Eq. (7.2.3), which requires knowledge of both u and v. The velocity components, u and v, may be defined in terms of a stream function, $\psi(x, y)$, where

$$u \equiv \frac{\partial \psi}{\partial y} \quad \text{and} \quad v \equiv -\frac{\partial \psi}{\partial x} \tag{7.2.4}$$

automatically satisfying Eq. (7.2.1). The solution thereafter requires that two new independent variables, f and η are defined, such that,

$$f(\eta) \equiv \frac{\psi}{u_\infty\sqrt{(\nu x/u_\infty)}} \tag{7.2.5}$$

and

$$\eta \equiv y\sqrt{(u_\infty/\nu x)} \tag{7.2.6}$$

The term η is called the *similarity variable* as the form of the boundary layer velocity profile, u/u_∞, remains *geometrically similar*, that is, parabolic, over

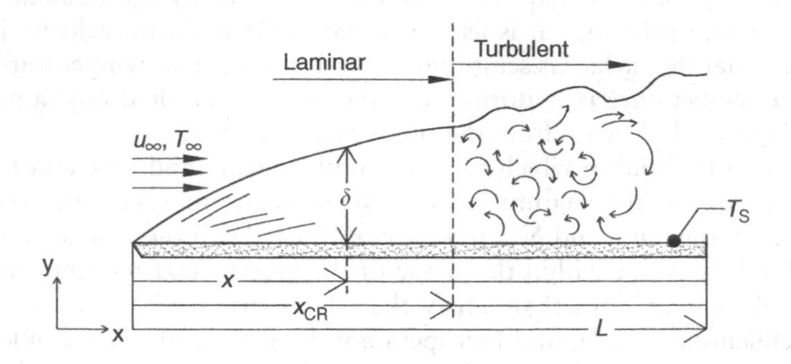

Figure 7.2.1 The flat plate in parallel flow.

the length of the plate, irrespective of the distance x from the leading edge. If, as shown in Fig. 7.2.1, y represents the distance perpendicular to the plate surface, and δ the boundary layer thickness, the above mentioned similarity may be expressed in the form given as

$$\frac{u}{u_\infty} = \phi\left(\frac{y}{\delta}\right) \tag{7.2.7}$$

From our earlier consideration of factors affecting boundary layer development, it is clear that, as the fluid kinematic viscosity, $\nu(\text{m}^2/\text{s})$, and distance, x (m), from the leading edge increase, the thickness of the *velocity* boundary layer will also increase. Thickness δ will however decrease for increasing free stream velocity $u_\infty(\text{m/s})$. If we assume that thickness varies as $(\nu x/u_\infty)^{1/2}$, Eq. (7.2.7) may be rewritten as

$$\frac{u}{u_\infty} = \phi(\eta) \tag{7.2.8}$$

Thus, the velocity profile may be defined solely in terms of the *similarity variable* η, which itself depends on x and y (Eq. (7.2.6)). From Eqs (7.2.4) to (7.2.6) we get, first, for the u component of velocity,

$$u = \frac{\partial\psi}{\partial\eta}\frac{\partial\eta}{\partial y} = u_\infty\sqrt{\frac{\nu x}{u_\infty}}\frac{\mathrm{d}f}{\mathrm{d}\eta}\sqrt{\frac{u_\infty}{\nu x}} = u_\infty\frac{\mathrm{d}f}{\mathrm{d}\eta} \tag{7.2.9}$$

and, second, for the v component,

$$v = -\frac{\partial\psi}{\partial x} = -\left(u_\infty\sqrt{\frac{\nu x}{u_\infty}}\frac{\partial f}{\partial x} + \frac{u_\infty}{2}\sqrt{\frac{\nu}{u_\infty x}}f\right)$$

which reduces to

$$v = \frac{1}{2}\sqrt{\frac{\nu u_\infty}{x}}\left(\eta\frac{\mathrm{d}f}{\mathrm{d}\eta} - f\right) \tag{7.2.10}$$

By differentiating the respective velocity components it can also be shown that

$$\frac{\partial u}{\partial x} = -\frac{u_\infty}{2x}\eta\frac{d^2 f}{d\eta^2} \tag{7.2.11}$$

$$\frac{\partial u}{\partial y} = u_\infty\sqrt{\frac{u_\infty}{\nu x}}\frac{d^2 f}{d\eta^2} \tag{7.2.12}$$

$$\frac{\partial^2 u}{\partial y^2} = \frac{u_\infty^2}{\nu x}\frac{d^3 f}{d\eta^3} \tag{7.2.13}$$

If we substitute these equations into the hydrodynamic boundary layer momentum equation, Eq. (7.2.2), we obtain the non-linear third-order ordinary differential equation as

$$2\frac{d^3 f}{d\eta^3} + f\frac{d^2 f}{d\eta^2} = 0 \tag{7.2.14}$$

In terms of the velocity components, u and v, the appropriate boundary conditions, or constraints, for Eq. (7.2.14) are, first for the fluid layer immediately adjacent to the surface, at $y = 0$,

$$u(x,0) = v(x,0) = 0 \tag{7.2.15}$$

and, where y is greater than or equal to the boundary layer thickness, which may be taken as $y = \infty$, or, $y \to \infty$.

$$u(x,\infty) = u_\infty \quad \text{or, alternatively} \quad \frac{\partial u}{\partial y} = 0 \tag{7.2.16}$$

These boundary conditions may also be expressed in terms of the similarity variables, such that

$$\frac{df}{d\eta} = f(0) = 0 \text{ at } \eta = 0 \quad \text{and} \quad \frac{df}{d\eta} = 1.0 \text{ at } \eta \to \infty \tag{7.2.17}$$

The solution to Eq. (7.2.14) may be afforded, subject to the constraints of Eq. (7.2.17), either through a series expansion, or by numerical integration. While Incropera and DeWitt (2002) present selected results from the expansion, a detailed account of the solution afforded by the integration method is given in Holman (1990). Both methods yield the same result for the hydrodynamic boundary layer thickness, given as

$$\delta = \frac{5x}{\mathrm{Re}_x^{1/2}} \tag{7.2.18}$$

The thickness of the hydrodynamic boundary layer is therefore proportional to the distance from the plate leading edge, and inversely proportional to the root of the Reynolds number.

From Eq. (7.2.12) the wall shear stress, τ_S, may be expressed as

$$\tau_S = \mu \left.\frac{\partial u}{\partial y}\right|_{y=0} = \mu u_\infty \sqrt{\frac{u_\infty}{\nu x}} \left.\frac{d^2 f}{d\eta^2}\right|_{\eta=0} \tag{7.2.19}$$

Solving Eq. (7.2.19) for the wall shear stress, yields

$$\tau_S = 0.332 u_\infty \sqrt{\frac{\rho \mu u_\infty}{x}} \tag{7.2.20}$$

and substituting Eq. (7.2.20) into the equation for the local friction coefficient, Eq. (6.3.3), gives a value for the local friction coefficient in terms of the Reynolds number,

$$C_{f,x} \equiv \frac{\tau_S}{\rho u_\infty^2/2} = 0.664 \mathrm{Re}_x^{1/2} \tag{7.2.21}$$

The local friction coefficient is an important parameter in defining surface frictional drag.

Thus far, the analysis presented has afforded a solution to the boundary layer equation, Eq. (7.2.2). Through this solution, we now concern ourselves with obtaining a solution to the energy equation, Eq. (7.2.3). As before, we now introduce another dimensionless variable, the dimensionless temperature, T^*, such that,

$$T^* \equiv \frac{T - T_S}{T_\infty - T_S} \tag{7.2.22}$$

and again assume a similarity solution, $T^* = T^*(\eta)$. Through making the appropriate substitutions, Eq. (7.2.3) reduces to,

$$\frac{d^2 T}{d\eta^2} + \frac{\mathrm{Pr}}{2} f \frac{dT^*}{d\eta} = 0 \tag{7.2.23}$$

The conditions that the temperature distribution must satisfy are that,

$$T = T_S \quad \text{at } y = 0 \tag{7.2.24a}$$

$$\frac{\partial T}{\partial y} = 0 \quad \text{at } y = \delta_t \tag{7.2.24b}$$

and

$$T = T_\infty \quad \text{at } y = \delta_t \tag{7.2.24c}$$

Applying the constraints outlined in Eqs (7.2.24a)–(7.2.24c), Kays and Crawford (1980) solved Eq. (7.2.23) by numerical integration for different values of the Prandtl number. They found that for $\Pr \geq 0.6$, the surface temperature gradient, $\mathrm{d}T^*/\mathrm{d}\eta|_{\eta=0}$ could be represented by

$$\left.\frac{\mathrm{d}T^*}{\mathrm{d}\eta}\right|_{\eta=0} = 0.332\,\Pr^{1/3} \tag{7.2.25}$$

Fortunately, most liquids and gases fall within this category, making Eq. (7.2.25) broadly applicable. However, as liquid metals have Prandtl numbers of the order of 0.01, the above relationship does not apply.

From Newtons law of cooling, the local convection heat transfer coefficient, h, may be expressed as,

$$h_x = \frac{q''_\mathrm{S}}{T_\mathrm{S} - T_\infty} = \frac{T_\infty - T_\mathrm{S}}{T_\mathrm{S} - T_\infty} k \left.\frac{\partial T^*}{\partial y}\right|_{y=0} \tag{7.2.26}$$

which may be reduced to

$$h_x = k\left(\frac{u_\infty}{\nu x}\right)^{1/2} \left.\frac{\mathrm{d}T}{\mathrm{d}\eta}\right|_{\eta=0} \tag{7.2.27}$$

Making the necessary substitutions and rearranging Eq. (7.2.27), the local Nusselt number is found to be of the form,

$$\mathrm{Nu}_x = \frac{h_x x}{k} = 0.332\mathrm{Re}^{1/2}\,\Pr^{1/3} \quad \text{for } \Pr \geq 0.6 \tag{7.2.28}$$

Also, from the solution to Eq. (7.2.23), the ratio of the velocity and thermal boundary layer thickness is given by,

$$\delta_\mathrm{T} = \frac{\delta}{\Pr^{1/3}} \tag{7.2.29}$$

Should the reader wish to further explore the effect of fluid properties and system conditions on boundary layer thickness, the parameters given in Example 2.5.2, and file Ex02-05-02.xls may be altered to suit.

Equations (7.2.18), (7.2.21), (7.2.28), and (7.2.29) may be used to compute the thickness of the hydrodynamic boundary layer, the local friction coefficient, the local Nusselt number, and the thickness of the thermal boundary layer respectively for laminar flow only. That is, they apply only where $0 < x < x_\mathrm{CR}$. If we define the average friction coefficient as,

$$\overline{C}_{\mathrm{f},x} \equiv \frac{\overline{\tau}_{\mathrm{S},x}}{\rho u_\infty^2/2} \tag{7.2.30}$$

where,

$$\overline{\tau}_{\mathrm{S},x} \equiv \frac{1}{x}\int_0^x \tau_{\mathrm{S},x}\,\mathrm{d}x \tag{7.2.31}$$

then, substituting the form of $\tau_{S,x}$ from Eq. (7.2.21) and performing the integration, the average friction coefficient may be obtained as,

$$\overline{C_{f,x}} = 1.328\mathrm{Re}_x{}^{-1/2} \tag{7.2.32}$$

Considering the average heat transfer coefficient to be of the form,

$$\bar{h}_x = \frac{1}{x}\int_0^x h\,\mathrm{d}x \tag{7.2.33}$$

and integrating and substituting from Eq. (7.2.28), the *average heat transfer coefficient* is found to be twice that of the local heat transfer coefficient. Hence,

$$\overline{\mathrm{Nu}_x} \equiv \frac{\bar{h}_x x}{k} = 0.664\mathrm{Re}_x{}^{1/2}\,\mathrm{Pr}^{1/3} \quad \text{for Pr} \geq 0.6 \tag{7.2.34}$$

Equations (7.2.32) and (7.2.34) show that, for laminar flow, both the average friction coefficient and the average heat transfer coefficient, from the leading edge to a point x on the surface, are *twice* the *local coefficients* at that point. This is shown schematically in Fig. 7.2.2. As detailed in Chapter 6, all fluid properties are taken for the film temperature.

As stated above, Eqs (7.2.28) and (7.2.34) are not applicable for fluids with a small Prandtl number. For low Prandtl numbers, the thermal boundary layer develops much more rapidly than the velocity boundary layer. In this instance it is not unreasonable to assume that the *velocity profile* is uniform within the thermal boundary layer. By solving the thermal boundary layer equation, based on this assumption, Kays and Crawford (1980) showed that,

$$\mathrm{Nu}_x = 0.565\mathrm{Pe}_x^{1/2} \quad \text{for Pr} \leq 0.05 \text{ and Pe} \geq 100 \tag{7.2.35}$$

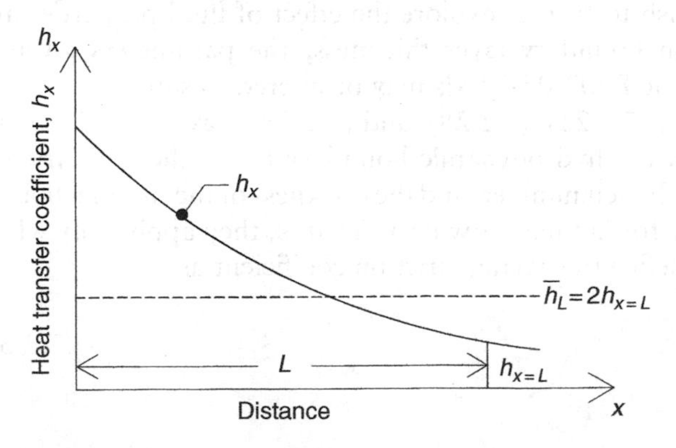

Figure 7.2.2 Variation of the local heat transfer coefficient hx with distance from the leading edge.

where Pe_x is the Peclet number, and is given by,

$$Pe_x = Re_x \, Pr \tag{7.2.36}$$

While the above relationships have been derived through an analytical approach, Churchill and Ozoe (1973) recommend the following regression for all values of the Prandtl number, for laminar flow.

$$\overline{Nu_L} = \frac{0.6774 \, Pr^{1/3} Re_L^{1/2}}{\left[1 + \left(0.0468 \, Pr^{2/3}\right)\right]^{1/4}} \tag{7.2.37}$$

with $Nu_{x=L} = \overline{Nu_L}/2$.

Example 7.2.1 Laminar parallel flow over a flat plate with *uniform surface temperature.*

Air at a temperature of 26°C flows over a flat plate of length 2.3 m by 0.2 m at a velocity of 4 m/s. If the plate has a uniform surface temperature of 44°C, find the rate of heat loss from the plate to the air for the first 0.9 m of the plate. Also, determine whether the flow is laminar over the entire length of the plate.

Analysis

In order to determine the plate heat loss, we must first define the fluid film temperature, being the average of the plate surface and fluid free stream temperatures. For the film temperature, the appropriate fluid properties may be obtained, and the Reynolds number calculated. For a flat plate, the *critical* Reynolds number, being that value above which flow becomes turbulent, is normally taken as 5×10^5. For the plate length considered, provided the Reynolds number calculated is less than the *critical* value, Eqs (7.2.28) and (7.2.34) may be used to calculate the local and average Nusselt numbers respectively. Thereafter, the average heat transfer coefficient and heat transfer rate may be defined.

For the fluid properties given, setting the Reynolds number to 5×10^5 allows the *critical length*, referred to as x_{CR}, or L_{CR}, to be found.

Solution

Open the workbook Ex07-02-01.xls. This workbook comprises two sheets, namely, *Compute*, and *Properties*. Select the *Compute* sheet by clicking on its tab and input the data given in **cells (B4:B9)**. The information required is detailed in **cells (A12 and B12)**.

Excel performs the calculations as follows:

- The mean fluid temperature is calculated in **cell B15**.
- For the given film temperature Excel then calculates the thermophysical properties from the data given in the Properties sheet, and posts the data in **cells (G8:N8)**.
- The Reynolds number, **cell B16**, is calculated for the fluid velocity and distance from the plate leading edge specified. As it is below the limit for transition (5×10^5) flow is laminar at this point.
- The local Nusselt number is calculated via Eq. (7.2.28), **cell B17**, followed by the average Nusselt number (Eq. (7.2.34)) **cell B18** for the part of the plate considered. Nu_{AV} is found to be 274.4.
- This yields a value for h_{AV} of $8.2\,W/m^2\,K$ **cell B19**, and thereafter, a value of 26.6 W for q, **cell B20**.

Discussion

To determine whether flow is laminar over the entire length of the plate, for the given fluid velocity, the transition point may be determined by setting the Reynolds number to 5×10^5, and solving for L. As can be seen from the value of 2.09 m obtained for L_{CR}, **cell B21**, fluid flow is laminar for the majority of the plate length.

The foregoing solutions and equations have been derived for *laminar* flow over a flat plate with a uniform surface temperature. For *uniform heat flux*, as the heat transfer rate is already known, there is often no need to find the average heat transfer coefficient. However, where knowledge of the local surface temperature is required, the regressions presented in Eqs (7.2.38)–(7.2.41) may be applied to afford a solution for the surface temperature for *laminar* flows.

$$Nu_x = 0.453 Re_x^{1/2} Pr^{1/3} \quad \text{for } Pr > 0.1 \tag{7.2.38}$$

For fluids with a low Prandtl number the regression is

$$Nu_x = 0.886 Re_x^{1/2} Pr^{1/3} \quad \text{for } Pr < 0.05 \tag{7.2.39}$$

As before, Churchill (1976) recommends the use of a single relationship, as given in Eq. (7.2.40), for all Prandtl numbers.

$$Nu_x = \frac{0.464 Re_x^{1/2} Pr^{1/3}}{\left[1 + (0.0207/Pr)^{1/3}\right]^{1/4}} \tag{7.2.40}$$

Solving for the Nusselt number allows evaluation of the heat transfer coefficient, from $Nu_x = h_x x/k$, and, hence, by Eq. (7.2.41), the local surface temperature at a distance x from the leading edge:

$$T_S = \frac{q_x}{h_x} + T_\infty \tag{7.2.41}$$

Example 7.2.2 Laminar parallel flow over a flat plate with uniform heat flux.

Air at a temperature of 15°C flows over a flat plate of length 1.8 m by 0.2 m at a velocity of 3 m/s. If the plate has a uniform heat flux of 50 W/m^2, find the plate surface temperature at points 0.3, 0.6, 0.9, 1.2, 1.5, and 1.8 m from the plate leading edge. Plot the relationship between distance from the leading edge and surface temperature, and state any assumptions that you have made.

Solution

Open the workbook Ex07-02-02.xls. This workbook comprises two sheets, namely, *Compute*, and *Properties*. Input the air temperature and velocity, together with the plate dimensions and heat flux data in **cells (B4:B9)**. Also, input the distances from the leading edge for the respective points on the plate surface for which the temperature is required in **cells (B7:G7)**. The information required is detailed in **cell A12**.

To provide a solution to the problem the user must make an initial guess for the temperature at each point. Thereafter, Excel calculates the film temperature, **cell B15**, based on this estimate, and defines the associated fluid properties. The Reynolds number is then calculated for each position considered.

The Nusselt number is calculated via Eq. (7.2.40), **cells (B17:G17)**, followed by the respective heat transfer coefficients, **cells (B18:G18)**. Finally, the heat flux q is calculated, **cells (B19:G19)**.

If the heat flux calculated for each position is sufficiently close to that specified, 50 W/m^2 for the present case, then the initial estimate for the temperature at that point is therefore also accepted. However, if the difference between the calculated and stipulated values for q is unacceptably large, an iterative procedure must be adopted. Excel's **Goal Seek** function is ideally suited to this task. While discussed in detail in Chapter 2, it is now applied to the present situation.

For the initial point of concern on the plate surface, namely, 0.3 m, the user selects the **Goal Seek** function in the manner previously described. In the 'Set Cell' box the user selects **cell B19**, then types in 50 in the 'To Value' box. In the 'By Changing Cell' box, the user selects **cell B20**. On 'clicking on' the OK button Excel performs the iterative routine until the calculated and defined values for q converge. This process is repeated for each of the points considered:

- The temperatures obtained are presented in cells **(B20:G20)**.
- To obtain a plot of the temperature profile, these results are copied to the **cells (D24:D29)** adjacent to the associated surface position details **cells (A24:A29)**.
- On selecting the 'create chart' icon from the uppermost toolbar, the user should select *XY*(Scatter) from the 'Chart type' menu, and follow the procedure presented. On completing the procedure the user should thereafter select an appropriate scale for each axis to afford an easier interpretation of plot provided.

Assumptions and discussion

As can be seen from careful examination of the Sheet *Compute*, only one film temperature, namely that obtained for the temperature difference at the point closest to the leading edge is given. All fluid properties used for subsequent calculation for all other points are based on this temperature. It has therefore been assumed that, as there is only marginal difference in the temperature across the plate the error thus induced will be minimal. Greater accuracy could, however, be achieved by considering the film temperature at each point.

We now consider the cases of uniform surface temperature and uniform heat flux for turbulent boundary layers.

7.2.2 *Turbulent flow across a flat plate*

We first consider the case of turbulent flow across a semi-infinite flat plate with a uniform surface temperature. For Reynolds number up to 10^7, Schlichting (1960) found that the local friction coefficient was well represented by

$$C_{f,x} = 0.0592\mathrm{Re}_x^{-1/5} \quad \text{for } \mathrm{Re}_x < 10^7 \tag{7.2.42}$$

Incropera and DeWitt (2002) state that this expression may also be used for Reynolds numbers up to 10^8 to within 15% accuracy. It has also been shown that, for turbulent flow, the velocity boundary layer thickness may be approximated from:

$$\delta = 0.37x\mathrm{Re}_x^{-1/5} \tag{7.2.43a}$$

From Eq. (7.2.43a) it can be seen that the boundary layer thickness increases with $x^{4/5}$ as opposed to $x^{1/2}$ for laminar flow (Eq. (7.2.18)). Hence, the growth of the turbulent boundary layer is much more rapid than that of the laminar layer. In turbulent flow, as boundary layer development is highly dependent on random fluctuations in fluid motion, and less so on molecular diffusion, the relative growth of the velocity and thermal boundary layers is effectively independent of the Prandtl number. Hence, for turbulent flow,

$$\delta \approx \delta_t \tag{7.2.43b}$$

For $0.6 < \Pr < 60$, the local Nusselt number may be obtained from Eqs (7.2.44a) and (7.2.44b).

$$\mathrm{Nu}_x = 0.0296\mathrm{Re}_x^{4/5}\,\Pr^{1/3} \quad \text{for } \mathrm{Re}_{\mathrm{CR}} < \mathrm{Re}_x < 10^7 \tag{7.2.44a}$$

and

$$\mathrm{Nu}_x = 1.596\mathrm{Re}_x(\ln \mathrm{Re}_x)^{-2.584}\,\Pr^{1/3} \quad \text{for } 10^7 < \mathrm{Re}_x < 10^9 \tag{7.2.44b}$$

Example 7.2.3 Turbulent flow over a flat plate with uniform surface temperature.

Air at a temperature of 13.5°C flows over a flat plate of dimensions 4.8 m long by 2.0 m wide at a velocity of 5.0 m/s. If the temperature of the plate surface is uniform at 52.3°C, determine the rate of heat transfer from the plate to the air.

Solution

Open the workbook Ex07-02-03.xls. This workbook comprises two worksheets, namely *Compute* and *Properties*. A table of the thermophysical properties of air is included in the *Properties* sheet.

Input the data given in **cells (B4:B8)**. As before, Excel again performs the necessary calculations. First, the film temperature T_f is calculated in **cell B15**. Excel then calculates the fluid properties accordingly, and returns the values to **cells (E8:L8)**. The Reynolds number is then calculated, **cell B16**. As this value is greater than the critical value of 5×10^5 for flat plates, the flow is therefore turbulent for some part of the plate.

The critical length L_{CR}, **cell E16**, is found to be **1.65 m**. As the plate length is only slightly greater than twice L_{CR}, and Re_L is less than 10^7, Eq. (7.2.48) is employed. The average Nusselt number is calculated as 2030.3 **cell B17**. Thereafter, the heat transfer coefficient is found to be 11.3 W/m^2 K, and, finally, q is determined as 4214.2 W, **cell B19**.

To find the total heat transfer rate for a plate where the free stream velocity and plate length are such that the boundary layer is turbulent for some part, we require to add the sum of the respective rates of heat transfer from the laminar and turbulent regions. Thus,

$$q = \int_0^{x_{\mathrm{CR}}} h_x(T_S - T_\infty)W\,dx + \int_{x_{\mathrm{CR}}}^{L} h_x(T_S - T_\infty)W\,dx \tag{7.2.45}$$

where W is the width of the plate.

Substituting the equation for the local Nusselt number for laminar flow, Eq. (7.2.28), for $0 < x < x_{CR}$, and, for the present discussion assuming $Re_L < 10^7$, Eq. (7.2.44a) for turbulent flow over the length x_{CR} to L, as shown in Fig. 7.2.1, and integrating, we obtain

$$q = 0.664\left(\frac{u_\infty}{\nu}\right)^{1/2} \Pr^{1/3} kW(T_S - T_\infty)^{1/2} + 0.037\left(\frac{u_\infty}{\nu}\right)^{4/5} \Pr^{1/3} kW(T_S - T_\infty)\left(L^{4/5} - x_{CR}^{4/5}\right) \quad (7.2.46)$$

which reduces to,

$$\overline{Nu_L} = \left[0.664 Re_{CR}^{1/2} + 0.037\left(Re_L^{4/5} - Re_{CR}^{4/5}\right)\right] \Pr^{1/3} \quad (7.2.47)$$

For $Re_{CR} = 5 \times 10^5$ Eq. (7.2.47) reduces to

$$\overline{Nu_L} = (0.037 Re_L^{4/5} - 871) \Pr^{1/3} \quad \text{for } 5 \times 10^5 < Re_L < 10^7 \quad (7.2.48)$$

For $10^7 < Re_L < 10^9$ and $\Pr \geq 0.6$, Eq. (7.2.44b) is used instead of Eq. (7.2.44a), and we get,

$$\overline{Nu_L} = [1.967 Re_L (\ln Re_L)^{-2.584} - 871] \Pr^{1/3} \quad (7.2.49)$$

Where the system characteristics are such that $x_{CR} \ll L$, Eq. (7.2.48) may be approximated as,

$$\overline{Nu_L} = 0.037 Re_L^{4/5} \Pr^{1/3} \quad (7.2.50)$$

and, similarly, for Eq. (7.2.49),

$$\overline{Nu_L} = 1.967 Re_L (\ln Re_L)^{-2.584} \Pr^{1/3} \quad (7.2.51)$$

Example 7.2.4 Determination of the critical length for different fluids at the same temperature and velocity.

Determine the critical length, for air, oil, and water, all at a temperature of 100°C and flowing at 4 m/s over an infinite flat plate with uniform surface temperature of 15°C. Comment on your findings.

Solution

Open the workbook Ex07-02-04.xls. This workbook contains four Sheets, namely, *Compute, PropAir, PropWat,* and *PropOil.* The thermophysical property tables for the three fluids in question are contained in the above mentioned respective *Prop* sheets.

Select Sheet *Compute* by clicking on its tab and insert the given data in cells **(B4:B8)**. For each of the fluids, Excel automatically computes the required fluid properties from the above mentioned tables, and performs all necessary calculations. The critical length is determined by setting the Reynolds number, **cell B15**, 5×10^5, and solving for L. x_{CR} is thus found to be, for air, 2.4 m, **cell B18**, for water, 0.06 m, **cell B20**, and for oil 11.8 m, **cell B22**.

Discussion

As the critical length is proportional to ν, ($x_{CR} = \mathrm{Re}_{CR}.\nu/V$), for the same fluid velocity and Reynolds number it is therefore the kinematic viscosity, ν, that determines the difference in critical length. For the present case, with the parameters defined, the critical length for engine oil is approximately 400 times that for water, which in turn is approximately one tenth that of air. At higher temperatures, for example 220°C, as the kinematic viscosity of oil is more highly dependent on temperature than that of water, the above oil/water critical length ratio reduces to approximately 50.

Example 7.2.5 Heat transfer between two streams of gas separated by a thin plate.

In this example we consider the exchange of heat between two streams of gas that are separated by a thin plate. This situation is common in many engineering applications such as air to air or exhaust gas heat exchangers.

Two opposing streams of air (counter flow) are separated by a thin metal plate of length 1.6 m. If the temperature and velocity of stream one are, 190°C and 28 m/s, respectively, what is the temperature of the plate 1.3 m from the leading edge, with respect to stream 1, if the *free stream temperature* and velocity of the second stream are, respectively, 30°C and 12 m/s? Assume that the thermal resistance of the plate is negligible compared to that of the fluid streams.

Analysis

From an electrical current circuit analogy, the heat transfer rate may be defined as,

$$q = \frac{T_{\infty 1} - T_{\infty 2}}{(1/h_1) + (1/h_2)}$$

and, as

$$h_1(T_{\infty 1} - T_p) = h_2(T_p - T_{\infty 2})$$

it therefore follows that

$$T_p = \frac{h_1 T_{\infty 1} + h_2 T_{\infty 2}}{h_1 + h_2} \tag{7.2.52}$$

However, as both h_1 and h_2 depend on T_p, an iterative solution is therefore required. We must first therefore assume a temperature for the plate at the point in question in order to determine the values of the above parameters. For both streams, we therefore define the respective stream fluid properties based on the assumed plate temperature, calculate the Reynolds number, followed by the Nusselt, and thence values for the respective surface heat transfer coefficients, h_1 and h_2. The accuracy of the initial guess is determined through comparison with the plate temperature, as calculated through the above equation. If the respective estimated and calculated values are not sufficiently close, a new assumed value, equal to the average of the original and calculated is chosen, and the procedure repeated until the two values converge.

Solution

Open the workbook Ex07-02-05.xls. This workbook comprises two Sheets, namely, *Compute* and *Properties*. Select the *Compute* sheet by clicking on its tab. Insert the data for the two streams in, first, **cells (B6:B8)** for stream 1, and **cells (D6:D8)** for stream 2. Note that as the temperature is required at a point 1.3 m from the plate leading edge with respect to stream 1, this distance, for the 1.6 m plate, equates to L_2 of 0.3 m for stream 2.

Insert your initial estimate for the plate surface temperature, for example, 150°C in **cell B23.** Excel performs all necessary calculations of fluid properties, Reynolds number, Nusselt number, the respective heat transfer coefficients, q, and the calculated plate temperature **cell D23.** For the present case, as the Reynolds number for stream 1 (1.15×10^6 in **cell B17**) is greater than the critical value for a flat plate, 5×10^5, Eq. (7.2.44a) is employed to calculate the Nusselt number. For stream 2, at a distance of 0.3 m from the plate leading edge, $Re_2 = 1.61 \times 10^5$, **cell D17**, and hence Eq. (7.2.28) is used.

Assuming an initial temperature of 150°C yields a calculated value of 159.7°C for the plate temperature, **cell D23**. The difference between these two values is unacceptable. In previous examples Excel's **Goal Seek** and **Solver** functions have been used to 'home in' on a solution. For the present case, a macro called *Iterate* has been developed. The user should select and run this macro via the 'Tools' menu. This macro performs 100 iterations per run, selecting the average of the assumed and calculated values as the input for each iterative run. For any reasonable initial assumed temperature (0–200°C) the values are seen to converge within one run of the aforementioned macro.

For the present case, the values converge at 159.6°C, **cells (B23 and D23)**.

Discussion

If an initial value of 0°C is assumed for the plate temperature, **cell B23**, the calculated value returned, prior to iteration, would be 161.9°C. Similarly, if **cell B23** were set to 200°C, the value returned would be 158.9°C. The computed temperature is therefore relatively insensitive to the initial temperature assumed. For the present case there is little gain in accuracy to be obtained through using the *Iterate* macro.

The value obtained for q, **cell B21**, relates only to the point on the plate considered. As the plate temperature, and hence film temperature and fluid properties will necessarily vary along the plate, the Reynolds and Nusselt numbers, and therefore also h and q, will vary for both fluid streams along the plate. To calculate the overall rate at which heat is transferred between the two fluids, the plate would have to be considered in sections, with the dimensionless groups and heat transfer rates calculated for each section. The local heat flux rates would then be summed to provide a measure of the total heat flux along the plate.

Where an application requires that the heat transfer rate is maximised for a given temperature difference and free stream velocity, a turbulent boundary layer may be established from the leading edge through a fine wire or an alternative *tripper* device being placed at the leading edge.

For fluids with a highly temperature dependent viscosity, Whitaker (1985), recommends the regressions given in Eqs (7.2.53a) and (7.2.53b) for the local Nusselt number,

$$\mathrm{Nu}_x = 0.332\,\mathrm{Re}_x^{1/2}\mathrm{Pr}^{1/3}\left(\frac{\mu_\infty}{\mu_S}\right)^{1/4} \quad \text{for } \mathrm{Re}_x < \mathrm{Re}_{CR} \qquad (7.2.53a)$$

$$\mathrm{Nu}_x = 0.0296\,\mathrm{Re}_x^{4/5}\mathrm{Pr}^{0.43}\left(\frac{\mu_\infty}{\mu_S}\right)^{1/4} \quad \text{for } \mathrm{Re}_x > \mathrm{Re}_{CR} \qquad (7.2.53b)$$

Setting $\mathrm{Re}_{CR} = 5\times10^5$, the average heat transfer coefficient may be obtained from

$$\mathrm{Nu}_L = 0.664\,\mathrm{Re}_x^{1/2}\mathrm{Pr}^{1/3}\left(\frac{\mu_\infty}{\mu_S}\right)^{1/4}, \quad \mathrm{Re}_L < 5\times10^5 \qquad (7.2.54a)$$

$$\mathrm{Nu}_L = (0.037\,\mathrm{Re}_L^{4/5} - 871)\,\mathrm{Pr}^{0.43}\left(\frac{\mu_\infty}{\mu_S}\right)^{1/4}, \quad \mathrm{Re}_L > 5\times10^5 \qquad (7.2.54b)$$

All the foregoing expressions for the Nusselt number are restricted to situations of uniform surface temperature or uniform heat flux. We now consider the special case, as shown in Fig. 7.2.3, where the aforementioned flat plate has an unheated starting length ($T_S = T_\infty$) upstream of a heated section ($T_S \neq T_\infty$). It can be seen that, as before, the velocity boundary layer will begin to develop from $x = 0$, while the thermal boundary layer will not become established until the fluid stream reaches the heated section of the

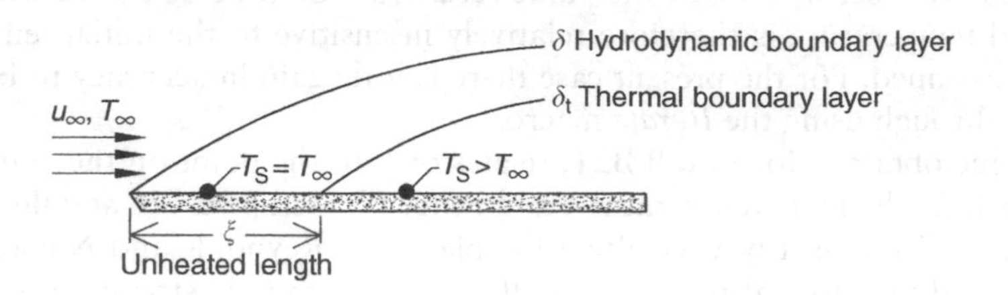

Figure 7.2.3 The flat plate in parallel flow with unheated starting length.

plate at $x = \xi$. There is therefore no heat transfer between the plate and the fluid stream for $x < \xi$.

The solution presented by Kays and Crawford (1980) assumes, as shown in Fig. 7.2.3, that the hydrodynamic boundary layer will always be thicker than the thermal boundary layer. For laminar boundary layers, for $x > \xi$, the local Nusselt number is given by

$$\mathrm{Nu}_x = \frac{0.332\,\mathrm{Re}_x^{1/2}\mathrm{Pr}^{1/3}}{[1-(\xi/x)^{3/4}]^{1/3}} \tag{7.2.55}$$

(Note: For $\xi = 0$, Eq. (7.2.55) reduces to Eq. (7.2.28)).

For turbulent flows, Kays and Crawford (1980) show that the local Nusselt number, for $x > \xi$, is

$$\mathrm{Nu}_x = \frac{0.0296\,\mathrm{Re}_x^{4/5}\mathrm{Pr}^{3/5}}{[1-(\xi/x)^{9/10}]^{1/9}} \tag{7.2.56}$$

For a uniform surface temperature and laminar flow from $x = \xi$ to $x = L$, the average value of the surface heat transfer coefficient is given by

$$h_\mathrm{L} = 2\frac{\lfloor 1-(\xi/L)^{3/4}\rfloor}{1-\xi/L}h_{x=L} \tag{7.2.57a}$$

If the boundary layer is turbulent,

$$h_\mathrm{L} = \frac{5\lfloor 1-(\xi/L)^{9/10}\rfloor}{4(1-\xi/L)}h_{x=L} \tag{7.2.57b}$$

If the boundary layer is partly laminar and partly turbulent then the average heat transfer coefficient

$$h_\mathrm{L} = \frac{1}{L-\xi}\int_{\xi}^{L} h(x)\,\mathrm{d}x \tag{7.2.58}$$

is found by integration by substituting Eqs (7.2.55) and (7.2.56) for the laminar and turbulent boundary layers, respectively.

7.2.3 Flow across cylinders and spheres

The boundary layers that develop as a result of flows perpendicular to the axes of cylinders and over spheres are quite distinct to those that form over the previously mentioned flat plate. With uniform flow over a flat plate, the pressure is uniform and there is no pressure gradient in the direction of flow. For flow perpendicular to the axis of a cylinder, as shown in Fig. 7.2.4, the fluid, with initial velocity V, is brought to rest at the *forward stagnation point*, with a subsequent rise in pressure. Beyond this point the pressure decreases as x, the streamline coordinate, increases, and the boundary layer develops under the influence of a favourable pressure gradient. The fluid accelerates away from the forward stagnation point, and the free stream velocity u_∞ increases, ($du_\infty/dx > 0$ when $dp/dx < 0$). However, the pressure will reach a minimum at some point, beyond which it will increase with increasing x (adverse pressure gradient), and the fluid will decelerate. As the fluid decelerates, the velocity gradient at the surface, $(\partial u/\partial y)|_{y=0}$, eventually becomes zero and the boundary layer detaches from the surface. The point at which the boundary layer detaches from the surface is termed the separation point. Flow in the region beyond the separation point, the *wake*, is characterised by vortex formation, and fluid motion is highly irregular.

The initial fluid velocity V, and hence the Reynolds number, strongly influences the position of the separation point. For cylinders, transition from laminar to turbulent flow occurs at approximately $Re_D = 2 \times 10^5$. For turbulent flow, as the momentum of the fluid in the boundary layer is greater than that of the laminar layer, separation is delayed, and the separation point moves downstream (Fig. 7.2.5a and b).

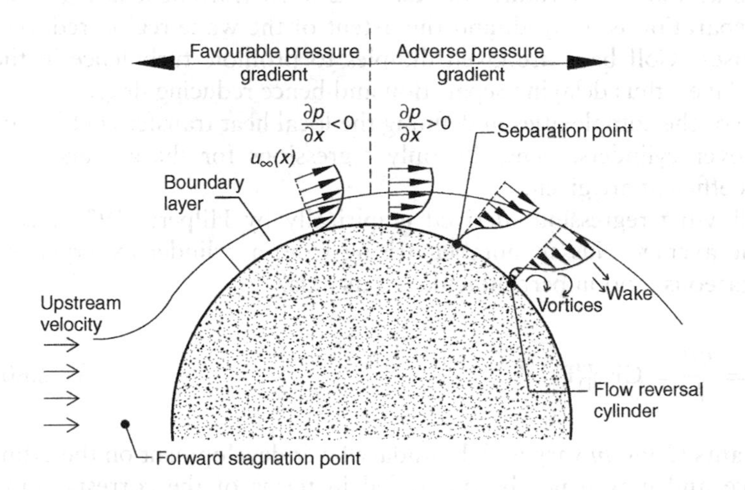

Figure 7.2.4 Velocity profile established over a cylinder in cross-flow.

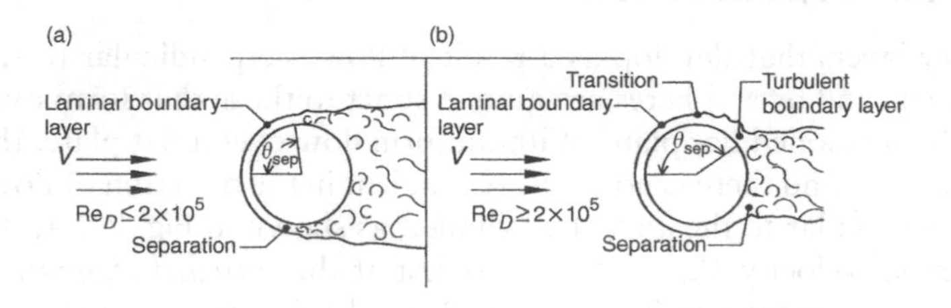

Figure 7.2.5 The effect of turbulence on the relative position of the separation point: (a) laminar flow; (b) turbulent flow.

The drag force F_D acting on the cylinder is also greatly influenced by the flow condition. As for flow over the flat plate, the drag force comprises a shear stress component (friction drag), but, in the present case, it also has a component deriving from the pressure differential in the flow direction that results from the formation of the wake. This latter component is termed the *form* or *pressure drag*. A dimensionless pressure drag may be defined as,

$$C_D \equiv \frac{F_D}{A_f\left(\rho V^2/2\right)} \tag{7.2.59}$$

where A_f is the cylinder frontal area, calculated as the area projected perpendicular to the free stream velocity. The drag coefficient is a function of the Reynolds number. For small Reynolds numbers, $Re_D < 2$, as the effects of separation are negligible, conditions are dominated by friction drag. However, as Re_D increases, the effect of separation increases, and hence form drag becomes more important. For $Re_D > 2 \times 10^5$(turbulent flow), however, as separation is delayed, and the extent of the wake region reduced, C_D decreases. Golf balls are given dimples to promote turbulence in the boundary layer, thus delaying separation and hence reducing drag.

Because of the complexities in defining the local heat transfer coefficients for flow over cylinders, generally only regressions for the average heat transfer coefficient are given.

The following regression, derived empirically by Hilpert (1933), represents the average Nusselt number arising from a cylinder experiencing external gaseous motion perpendicular to its axis.

$$\overline{Nu_D} = \frac{\bar{h}d}{k} = CRe_D^m Pr^{1/3} \tag{7.2.60}$$

The constants C and m vary with boundary layer development on the cylinder surface and may hence be quantified in terms of the corresponding Reynolds number groupings shown in Table 7.2.1.

Table 7.2.1 Constants for Eq. (7.2.60) (circular cylinder in a cross-flow of gas)

Re_D	C	m
0.4–4	0.989	0.330
4–40	0.911	0.385
40–4000	0.683	0.466
4000–40000	0.193	0.618
40000–400000	0.027	0.805

Example 7.2.6 Heat loss from a cylinder in cross-flow.

An industrial process requires a variable supply of water at a fixed temperature of 32°C. Part of the feedback control system comprises a hot wire anemometer for fluid velocity measurement. The device comprises an electrically heated platinum wire 0.2 mm in diameter, and 6 mm long, maintained at a constant temperature of 65°C immersed in the fluid stream. (The resistivity of platinum is 0.17 μΩm.) As the heat transfer rate from the wire to the fluid varies with fluid velocity, to maintain a *constant surface temperature*, the wire supply voltage necessarily varies. Through calibrating the voltage as a function of velocity, the instrument can be used to measure the velocity of fluids. For the case of water, determine the supply voltage required for fluid velocities of 0.5, 1, 2, and 5 m/s. Assume that all conditions are steady state and that wire end effects are negligible.

Analysis

The cooler fluid will remove heat from the wire at a rate defined by, among other factors, its velocity, the respective dimensionless groupings, their inter-relationships, and the difference in the fluid and wire temperatures. To maintain the wire surface temperature, the electrical power (heat) input to the wire must equal that lost to the fluid. Hence, as the heat lost to the fluid may be expressed as

$$q = hA(T_S - T_W)$$

and the electrical power input required to maintain the wire surface temperature obtained from

$$P(\text{or } q) = \frac{V^2}{R}$$

the voltage required to maintain the wire surface temperature for a given flow rate may therefore be expressed as

$$V = \sqrt{hA(T_S - T_W)R}$$

Solution

Open the workbook Ex07-02-06.xls and select worksheet *Compute*. Insert the values given in the above example in the appropriate cells: namely **cell B4** for the wire surface temperature, **cell B5** for the fluid temperature, and so on, as labelled in column **'A'**.

Values for the fluid film temperature and resistance of the wire (for the dimensions specified) are returned in the *Calculations (1)* cells, namely **cells (B18 and B19)**, respectively. For the defined film temperature in **cell B16** fluid property values are obtained from the *Properties* worksheet and returned to **cells (F6:T6)** in the *Compute* worksheet. These values are then used to compute the respective values of the Reynolds numbers, as shown in Table 2 **cells (G14:J14)** for the different velocities considered.

The appropriate values of 'C' and 'm' for application in Eq. (7.2.60) are then selected from Table 7.2.1, using the **VLookup** function, based on the calculated Reynolds number, and a value returned for the average Nusselt number in **cells (G20:J20)**. A value for the surface heat transfer coefficient is calculated, **cells (G21:J21)**, allowing an evaluation of the total heat loss, q, from the wire, for each velocity considered, **cells (G22:J22)**.

To maintain a constant surface temperature, as described above, the heat input must balance the heat loss. Hence, from the electrical power equation $P = V^2/R$, $V = (PR)^{0.5}$, a measure of the voltage required for each of the fluid velocities defined may be obtained.

Discussion

A plot of the data obtained is embedded in the *Compute* worksheet. Through employing Excel's trend-line fitting facility, it can be seen that, through the R^2 value obtained, the data correlates well with the second order polynomial curve shown. From these initial results it would therefore appear that a function of the form shown on the chart would be required in calibrating the instrument for the measurement of water velocity.

In instances wherein non-circular cylinders encounter external, gaseous, transverse flow, Eq. (7.2.60) may still be employed. However, the characteristic length and constants associated with Eq. (7.2.60) are now given in Table 7.2.2 for such cylinders.

Irrespective of the shape of the cross-section of the cylinder in question, all thermophysical properties should be determined at the film temperature T_f.

Zhukauskas (1972) suggests the following regression for the cylinder in cross-flow,

$$\overline{\mathrm{Nu}_\mathrm{D}} = \frac{\bar{h}d}{k} = \mathrm{C}\mathrm{Re}_\mathrm{D}^m\mathrm{Pr}^n\left(\frac{\mathrm{Pr}}{\mathrm{Pr}_\mathrm{S}}\right)^{1/4}, \quad 0.7 < \mathrm{Pr} < 500, \;\; 1 < \mathrm{Re}_\mathrm{D} < 10^6 \tag{7.2.61}$$

Table 7.2.2 Constants for Eq. (7.2.60) (non-circular cylinder in a cross-flow of a gas)

	Re_D	C	m
Square (Flow →, D)	5×10^3 to 10^5	0.246	0.588
Square (Flow →, D)	5×10^3 to 10^5	0.102	0.675
Hexagon (Flow →, D)	5×10^3 to 1.95×10^4	0.160	0.638
	1.95×10^4 to 10^5	0.0385	0.782
Hexagon (Flow →, D)	5×10^3 to 10^5	0.153	0.638
Vertical plate (Flow →, D)	4×10^3 to 1.5×10^4	0.228	0.731

Table 7.2.3 Constants for Eq. (7.2.61) (circular cylinder in a cross-flow of a gas)

Re_D	C	m
1–40	0.75	0.4
40–4000	0.51	0.5
10^3 to 2×10^5	0.26	0.6
2×10^5 to 10^6	0.076	0.7

In this case however, all thermophysical properties are obtained at T_∞, with the exception of Pr_S which is obtained at T_S. The appropriate constants C and m are presented in Table 7.2.3, whilst $n = 0.37$ if $Pr \leq 10$ and $n = 0.36$ if $Pr > 10$.

The relation given in Eq. (7.2.62), reported by Churchill and Bernstein (1977), is applicable across both the entire range of data required for boundary layer analyses and the majority of Pr. Again, thermophysical properties are to be obtained at the film temperature.

$$\overline{Nu_D} = 0.3 + \frac{0.62 Re_D^{1/2} Pr^{1/3}}{[1 + (0.4/Pr)^{2/3}]^{1/4}} \left[1 + \left(\frac{Re_D}{282000}\right)^{5/8}\right]^{4/5} \qquad (7.2.62)$$

At this point the attention of the reader should be drawn to the fact that none of the preceding regressions are definitive, that is, they provide results to a reasonable degree of accuracy over a wide range of easily influenced conditions. Indeed, Incropera and De Witt (2002), report that for the majority of engineering applications one should expect accuracy in predictions to be of the order of ±25%.

A heat transfer relation for examining the effect of gaseous cross-flow on a sphere has been developed by Whitaker (1972) such that,

$$\overline{Nu_D} = 2 + (0.4Re_D^{1/2} + 0.06Re_D^{2/3})Pr^{0.4}\left(\frac{\mu}{\mu_S}\right)^{1/4} \quad (7.2.63)$$

The above regression is reported as having an accuracy of ±30% for the following regimes; $0.71 < Pr < 380$ and $3.5 < Re_D < 7.6 \times 10^4$. In this case, thermophysical properties are determined at T_∞, with the exception of μ_S which is obtained at T_S.

Example 7.2.7 Heat gain to a sphere in cross-flow.

In this example we consider the case of heat gain to a sphere with constant surface temperature in a gaseous cross-flow.

An underwater echo sounder is housed in a small spherical pod of diameter 110 mm. To maintain the equipment at a safe operating temperature when the pod is operating in deep water, the device is fitted with a 350 W heater. The pressure controlled safety switch which normally automatically deactivates the heater when the pod is brought to the surface fails. If the pod is raised above the surface into air at 15°C flowing at a velocity of 3.5 m/s, assuming steady state conditions, calculate the pod surface temperature. Comment on your findings.

Analysis

As with previous examples, the solution requires an iterative approach. For the present case of a sphere however, as defined by Eq. (7.2.63), fluid properties are determined for the air temperature. μ_S is determined at the sphere surface temperature. We therefore do not need to calculate the film temperature.

We must first assume a value for the sphere surface temperature and calculate q accordingly. If the calculated value of q is sufficiently close to that specified, the initial assumed value of the sphere surface temperature may be considered acceptable. Otherwise, we require to either increase or reduce our assumed temperature accordingly.

Solution

Open the workbook Ex07-02-07.xls. This workbook comprises two worksheets, namely, *Properties*, and *Compute*. Select the *Compute* worksheet by clicking on its tab and insert the given data in **cells (B4:B7)**. Insert an assumed value for the sphere surface temperature T_{SP} in **cell B18**. Excel performs all the necessary calculations.

The Reynolds number is calculated based on the sphere diameter, air velocity, and the kinematic viscosity ν for the air film temperature. The ratio μ/μ_S, **cell B14,** is calculated with μ_S based on the assumed sphere surface temperature. The Nusselt number is calculated via Eq. (7.2.63), followed by h, and finally q, **cell B17**. As outlined above, if the value of q is sufficiently close to that defined in the problem, we need proceed no further. However, if, for example, our initial estimate for T_{SP} was 100°C, providing a calculated value for q of 73.6 W, we require to use Excel's **Goal Seek** function.

Setting the target value of the 'To Value' box to 350 for **cell B17**, by changing **cell B19** returns a value of 469.7°C for T_{SP}.

Discussion

The spheres surface temperature, and consequently the temperature within it, may be so high as to cause damage to the pods components. If switch failure is a common problem it may be financially worthwhile to alter the device's design to incorporate a more robust method of heater switching.

7.2.4 *Flow across tube banks*

There are a great many applications whereby the engineer may encounter the use of a bank of tubes as a device to implement heat transfer between separate fluids. Knowledge of the heat transfer coefficient is required, for example, in determining the size of a heater battery for an air-handling unit, or the off-air temperature from the evaporator of a roof-mounted chiller.

In most situations it is generally the case that one fluid passes across the tube bank, thus influencing the thermal characteristics of a second fluid flowing through the tubes at a different temperature. (Example 10.4.1 considers the combined effects of conduction, radiation, and forced convection on the heat transfer rate for a tube bank in cross-flow.) A schematic representation of a tube bank in cross-flow is shown in Fig. 7.2.6. The tubes contained within the bank may be geometrically configured in such a manner that they are either aligned with, or staggered against, the direction of external fluid flow. Either arrangement can be described in terms of the tube diameter D, the transverse pitch S_T and the longitudinal pitch S_L as shown in Figs. 7.2.7a and b (with the diagonal pitch S_D in Fig. 7.2.7b given by $(S_L^2 + (S_T/2)^2)^{0.5}$.

In a bank of tubes, the heat transfer coefficient corresponding to a specific tube is proportional to that tube's position within the bank. In the case of a tube positioned in the first row, the corresponding heat transfer coefficient may be approximated as that of a single tube in cross-flow. The tubes positioned in the first rows serve as to increase the turbulence of the external

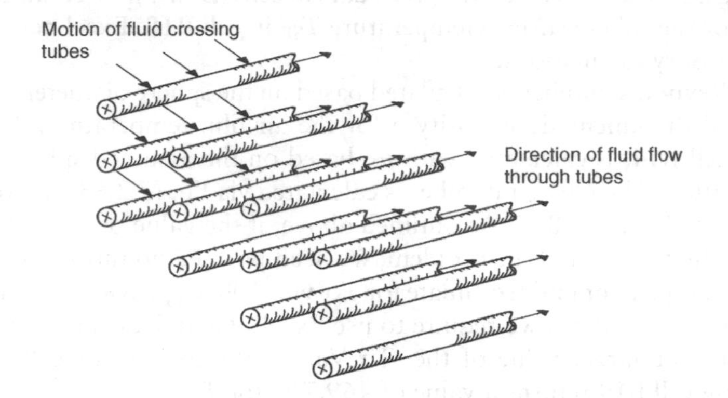

Figure 7.2.6 A bank of tubes in cross-flow.

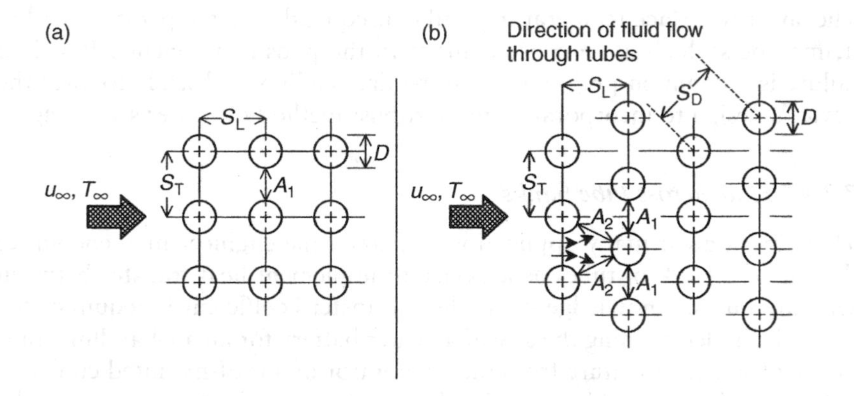

Figure 7.2.7 Arrangement of tubes in bank: (a) aligned and (b) staggered, and the dimensions required for determining cross-flow flow rates.

fluid motion, thus also increasing the heat transfer coefficients of tubes in the following rows. It has been shown however that beyond the fourth or fifth rows the convection coefficient reaches a state of equilibrium as heat transfer conditions stabilise. For the majority of engineering applications it is of interest to consider the average convection heat transfer coefficient across a tube bank rather than at specific points. The forthcoming regressions for predicting this average value are therefore functions of the maximum Reynolds number occurring in the tube bank.

The maximum Reynolds number, Re_{Dmax}, will occur at that point in the tube arrangement where maximum fluid velocity, V_{max}, occurs, and hence

$$Re_{Dmax} \equiv \frac{\rho V_{max} D}{\mu} \tag{7.2.64}$$

The maximum velocity is dependent upon the configuration of the tube bank and, for an aligned series of tubes, can be determined as

$$V_{max} = V\frac{S_T}{S_T - D} \tag{7.2.65}$$

However, in a staggered tube bank V_{max} can occur in either the transverse plane A_1 or the diagonal plane A_2, as shown in Fig. 7.2.7b. If V_{max} occurs in plane A_1 of a staggered arrangement, then Eq. (7.2.65) may again be employed. However, V_{max} will occur in plane A_2 if the staggered tube arrangement is such that,

$$S_D < \frac{S_T + D}{2} \tag{7.2.66}$$

and can be determined by

$$V_{max} = V\frac{S_T}{2(S_D - D)} \tag{7.2.67}$$

For a tube bundle containing 10 or more rows ($N_L \geq 10$), Grimison (1937) presented the following relationship describing the average convection coefficient due to air in cross-flow,

$$\overline{\mathrm{Nu}_D} = C_1 \mathrm{Re}_{D,max}^m, \quad 2000 < \mathrm{Re}_{Dmax} < 40000, \quad \mathrm{Pr} = 0.7 \tag{7.2.68}$$

The constants 'C_1' and 'm' are functions of the geometric arrangement of the tube bundle and are given accordingly in Table 7.2.4.

For fluids other than air in external contact with a tube bundle arranged such that $N_L \geq 10$, the factor $1.13\mathrm{Pr}^{1/3}$ is inserted into Eq. (7.2.68) such that,

$$\overline{\mathrm{Nu}_D} = 1.13 C_1 \mathrm{Re}_{D,max}^m \mathrm{Pr}^{1/3}, \quad 2000 < \mathrm{Re}_{Dmax} < 40000, \quad \mathrm{Pr} \geq 0.7 \tag{7.2.69}$$

Again, the values given in Table 7.2.4 for constants 'C_1' and 'm' apply. In the case of both the above equations, all thermophysical properties are determined at the film temperature.

To estimate the average convection heat transfer coefficient for tube bundle arrangements where $N_L < 10$, a factor, 'C_2', is applied, as shown as

$$\overline{\mathrm{Nu}_D}\Big|_{(N_L<10)} = C_2 \overline{\mathrm{Nu}_D}\Big|_{(N_L \geq 10)} \tag{7.2.70}$$

Values of the additional correction factor 'C_2' are given in Table 7.2.5.

Table 7.2.4 Constants for Eqs (7.2.68) and (7.2.69) for air flow across a tube bank of $N_L \geq 10$

S_L/D	S_T/D							
	1.25		1.50		2.0		3.0	
	C_1	m	C_1	m	C_1	m	C_1	m
Aligned								
1.25	0.348	0.592	0.275	0.608	0.100	0.704	0.0633	0.752
1.50	0.367	0.586	0.250	0.620	0.101	0.702	0.0678	0.744
2.00	0.418	0.570	0.299	0.602	0.229	0.632	0.198	0.648
3.00	0.290	0.601	0.357	0.584	0.374	0.581	0.286	0.608
Staggered								
0.600	—	—	—	—	—	—	0.213	0.636
0.900	—	—	—	—	0.446	0.571	0.401	0.581
1.000	—	—	0.497	0.558	—	—	—	—
1.125	—	—	—	—	0.478	0.565	0.518	0.560
1.250	0.518	0.556	0.505	0.554	0.519	0.556	0.522	0.562
1.500	0.451	0.568	0.460	0.562	0.452	0.568	0.488	0.568
2.000	0.404	0.572	0.416	0.568	0.482	0.556	0.449	0.570
3.000	0.310	0.592	0.356	0.580	0.440	0.562	0.428	0.574

Table 7.2.5 C_2 correction factor values for Eq. (7.2.70) for air flow across a tube bank of $N_L < 10$

N_L	1	2	3	4	5	6	7	8	9
Staggered	0.64	0.80	0.87	0.90	0.92	0.94	0.96	0.98	0.99
Aligned	0.68	0.75	0.83	0.89	0.92	0.95	0.97	0.98	0.99

Zhukauskas (1972) provides the following regression for a fluid in cross-flow with a bank of 20 or more tubes.

$$\mathrm{Nu_D} = C\mathrm{Re}_{\mathrm{D,max}}^{m}\mathrm{Pr}^{0.36}\left(\frac{\mathrm{Pr}}{\mathrm{Pr_S}}\right)^{1/4}$$

$$\text{where} \begin{bmatrix} N_L \geq 20 \\ 1000 < \mathrm{Re_{D,max}} < 2 \times 10^6 \\ 0.7 < \mathrm{Pr} < 500 \end{bmatrix} \tag{7.2.71}$$

In this case, all thermophysical properties with the exception of $\mathrm{Pr_S}$ are taken at the arithmetic mean of the fluid inlet and outlet temperatures. The constants ‘C’ and ‘m’ are not only categorised in terms of tube geometry, but also in terms of the corresponding Reynolds number, as shown in Table 7.2.6.

Table 7.2.6 Constants for Eq. (7.2.71) for fluids in cross-flow with a tube bank

Tube bank geometry	$Re_{D,max}$	C_1	*m*
Aligned	10 to 10^2	0.8	0.40
Staggered	10 to 10^2	0.9	0.40
Aligned	10^2 to 10^3	Treat as a single	
Staggered	10^2 to 10^3	cylinder	
Aligned ($S_T/S_L > 0.7$)*	10^3 to 2×10^5	0.27	0.63
Staggered ($S_T/S_L < 2$)	10^3 to 2×10^5	$0.35(S_T/S_L)^{1/5}$	0.60
Staggered ($S_T/S_L > 2$)	10^3 to 2×10^5	0.4	0.60
Aligned	2×10^5 to 2×10^6	0.021	0.84
Staggered	2×10^5 to 2×10^6	0.022	0.84

Note
* For $S_T/S_L < 0.7$, heat transfer is inefficient and aligned tubes should not be used.

Table 7.2.7 C_2 correction factor values for Eq. (7.2.72) for air flow across a tube bank of $N_L < 20(Re_D > 10^3)$

N_L	1	2	3	4	5	7	10	13	16
Staggered	0.7	0.8	0.86	0.9	0.92	0.95	0.97	0.98	0.99
Aligned	0.64	0.76	0.84	0.89	0.92	0.95	0.97	0.98	0.99

For tube bundle arrangements containing less than twenty rows the correction factor given below should be applied.

$$\overline{Nu}_D \Big|(N_L < 20) = C_2 \overline{Nu}_D \Big|_{(N_L \geq 20)} \tag{7.2.72}$$

Values of 'C_2' are given in Table 7.2.7 according to tube bundle arrangement and geometry.

As stated previously, there are a great many applications whereby the engineer may encounter the use of a bank of tubes as a device to implement heat transfer between separate fluids. In many tube bundle applications the surface temperature is greater than that of the free stream fluid. However, for the case of the water tube boiler presented in Example 10.2.1, hot combustion gases are used to heat the fluid within the tubes. Experiments have shown that the temperature difference between the two fluids falls most rapidly as the free stream fluid passes over the first rows in the tube bank. This implies that the rate of heat transfer is greatest at this station.

Here, the difference in temperature between the tube surface and that of the free stream is greatest. While greater convection coefficients are realised further downstream, generally reaching a maximum by about the fifth row, in respect of the heat transfer rate, the lower initial values are more than compensated by the larger temperature difference.

As the fluid passes across the tube bundle, it can undergo such a large change in temperature that if one were to assume the temperature difference in Newton's Law of Cooling to be constant at $\Delta T = T_S - T_i$ then the results generated could be greatly over-predictive. Therefore, in defining the rate of heat transfer for tube bundles and heat exchangers where the temperature of a fluid is influenced by a surface at a different, constant temperature, the relationship known as the *log-mean temperature difference* is used.

If one considers the case of a constant tube surface temperature, the log-mean temperature difference may be expressed as

$$\Delta T_{lm} = \frac{(T_S - T_i) - (T_S - T_o)}{\ln\left[(T_S - T_i)/(T_S - T_o)\right]} \tag{7.2.73}$$

where T_i and T_o are the inlet and outlet temperature of the fluid as it enters and leaves the bank. To obtain a measure of ΔT_{lm}, the outlet temperature, T_o, may be found from

$$\frac{T_S - T_o}{T_S - T_1} = \exp\left(\frac{-\pi D N \bar{h}}{\rho V N_T S_T c_p}\right) \tag{7.2.74}$$

where N is the total number of tubes in the bundle, and N_T is the number of tubes in the transverse plane. With knowledge of ΔT_{lm} the heat transfer rate per unit length of the tubes in the bundle may be calculated by

$$q = N(\bar{h}\pi D \Delta T_{lm}) \tag{7.2.75}$$

Example 7.2.8 The effect of tube spacing on the heat transfer rate from a tube bank heater.

In the present example we consider the effect of altering the tube spacing on the overall rate of heat transfer from a bank of tubes that are used to preheat ventilation air.

In air duct heaters banks of cylindrical electrical heating elements are often used to preheat ventilation air. One such duct heater comprises a bank of 12 tubes with $N_L = 3$, and $N_T = 4$. Both the transverse and longitudinal pitches are 24 mm. Air at atmospheric pressure, and an initial temperature, and velocity of 25°C and 12 m/s, moves in cross-flow over the heater bank. If the elements are 12 mm in diameter and 250 mm long, and have a surface temperature of 350°C, determine the total rate of heat transfer to the air, and the temperature of the air as it leaves the tube bank. If both the longitudinal and transverse pitch were increased to 36 mm what would be the effect on the rate of heat transfer and the air exit temperature?

Analysis

The total rate of heat transfer may be obtained from Eq. (7.2.75), with the log-mean temperature difference defined through Eq. (7.2.73). However, to define these respective values, we must make an initial estimate of the air exit temperature, and thereafter employ an iterative method to provide a solution.

On assuming an exit temperature, fluid properties may be obtained for the average temperature, thus allowing an estimate of the Reynolds number. For the present case of an aligned series of tubes, the Reynolds number is calculated based on the maximum fluid velocity, Eq. (7.2.65), and is therefore itself a function of the tube bank configuration. The Nusselt number is obtained via Eq. (7.2.69), with the appropriate values for the constants 'C_1' and 'm' taken from Table 7.2.4. As $N_L < 10$, an additional correction factor 'C_2', obtained from Table 7.2.5, is applied, as shown in Eq. (7.2.70). Thereafter, the surface heat transfer coefficient, and then the rate of heat transfer, may be calculated. In respect of the value obtained for the heat transfer rate, both h and ΔT_{lm}, as terms included in Eq. (7.2.75), are calculated based on the assumed exit temperature. The exit temperature may also be calculated through Eq. (7.2.74), and compared with the original assumed value. If required, a new assumed value, equal to the average of the calculated and original assumed values, is applied, and the procedure repeated until the two respective values of the assumed and calculated exit temperature are seen to converge.

Solution

Open the workbook Ex07-02-08.xls and insert the data given in the example detailed above in **cells (C4:C12)** of worksheet *Compute*. Also, in **cell B28**, insert a value for your initial estimate for the exit temperature. Excel performs all the necessary calculations.

Fluid properties are returned to the *Compute* worksheet for air at the inlet temperature, the heater element surface temperature, and also the mean fluid temperature, as based on the initial assumed exit temperature. Values of S_T/D and S_L/D are given in **cells (B18 and B19)** to allow the appropriate selection of the constant values 'C_1' and 'm' from Table 7.2.4. As detailed in the above analysis, the fluid maximum velocity is calculated, **cell B21**, followed by the Reynolds number, **cell B22**.

The user must then select the appropriate values of 'C_1' and 'm', obtained from Table 7.2.6, and insert them in **cells (D21 and E21)** respectively, to allow calculation of the Nusselt number. For the present case, assuming an initial value of 40°C for the exit temperature, **cell B28,** a value of 96.3, **cell B23**, is returned for the Nusselt number. Thereafter, values of 214.4 W/m^{2}K, 317.4°C, 7699.2 W, and 47.4°C, are returned for h, ΔT_{lm}, q, and T_o in **cells (B24, B25, B26, and B27)**, respectively.

It will be immediately apparent that the initial estimate for the exit temperature is too low, and that the values obtained for the above mentioned parameters are subject to errors greater than those implied through use of the regression given in Eq. (7.2.69). In addition to the functions embedded in the *Compute* and *Properties* worksheets, the workbook Ex07-02-08.xls comprises a 'macro', called *Iterate*. The user should access and run the macro either via the 'Tools' menu or by simultaneously pressing the 'Alt' and 'F8' keys. The values in **cells (B27 and B28)** will be seen to converge, thus affording a more accurate solution to the problem posed.

Running this macro yields values of 213.6 W/m^2 K, 313.7°C, and 7577.1 W for $h, \Delta T_{lm}, q$ **cells (B24, B25,** and **B26)**, and 47.3°C for T_o in both **cells (B27** and **B28)**.

In considering the effect of increasing both the longitudinal and transverse pitch, the above procedure is repeated.

On changing the values for both S_T and S_L to 0.036 m, appropriate values of 0.286 and 0.608 should be inserted into the appropriate **cells** for 'C_1', and 'm', respectively. 'C_2' remains unchanged. Selecting an initial estimate for the exit temperature, and thereafter running the *Iterate* macro, Excel returns values of 178.1 W/m^2 K, 318.7°C, 6417.6 W for $h, \Delta T_{lm}, q$ **cells (B24, B25,** and **B26)**, and 37.6°C for T_o in both **cells (B27** and **B28)**.

Discussion

Through increasing both the vertical and horizontal distance between adjacent tubes, the maximum air velocity is reduced from 24 to 18 m/s, **cell B21**. Hence, the Reynolds number is reduced, reducing the rate of heat transfer, and, consequently, the exit temperature. As the exit temperature is reduced, so ΔT_{lm} is seen to increase slightly. While the heat transfer rate is reduced, the pressure drop across the heater bank will also be slightly reduced.

Example 7.2.9 The effect of tube number and spacing on the condensation rate of a tube bank condenser.

Steam at a pressure of 2 bar and saturation temperature of 120.2°C (h_{fg} = 2183 kJ/kg), is passed through a tube bank condenser. A fan blows air at a temperature of 18°C and velocity 5 m/s through the tube bank which has the following characteristics: Tubes are aligned, with, $S_T = S_L = 0.05$ m, $N_T = N_L = 20$, tube diameter = 0.02 m, and length = 1.25 m. Calculate the heat transfer rate from the tubes, the air outlet temperature, and the rate of condensation.

If we require to increase the rate of condensation but cannot increase the overall size of the condenser, calculate the effect on the rate of condensation by, firstly, increasing the number of tubes in the lateral plane (N_L) to 25 while reducing S_L to 0.04 m, and, secondly, for the original condenser configuration, increasing the initial air velocity by 25%.

Analysis

In the present example we consider the application of Eq. (7.2.71). As for the previous example, the maximum velocity is again calculated via Eq. (7.2.65), followed by the Reynolds number. Based on the Reynolds number, the appropriate constants 'C_1' and 'm', for application in Eq. (7.2.71), are selected from Table 7.2.6.

Equation (7.2.71) requires knowledge of the air outlet temperature in order to define the fluid properties for the mean fluid temperature. We therefore must make an initial estimate of the outlet temperature to define the necessary fluid property data. After calculating h and ΔT_{lm}, a computed value for the outlet temperature may thereafter be obtained via Eq. (7.2.74) and compared with the initial estimate. As previously described, an iterative procedure is then applied until the estimated and calculated values for the air outlet temperature converge. At each successive iterative step, as the input/estimated value for the outlet temperature changes, so the related fluid properties must also be recalculated.

The rate of heat transfer may be calculated via Eq. (7.2.75), and, assuming the temperature of the fluid in the tubes remains constant, the condensation rate may be determined from the following equation

$$\dot{m} = q/h_{\mathrm{fg}}$$

Solution

Open the workbook Ex07-02-09.xls and enable the 'macro', when prompted. The workbook contains two worksheets, namely, *Compute* and *Properties*. As with many of the previous examples, the user need only concern themselves with the *Compute* worksheet. Input the given data for the first scenario in **cells (C4:C12)**. As stated in the above *Analysis*, we require to make an initial estimate of the air outlet temperature. This should be inserted in **cell B24**.

Excel automatically calculates the maximum velocity and Reynolds number. The user is required to insert the appropriate values for 'C_1' and 'm', taken from Table 7.2.6, in **cells (D20 and E20)**, respectively. Values for the Nusselt number (Eq. (7.2.71)), and the surface convective heat transfer coefficient are then returned in **cells (B20 and B21)**. The log-mean temperature difference, rate of heat transfer, and outlet temperature are given in **cells (B22, B23,** and **B25)**. On completion of inputting all required data, the user should run the 'macro' *Iterate* via the 'Tools' menu. Presently, this 'macro' performs one hundred iterations per run. If the 'Try' and 'Computed' temperature values do not converge on the first run, the macro should be re-run until convergence is observed. The results obtained for each of the three scenarios considered are given in Table 7.2.8. On running the 'macro', in addition to both the 'Try' and 'Computed' values of the outlet temperature changing, the values for Re, Nu, h, q, ΔT_{lm} will also be seen to change.

Table 7.2.8 Results for Example 7.2.9

	$S_T = S_L = 0.05\,m$ $N_T = N_L = 20$ $V = 5\,m/s$	$S_T = 0.05\,m, S_L = 0.04\,m$ $N_T = 20, N_L = 25$ $V = 5\,m/s$	$S_T = S_L = 0.05\,m$ $N_T = N_L = 20$ $V = 6.25\,m/s$
V_{max}(m/s)	8.33	8.33	10.42
Re	9.85×10^3	9.64×10^3	1.24×10^4
Nu_D	78.49	77.40	90.77
h_D(W/m^2 K)	106.20	105.78	122.40
ΔT_{lm}(°C)	81.54	77.14	82.99
q(kW)	435	513	511
T_o(°C)	56.34	63.64	53.85
$\dot{m}$(kg/s)	0.1994	0.2348	0.2339

Discussion

For the initial case of $S_T = S_L = 0.05\,m$, $N_T = N_L = 20$, and $V = 5\,m/s$, as shown in Table 7.2.8, the condensation rate is found to be 0.199 kg/s. Compared to increasing the air inlet velocity to 6.25 m/s, increasing N_L to 25 will have a slightly greater effect on the rate of condensation. However, in respect of retrofitting, it would most likely be cheaper and easier to install an up-rated fan than alter the tube bank. However, due to the increased power demand of the up-rated fan, system running costs would increase.

The iterative method of solution necessarily employed in the present example once again demonstrates the value of using Excel. As outlined above, for each iterative step a number of calculations, including the redefinition of fluid property values, are required. Manually, this would be a laborious task.

In the present example the user is required to insert the appropriate values for the constants 'C_1' and 'm'. Care should be taken to ensure that, after each run of the *Iterate* 'macro', over which the Reynolds number will vary, the values initially chosen are still appropriate. The user may wish to further develop this example by incorporating Table 7.2.6 in the workbook in the form of a **Lookup** table linked to the calculated Reynolds number.

7.3 Internal flow

7.3.1 Introduction

When the motion of a fluid is limited by an adjacent surface, such as the flow of water through a pipe, in contrast with external flows, boundary layers are unable to develop fully. As boundary layer development is restricted, alternative methodologies to those presented in Section 7.2 are required for solving convection heat transfer problems under such conditions. As we have already seen, boundary layer development is dependent upon fluid velocity.

We shall therefore consider the velocity effects (hydrodynamic) upon boundary layer development, as well as the effects of any thermal variance in the fluid throughout its flow path.

7.3.2 *Hydrodynamic considerations*

In our discussion of external flow conditions, we were primarily concerned with either laminar or turbulent flows when trying to solve convection heat transfer problems. For the present case however, we must also bear in mind the effects of the entrance and *fully developed flow* regions on the heat transfer coefficient.

We begin our analysis of internal flow by considering the case of laminar fluid flow through a circular pipe of radius r_o, as shown in Fig. 7.3.1. Further, we assume that, at inlet, the fluid is flowing at a uniform velocity. We know, as described in Chapter 6, that at the fluid and the tube wall internal surface interface shear forces (viscosity) influence boundary layer development as the distance, x from the pipe entrance increases. This development continues until convergence occurs along the tube's centre line due to an ever-decreasing inviscid flow region. This process continues across the entire cross-sectional area of the tube until the velocity profile of the fluid remains constant, irrespective of x. Beyond this point, as shown in the above mentioned figure, the flow may be described as fully developed. The distance from the fluid's point of entry to the tube to the point where this condition is seen to occur is referred to as the *hydrodynamic entry length*. In the case of laminar flow, the fully developed velocity profile of a fluid flowing in a circular tube is parabolic. However, turbulent flow in a similar tube would result in a fully developed velocity profile that was much flatter as a result of turbulent mixing in the radial direction.

Recognition of the extent of the entry region is highly important when dealing with internal flow conditions. For fluid flow in a *circular tube*, the

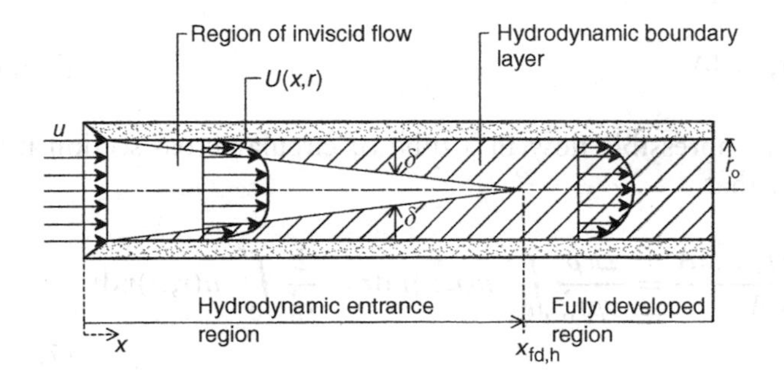

Figure 7.3.1 Development of a laminar hydrodynamic boundary layer in a circular tube of radius r_o.

Reynolds number is defined as

$$\mathrm{Re_D} \equiv \frac{\rho u_{\mathrm{m}} D}{\mu} \tag{7.3.1}$$

where u_{m} is the mean fluid velocity across the tube's cross-section and D is the diameter of the tube. For turbulent flow, the critical Reynolds number is 2300. For laminar flow, as shown in Fig. 7.3.1, the hydrodynamic entry length may be obtained from

$$\left(\frac{x_{\mathrm{fd,h}}}{D}\right)_{\mathrm{lam}} \approx 0.05\,\mathrm{Re_D} \tag{7.3.2}$$

This is based on the assumption that flow enters the tube with a uniform velocity profile and hence the boundary layer develops at the same rate at all points around the tube circumference.

As has previously been discussed, the velocity of a fluid moving within the confines of an enveloping surface does not remain constant across the cross-section of the body in which it is contained. We therefore use the mean velocity, u_{m}, of the fluid as a guide to providing an insight into boundary layer development. The mean velocity can also be used to determine other fluid characteristics, such as mass flow rate, as shown below:

$$\dot{m} = \rho u_{\mathrm{m}} A_{\mathrm{C}} \tag{7.3.3}$$

where $\dot{m}$ is the mass flow rate, ρ the fluid density, and A_{C} the cross-sectional area of the enclosure. In the case of steady *incompressible flow* in a tube of uniform cross-sectional area $\dot{m}$ and u_{m} remain constant, regardless of the hydrodynamic entrance regions length. It therefore follows that, for similar flow in a circular tube ($A_{\mathrm{C}} = \pi D^2/4$), the Reynolds number reduces to,

$$\mathrm{Re_D} = \frac{4\dot{m}}{\pi D \mu} \tag{7.3.4}$$

Integration of the mass flux, ρu, over the cross-section is another method that may be used to quantify mass flow rate. Thus

$$\dot{m} = \int_A \rho u(r,x)\,\mathrm{d}A \tag{7.3.5}$$

In the case of incompressible flow in a tube of circular cross-section it is therefore the case that,

$$u_{\mathrm{m}} = \frac{\int_A \rho u(r,x)\,\mathrm{d}A}{\rho A} = \frac{2\pi\rho}{\pi\rho r_{\mathrm{o}}^2}\int_0^{r_{\mathrm{o}}} u(r,x) r\,\mathrm{d}r = \frac{2}{r_{\mathrm{o}}^2}\int_0^{r_{\mathrm{o}}} u(r,x) r\,\mathrm{d}r \tag{7.3.6}$$

Therefore, the mean fluid velocity at any axial location x can now be determined from information concerning the velocity profile, $u(r)$, at that location. This may then be used, as will be discussed later, to determine the convective heat transfer coefficient for the hydrodynamic entry length. Presently, however, the velocity profile of the fully developed region of laminar flow is considered.

7.3.3 *Velocity profile of the fully developed region of laminar flow*

As has previously been stated, in the fully developed region of laminar flow, the velocity profile incurred remains unchanged for further increases in x. Alternatively, it can be said that both the radial velocity component (v) and the gradient of the axial velocity flow ($\mathrm{d}u/\mathrm{d}x$) equate to zero at all points in this region, that is,

$$v = 0 \quad \text{and} \quad \frac{\partial u}{\partial x} = 0 \tag{7.3.7}$$

The velocity component therefore is only dependent on r, thus, $u(x, r) = u(r)$.

Solution of the appropriate x-momentum equation will demonstrate this radial dependence of the axial velocity. In doing so, one must first be cognisant with the condition that the net momentum flux in the aforementioned region is also zero at all points of concern. The requirement for conservation of momentum therefore reduces to a simple balance between the shear and pressure forces induced by the fluid flow. If one considers the force balance occurring on the annular differential element shown in Fig. 7.3.2, it can be reported that

$$\tau_r(2\pi r\,\mathrm{d}x) - \left\{\tau_r(2\pi r\,\mathrm{d}x) + \frac{\mathrm{d}}{\mathrm{d}r}[\tau_r(2\pi r\,\mathrm{d}x)]\,\mathrm{d}r\right\} + p(2\pi r\,\mathrm{d}r) - \left\{p(2\pi r\,\mathrm{d}r) + \frac{\mathrm{d}}{\mathrm{d}x}[p(2\pi r\,\mathrm{d}r)]\,\mathrm{d}x\right\} = 0 \tag{7.3.8}$$

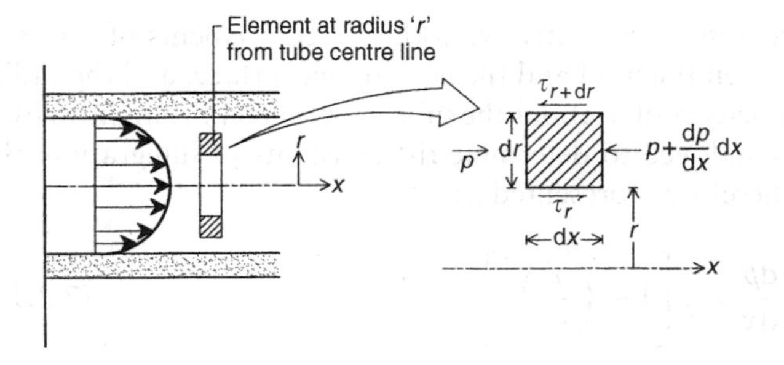

Figure 7.3.2 Balance of forces on an annular differential element for laminar, fully developed flow in a circular tube.

This may be reduced to:

$$-\frac{\mathrm{d}}{\mathrm{d}r}(r\tau_r) = r\frac{\mathrm{d}p}{\mathrm{d}x} \tag{7.3.9}$$

In the present case, the single relevant sheer stress component acting upon the differential element under analysis is given via the following modified version of Eq. (6.3.1):

$$\tau_r = -\mu\frac{\mathrm{d}u}{\mathrm{d}r} \quad \text{since} \quad y = r_\mathrm{o} - r \tag{7.3.10}$$

The force balance on the differential element can therefore also now be reduced to the form

$$\frac{\mu}{r}\frac{\mathrm{d}}{\mathrm{d}r}\left(r\frac{\mathrm{d}u}{\mathrm{d}r}\right) = \frac{\mathrm{d}p}{\mathrm{d}x} \tag{7.3.11}$$

Equation (7.3.11) can be solved through a process of double-integration since the axial pressure gradient is independent of r. Thus, first,

$$r\frac{\mathrm{d}u}{\mathrm{d}r} = \frac{1}{\mu}\left(\frac{\mathrm{d}p}{\mathrm{d}x}\right)\frac{r^2}{2} + c_1 \tag{7.3.12}$$

and second,

$$u(r) = \frac{1}{\mu}\left(\frac{\mathrm{d}p}{\mathrm{d}x}\right)\frac{r^2}{4} + c_1 \ln r + c_2 \tag{7.3.13}$$

The constants of integration, c_1 and c_2, may be determined by limiting the boundary conditions of solution to

$$u(r) = 0 \quad \text{and} \quad \left.\frac{\partial u}{\partial r}\right|_{r=0} = 0 \tag{7.3.14}$$

These conditions represent, respectively, both the requirements of zero slip at the interface between the fluid and the enveloping surface, and the radial symmetry of the velocity profile about the tube's centreline. From the mathematical operation employed to determine the constants of integration, the velocity profile is therefore represented as

$$u(r) = -\frac{1}{4\mu}\left(\frac{\mathrm{d}p}{\mathrm{d}x}\right)r_\mathrm{o}^2\left[1 - \left(\frac{r}{r_\mathrm{o}}\right)^2\right] \tag{7.3.15}$$

From the form of Eq. (7.3.15) it can be seen that in the fully developed region, the velocity profile is parabolic and the pressure gradient is always negative. Substitution of Eq. (7.3.15) in Eq. (7.3.6) allows determination of the mean velocity of the fluid via a process of integration, thus

$$u_m = -\frac{r_o^2}{8\mu}\frac{dp}{dx} \tag{7.3.16}$$

The velocity profile in the fully developed region may now be determined through further substitution of Eq. (7.3.16) into Eq. (7.3.15) such that

$$\frac{u(r)}{u_m} = 2\left[1 - \left(\frac{r}{r_o}\right)^2\right] \tag{7.3.17}$$

We have previously shown that u_m can be determined from mass flow rate. This procedure detailed above can therefore be applied to determine the pressure gradient.

7.3.4 *Friction factor and pressure gradient in the fully developed region of laminar flow*

There are many engineering applications whereby details of the pump or fan output that are necessary to overcome the pressure drop and maintain an internal flow are required. Such information is essential for specifying optimal plant selection and also to maintain its efficient operation. For example, incorrect fan or pump selection can result in system over-performance, economic concerns in terms of both initial and running costs, the possible creation of situations which are both hazardous to life and property and long-running contractual disputes between the designer and the client over this matter. The practicing engineer will turn to product-specific data prepared by the manufacturer when selecting items of plant to be incorporated into a design. This information is often presented in the form of tables, graphs, or ready-reckoners and is often the result of much research and development on behalf of the manufacturer. Rules-of-thumb are often used to provide initial foundation for these selections. However, in order to understand how the manufacturer has produced their technical literature, the engineer must be fully conversant with the principles involved in the production of such data. This section therefore aims to provide an overview of the above mentioned principles.

The Moody friction factor is a term often encountered in engineering when determining system pressure drop. This is a dimensionless parameter, referred to in some texts as the Darcy friction factor, and is defined as

$$f \equiv \frac{-(dp/dx)D}{\rho u_m^2/2} \tag{7.3.18}$$

It is important not to confuse the above with the friction coefficient, or the Fanning friction factor as it is often called, which can be found from

$$C_f = \frac{\tau_S}{\rho u_m^2/2} \tag{7.3.19}$$

From the shear stress equation, Eq. (6.3.2), $\tau_S = -\mu(du/dr)_{r=r_o}$, it therefore follows from Eq. (7.3.18) that

$$C_f = \frac{f}{4} \tag{7.3.20}$$

To determine the friction factor associated with laminar flow in the fully developed region, Eqs (7.3.1) and (7.3.16) are substituted into Eq. (7.3.18) to give

$$f = \frac{64}{Re_D} \tag{7.3.21}$$

However, in the case of turbulent flow, the analysis becomes more complex. As a result of the variety of situations that may be encountered in this type of flow, the engineer now relies on experimental rather than numerical results. These experimental results, for a wide variety of Reynolds numbers, are often presented in a Moody diagram (Moody, 1944). Note that the Colebrook model for the friction factor (Eq. (2.9.3)) represents the complete Moody diagram. A comparison of friction factors obtained via the Colebrook and Swamee and Jain equations, read from a Moody chart, and also from Eqs (7.3.22a) and (7.3.22b) and Eq. (7.3.63), is presented in Section 7.4. Considering the Colebrook equation, in addition to the Reynolds number, the friction factor is now also seen to be a function of the condition of the tube surface, a factor whose implications are negligible in the case of laminar flow. The magnitude of the friction factor is proportional to the tube surface roughness 'ε'. For smooth surfaces such as uPVC pipe 'f' is small compared to the values obtained when considering materials, such as cast iron, which have a markedly rougher surface. The surface roughness parameter is a measurement of the abrasive encroachment from the mean internal surface of the tube towards its central axis. The dimensionless parameter ε/D is used in Moody diagrams to represent different material types through which the turbulent flow may occur at a variety of Reynolds numbers.

For smooth surfaces the following correlations can be used to approximate the friction factor for both laminar and turbulent flows.

$$f = 0.316\,Re_D^{-1/4}, \quad Re_D \leq 2 \times 10^4 \tag{7.3.22a}$$

$$f = 0.184\,Re_D^{-1/5}, \quad Re_D \geq 2 \times 10^4 \tag{7.3.22b}$$

The Moody friction factor is shown to be constant in the fully developed region. It therefore follows that the pressure gradient dp/dx is also a constant value in this region. Let us consider the situation where, for *fully developed flow*, the pressure drop along an axial line from position x_1 to x_2 may be determined as $p_1 - p_2$. Therefore, from Eq. (7.3.18) it follows that

$$\Delta p = -\int_{p_1}^{p_2} \mathrm{d}p = f\frac{\rho u_\mathrm{m}^2}{2D}\int_{x_1}^{x_2} \mathrm{d}x = f\frac{\rho u_\mathrm{m}^2}{2D}(x_2 - x_1) \tag{7.3.23}$$

In this case, f may be obtained from either Eq. (7.3.21) for laminar flow or Eqs (7.3.22a) and (7.3.22b) for laminar and turbulent flow in smooth tubes respectively. The power required to overcome the resulting resistance to flow may be calculated as the product of the pressure drop and the volumetric flow rate, thus,

$$P = (\Delta p)\dot{V} \tag{7.3.24}$$

For an incompressible fluid, as previously discussed, the volumetric flow rate may be expressed as

$$\dot{V} = \dot{m}/\rho \tag{7.3.25}$$

As was discussed in Section 7.2.2, a tripper device may be used to promote turbulence and increase the convection coefficient. Similarly, the roughness of a surface can also affect the rate of heat transfer. Hence, there exists a relationship between the friction factor 'f', and the surface heat transfer coefficient 'h'. As will be seen later in this section (Eqs (7.3.63) and (7.3.65)), for the conditions specified, the Nusselt number, and hence h, may be determined through consideration of 'f'.

7.3.5 *Thermal effects*

Thus far we have provided the groundwork to allow analysis of the fluid dynamics of the internal flow scenario. We now turn our attention to the implications of the thermal effects on such fluid flow with the aid of Fig. 7.3.3. This figure details the development of a thermal boundary layer, established as a result of the effects of a convective heat transfer process. This process is induced through the presence of a temperature differential between the internal fluid and the tube wall. Considering the case where the fluid entering the tube is at a uniform temperature $T(r, 0)$ which is less than that of the surface temperature T_S of the tube, a *thermally fully developed boundary condition* will be achieved if the tube surface condition is controlled either through imposition of a uniform temperature (T_S = constant) or uniform heat flux (q_S = constant).

In practice, the above condition of a constant surface temperature may be realised where flue gases condense onto the surface of the secondary heat exchanger in a condensing boiler. Current carrying insulating tape, overlaid upon external water services pipework as a means of frost-protection or the

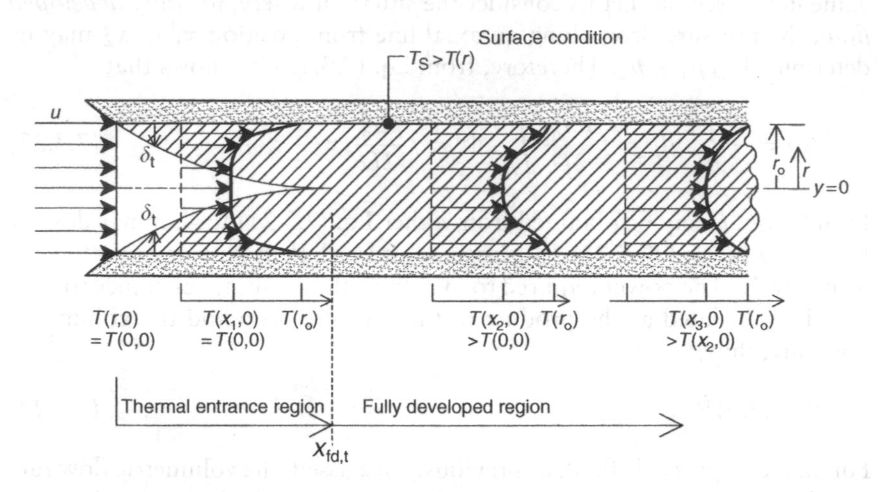

Figure 7.3.3 Development of the thermal boundary layer in a heated circular tube.

direct solar radiant energy falling on the absorber tubes of a solar collector on a warm, sunny day may provide a constant uniform heat flux.

The fully developed temperature profile varies in shape depending on whether or not a uniform surface temperature or uniform heat flux is sustained. For heat transfer to the internal fluid, the degree to which the fluid temperature exceeds the entry temperature increases with x, regardless of which type of surface condition prevails.

As reported by Kays and Crawford (1980), for laminar flow, the thermal entry length can be found from the following relationship,

$$\left(\frac{x_{fd,t}}{D}\right)_{lam} \approx 0.05\,\mathrm{Re_D\,Pr} \tag{7.3.26}$$

When compared to the hydrodynamic entry length (Eq. (7.3.2)) it is seen that if $\mathrm{Pr} > 1$ development of the thermal boundary layer occurs at a lesser rate than the velocity boundary layer, that is, $x_{fd,h} < x_{fd,t}$. Similarly the converse is true if $\mathrm{Pr} < 1$. In the case of fluids that have comparatively large Prandtl numbers, such as oils and organic compounds including glycol and glycerine, the hydrodynamic entry length is considerably shorter than that of its thermal counterpart. To fully understand the conditions in the fully developed thermal boundary layer it is first necessary to introduce the concept of a *mean temperature*, and the appropriate form of Newton's law of cooling.

7.3.6 *The mean temperature and Newton's law of cooling*

The mean or *bulk* temperature of the fluid at a specific cross-section within the tube is defined in terms of the thermal energy transported by the fluid as it passes the cross-section. This rate, q, may be determined by integrating the product of the mass flux (ρu), and the internal energy per unit mass ($c_V T$) over the cross-section. Thus,

$$q = \int_{A_C} \rho u c_V T \, dA_C \tag{7.3.27}$$

Hence, if the mean temperature is defined such that,

$$T_m = \frac{\int_{A_C} \rho u c_V T \, dA_C}{\dot{m} c_V} \tag{7.3.28}$$

we get

$$q \equiv \dot{m} c_V T_m \tag{7.3.29}$$

For incompressible flow in a circular tube, with constant c_V, from Eqs (7.3.3) and (7.3.28), we get,

$$T_m = \frac{2}{u_m r_o^2} \int_0^{r_o} u T r \, dr \tag{7.3.30}$$

While the mean temperature plays a similar role in internal flows as the free stream temperature does in external flows, unlike the free stream temperature, T_m varies along the tube length if heat transfer is occurring. If heat transfer is from the surface to the fluid, T_m will increase along the tube length, while it will decrease with x if heat transfer is from the fluid to the surface.

Considering the definition of the mean temperature, Newton's law of cooling may be expressed as,

$$q_S = h(T_S - T_m) \tag{7.3.31}$$

As the fluid mean temperature varies with x, the radial temperature gradient will also vary. Consequently the temperature profile $T(r)$ continually varies implying that fully developed conditions are never achieved. However, if we consider working with a dimensionless temperature difference of the form,

$$T_D = \frac{(T_S - T_r)}{(T_S - T_m)} \tag{7.3.32}$$

we may plot T_D against r for all positions along the tube length. We can see that, after a certain point along the tube, the *relative shape* of the temperature profile remains the same, and the flow may therefore be said to be *thermally fully developed.*

7.3.7 *Fully developed conditions*

The requirement for fully developed thermal conditions may be expressed as,

$$\frac{\partial}{\partial x}\left[\frac{T_S(x) - T(r,x)}{T_S(x) - T_m(x)}\right]_{fd,t} = 0 \qquad (7.3.33)$$

The condition given by Eq. (7.3.33) will eventually be met if the tube experiences either a uniform surface temperature or a uniform heat flux.

As the temperature ratio given in Eq. (7.3.32) is independent of x, the derivative of this ratio with respect to the tube radius r will also be independent of x. Hence,

$$\frac{\partial}{\partial r}\left(\frac{T_S - T}{T_S - T_m}\right)\bigg|_{r=r_o} = \frac{-\partial T/\partial r|_{r=r_o}}{T_S - T_m} \qquad (7.3.34)$$

Substituting for $\partial T/\partial r$ from Fourier's law, which, in respect of Fig. 7.3.3 is of the form,

$$q_S = -k\left.\frac{\partial T}{\partial y}\right|_{y=0} = k\left.\frac{\partial T}{\partial r}\right|_{r=0} \qquad (7.3.35)$$

it can be shown, via Eqs (7.3.35) and (7.3.31), that

$$\frac{h}{k} \neq f(x) \qquad (7.3.36)$$

Therefore, for fully developed flow for a fluid with constant properties the local convection coefficient is constant and independent of x. However, because the thermal boundary layer thickness varies with x along the entrance length, the heat transfer coefficient also varies with x along this length, being a maximum at $x = 0$ as the thermal boundary layer thickness is zero at this point. For the case of constant surface temperature, as the rate of heat transfer between the fluid and duct walls varies along the duct length, the value of $\partial T/\partial x$ is dependent on the radial coordinate.

As the flow is enclosed, an energy balance can be used to show how both the mean temperature T_m and the total convection heat transfer q_{conv} vary with the position along the tube. The main factors affecting the rate of change in convection heat transfer are those that derive from the transfer of thermal energy between the fluid and the enclosure walls, and the flow work that is performed in moving the fluid. The flow work may be expressed as the product of the fluid pressure p, and its specific volume v. Applying conservation of energy to the control volume shown in Fig. 7.3.4 we obtain

$$\mathrm{d}q_{conv} + \dot{m}(c_v T_m + pv) - \left[\dot{m}(c_v T_m + pv) + \dot{m}\frac{\mathrm{d}(c_v T_m + pv)}{\mathrm{d}x}\,\mathrm{d}x\right] = 0 \qquad (7.3.37)$$

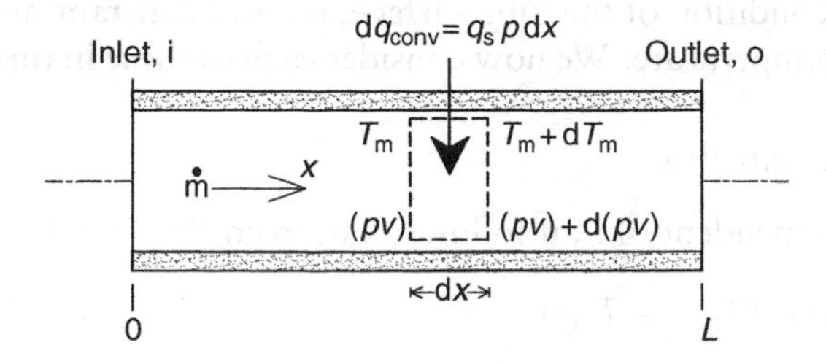

Figure 7.3.4 Elemental control volume for tube internal flow.

which reduces to

$$dq_{conv} = \dot{m}\,d(c_v T_m + pv) \tag{7.3.38}$$

Thus, the rate of convection heat transfer is the sum of the rate of change in the fluid thermal energy plus the net rate at which work is done in moving the fluid through the above mentioned control volume. If the fluid is assumed to be an ideal gas, with $pv = RT_m$ and $c_p = c_v + R$ with c_p constant, Eq. (7.3.38) reduces to,

$$dq_{conv} = \dot{m}c_p\,dT_m \tag{7.3.39}$$

For the entire tube, Eq. (7.3.39) may be integrated between the inlet and outlet to give,

$$q_{conv} = \dot{m}c_p(T_{m,o} - T_{m,i}) \tag{7.3.40}$$

For incompressible liquids, with $c_v = c_p$, and the specific volume, v, typically very small, the term $d(pv)$ in Eq. (7.3.38) is much smaller than $d(c_v T_m)$, and Eqs (7.3.39) and (7.3.40) may therefore also be used in this instance. However, where the pressure gradient is large, resulting from high fluid velocities, rough internal surfaces or narrow tubes, Eq. (7.3.38) may be integrated from inlet to outlet to obtain,

$$q_{conv} = \dot{m}\lfloor c_v(T_{m,o} - T_{m,i}) + (pv)_o - (pv)_i \rfloor \tag{7.3.41}$$

If we consider that the rate of heat transfer to the differential element of Fig. 7.3.4 may be expressed as,

$$dq_{conv} = q_S P\,dx \tag{7.3.42}$$

where P is the perimeter, $P = \pi D$ for a tube, substituting for Eq. (7.3.31) into Eq. (7.3.39), we get,

$$\frac{dT_m}{dx} = \frac{q_S P}{\dot{m}c_p} = \frac{P}{\dot{m}c_p} h(T_S - T_m) \tag{7.3.43}$$

The solution to Eq. (7.3.43) for the mean fluid temperature at point x depends on the thermal condition of the tube surface, namely constant heat flux or constant surface temperature. We now consider each of these in turn.

7.3.8 Constant surface heat flux

As q_S is constant and independent of x, it follows that, from Eq. (7.3.40)

$$q_{conv} = q_S(P \cdot L) = \dot{m}c_p(T_{m,o} - T_{m,i}) \tag{7.3.44}$$

where L is the tube length. Also, the right-hand term of Eq. (7.3.43) is constant and,

$$\frac{dT_m}{dx} = \frac{q_S P}{\dot{m}c_p} \tag{7.3.45}$$

If we integrate from $x = 0$, then

$$T_m(x) = T_{m,i} + \frac{q_S P}{\dot{m}c_p}x \tag{7.3.46}$$

If the tube perimeter is constant along its length, as is most often the case, as q_S is a constant, the mean temperature will vary linearly along the tube length.

7.3.9 Constant surface temperature

If we define $\Delta T = T_S - T_m$, Eq. (7.3.43) may be rewritten as,

$$\frac{dT_m}{dx} = -\frac{d(\Delta T)}{dx} = \frac{P}{\dot{m}c_p}\Delta T \tag{7.3.47}$$

By rearranging and integrating along the length of the tube we get,

$$\int_{\Delta T_i}^{\Delta T_o} \frac{d(\Delta T)}{\Delta T} = -\frac{P}{\dot{m}c_p}\int_0^L h\, dx \tag{7.3.48}$$

which reduces to,

$$\ln\frac{\Delta T_o}{\Delta T_i} = -\frac{PL}{\dot{m}c_p}\bar{h}_L \tag{7.3.49}$$

and rearranging gives,

$$\frac{\Delta T_o}{\Delta T_i} = \frac{T_S - T_{m,o}}{T_S - T_{m,i}} = \exp\left(-\frac{PL}{\dot{m}c_p}\bar{h}_L\right) \tag{7.3.50}$$

Therefore, the temperature difference $T_S - T_m$ varies exponentially along the tube length. For heat transfer to the fluid, for constant surface temperature

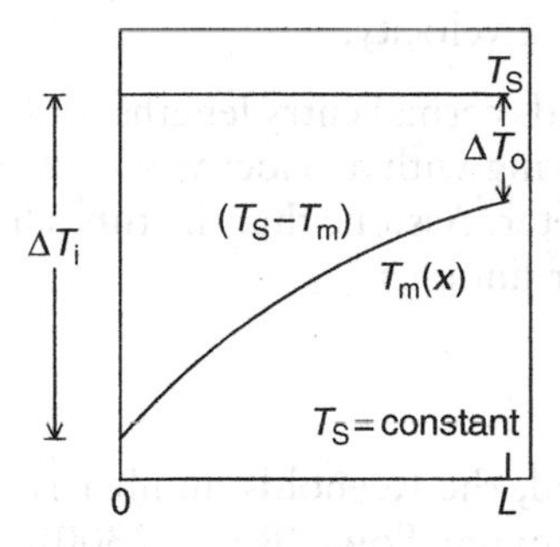

Figure 7.3.5 Axial temperature variation for heat transfer in a tube with constant surface temperature.

the profile of the mean fluid temperature will be as shown in Fig. 7.3.5. In defining the total heat transfer rate it can be shown that,

$$q_{\mathrm{conv}} = \bar{h}PL\Delta T_{\mathrm{lm}} \tag{7.3.51}$$

where ΔT_{lm} is the log-mean temperature, and is given by

$$\Delta T_{\mathrm{lm}} = \frac{\Delta T_{\mathrm{o}} - \Delta T_{\mathrm{i}}}{\ln(\Delta T_{\mathrm{o}}/\Delta T_{\mathrm{i}})} \tag{7.3.52}$$

7.3.10 *Laminar flow in circular tubes*

As for external flows, the correlations given for the convective heat transfer coefficients in internal flows are also given in terms of the Nusselt, Reynolds, and Prandtl numbers. For laminar flows Shah and Bhatti (1987) showed that the hydrodynamic entry length was given by

$$\frac{L_{\mathrm{e}}}{D} \approx 0.0565\,\mathrm{Re}_{\mathrm{D}} \tag{7.3.53}$$

and that, for a uniform surface temperature and uniform heat flux respectively, the thermal entry length could be obtained from

$$\frac{L_{\mathrm{e,th}}}{D} = 0.037\,\mathrm{Re}_{\mathrm{D}}\,\mathrm{Pr} \quad \text{(uniform surface temperature)} \tag{7.3.54}$$

$$\frac{L_{\mathrm{e,th}}}{D} = 0.053\,\mathrm{Re}_{\mathrm{D}}\,\mathrm{Pr} \quad \text{(uniform heat flux)} \tag{7.3.55}$$

We now consider the effect of fluid properties on the hydrodynamic and thermal entry length for laminar flow through Example 7.3.1.

Example 7.3.1 Comparison of the hydrodynamic and thermal entry lengths for oil and water at the same temperature and velocity.

Calculate and compare the hydrodynamic and thermal entry lengths for both water and oil at a temperature of 40°C, flowing with a velocity of 0.05 m/s through similar tubes each of 0.025 m diameter. Assume that the tubes have a uniform surface temperature. Discuss your findings.

Analysis

For both fluids, for the temperature specified, the Reynolds number is calculated and fluid properties defined. For laminar flow, (Re < 2300), the respective entry lengths for both fluids are calculated through Eqs (7.3.53) and (7.3.54).

Solution

Open the workbook Ex07-03-01.xls. This workbook comprises three worksheets, namely, *Propoil, Propwater*, and *Compute*. The thermophysical properties for oil and water are given in the respective worksheets. Input the data given, temperature, velocity, and pipe diameter, in **cells (B5:B7)**. Excel performs all necessary calculations, and returns the required results in **cells (B16:C18)**.

Discussion

While the density of the two fluids is of a similar magnitude, the dynamic viscosity μ_f of oil is approximately 300 times that of water. Consequently, there is a similar variation in the respective Reynolds numbers. For the conditions specified, the flow for both fluids is laminar, with the Reynolds number for water calculated as 1893.7, compared to 5.0 for oil. An increase in velocity to greater than 0.06 m/s would cause flow to become turbulent for water. Turbulence would not occur until much higher velocities for oil (approximately 23 m/s).

For the initial condition of a velocity of 0.05 m/s the respective hydrodynamic entry lengths are calculated as 2.7 and 0.007 m for water and oil, respectively. At the temperature specified, the Prandtl number for oil, **cell K8**, is approximately 700 times that of water. The thermal entry length for oil is therefore greater than that for water. As stated previously, the thermophysical properties for water and oil are given in the respective sheets. Through comparison of these data it is clear that, for the case of oil, those thermophysical properties that influence the hydrodynamic and thermal entry lengths are more temperature dependent. For a velocity of 0.015 m/s and a temperature of 150°C, the Reynolds number for water is approximately 30 times that of oil as compared to the 300-fold difference observed for the initial conditions.

In addition, through a theoretical analysis, it can be shown that, for fully developed flow, for each of the surface conditions of a uniform surface temperature and a uniform heat flux, the respective Nusselt numbers are constant. Thus

$$\mathrm{Nu_D} = 3.66 \quad \text{(uniform surface temperature)} \tag{7.3.56a}$$

$$\mathrm{Nu_D} = 4.36 \quad \text{(uniform heat flux)} \tag{7.3.56b}$$

In the entry region however, due to the variation in the thermal boundary layer thickness, the respective Nusselt numbers are greater than those given above, and vary with x over the entry length. For a constant surface temperature, the average heat transfer coefficient in the entrance region may be calculated, subject to the constraints detailed, using the correlation presented by Sieder and Tate (1936)

$$\overline{\mathrm{Nu_D}} = 1.86\left(\frac{\mathrm{Re_d\,Pr}}{L/D}\right)^{1/3}\left(\frac{\mu}{\mu_S}\right)^{0.14}$$

$$\text{where}\ \left\{\begin{array}{c} 0.48 < \mathrm{Pr} < 16700 \\ 0.0044 < \left(\frac{\mu}{\mu_S}\right) < 9.75 \\ L/D < 8(\mu/\mu_S)^{0.42}/(\mathrm{Re_D\,Pr}) \end{array}\right\} \tag{7.3.57}$$

The fluid temperature for evaluating properties is the mean of the tube inlet and exit ($T_{m,i}$ and $T_{m,o}$) average temperatures. For $L/D > 8(\mu/\mu_S)^{0.42}/(\mathrm{Re_D\,Pr})$ the fully developed region predominates, and the correlation for fully developed laminar flow given in Eq. (7.3.56a), with fluid properties again evaluated at the bulk temperature, is applicable.

For non-circular ducts the entrance length may be calculated by replacing the diameter D with the *hydraulic mean diameter*. This is defined as four times the ratio of the duct cross-sectional area divided by the internal perimeter of the duct, thus

$$D_h = 4A_C/P \tag{7.3.58}$$

For non-circular ducts, for turbulent flow, it is reasonable to use the regressions presented in the following section. However, for laminar flow, due to a larger variance in the convective heat transfer coefficient around the duct perimeter the preceding regressions, namely Eqs (7.3.56a) and (7.3.56b), are not valid. Table 7.3.1 presents details of the Nusselt numbers and friction factors for fully developed laminar flow in ducts of varying cross-sections.

7.3.11 *Turbulent flow in circular tubes*

For turbulent flows the hydrodynamic entry length can be found from the formula of Zhi-qing (1982):

$$\frac{x_{fd,h}}{D} = 1.359\,\mathrm{Re_D^{1/4}} \tag{7.3.59}$$

Table 7.3.1 Nusselt numbers and friction factors for fully developed laminar flow in ducts of varying cross sections

Cross-section	*B/A*	$Nu_D \equiv h D_h/k$		
		Uniform q''_S	*Uniform* T_S	$f\,Re_{Dh}$
Circle	—	4.36	3.66	64
Rectangle A × B	1.0	3.61	2.98	57
Rectangle A × B	1.43	3.73	3.08	59
Rectangle A × B	2.0	4.12	3.39	62
Rectangle A × B	3.0	4.79	3.96	69
Rectangle A × B	4.0	5.33	4.44	73
Rectangle A × B	8.0	6.49	5.60	82
Parallel plates	∞	8.23	7.54	96
Triangle	—	3.11	2.47	53

For most applications Re_D is less than 250000, and $L_{fd,h} < 30D$. Suryanarayana (1995), through a review of the literature, states that we may assume that, for a given Reynolds number, the respective Nusselt numbers, and hence the *heat transfer coefficients*, are constant for all $x/D > 10$. For example, for $Re = 10^5, Nu_{x,T} \approx 200$, and, for $Re = 10^4, Nu_{x,T} \approx 50$ for $x/D > 10$. However, the *Dittus–Boelter equations*, presented by Whitaker (1972), and given below, have been confirmed experimentally for the range of conditions specified:

$$Nu_D = 0.023\,Re_D^{4/5}\,Pr^n \quad \begin{bmatrix} n = 0.4 \text{ for heating} \\ n = 0.3 \text{ for cooling} \end{bmatrix} \quad \begin{bmatrix} 0.7 \le Pr \le 160 \\ Re_D \ge 10000 \\ L/D \ge 10 \end{bmatrix} \tag{7.3.60}$$

Incropera and DeWitt (2002) state that where there are large variations in temperature, and hence also in fluid properties, the following relationship, attributed to Sieder and Tate (1936), should be used in preference to the Dittus–Boelter equation above:

$$Nu = 0.027\,Re_D^{4/5}\,Pr^{1/3}\left(\frac{\mu}{\mu_S}\right)^{0.14}, \quad \begin{bmatrix} 0.7 \le Pr \le 16700 \\ Re_D \ge 10000 \\ L/D \ge 10 \end{bmatrix} \qquad (7.3.61)$$

Experimentation has shown that, for turbulent flows, the Nusselt numbers for uniform surface temperature and uniform heat flux are equal in the fully developed region, and only slightly higher for uniform heat flux in the developing region. All regressions for turbulent flows are therefore used for both uniform surface temperature and uniform heat flux. For fully developed turbulent flow in *smooth circular pipes*, Suryanarayana (1995) recommends the use of the regressions presented by Gnielinsky (1976, 1990) as given in

$$Nu_D = 0.0214(Re_D^{4/5} - 100)Pr^{2/5}\left[1 + \left(\frac{D}{L}\right)^{2/3}\right]$$

$$\text{where} \begin{bmatrix} 0.5 < Pr < 1.5 \\ 2300 < Re_D < 10^6 \\ 0 < D/L < 1 \end{bmatrix} \qquad (7.3.62a)$$

$$Nu_D = 0.012(Re_D^{0.87} - 280)Pr^{2/5}\left[1 + \left(\frac{D}{L}\right)^{2/3}\right]$$

$$\text{where} \begin{bmatrix} 1.5 < Pr < 500 \\ 2300 < Re_D < 10^6 \\ 0 < D/L < 1 \end{bmatrix} \qquad (7.3.62b)$$

The above equations themselves are approximations to the following equation, also by Gnielinsky (1976, 1990).

$$Nu_D = \frac{(f/8)(Re_D - 1000)\,Pr}{1 + 12.7(f/8)^{0.5}(Pr^{2/3} - 1)}\left[1 + \left(\frac{D}{L}\right)^{2/3}\right]$$

$$\text{where} \begin{bmatrix} f = (0.79 \ln Re_D - 1.64)^{-2} \\ 0 < D/L < 1, \quad 3000 < Re_D < 10^6 \\ 0.6 < Pr < 2000 \end{bmatrix} \qquad (7.3.63)$$

and fluid properties are, as previously, obtained for the fluid bulk temperature. To account for variations in the fluid properties that result from temperature, the above author recommends that the Nusselt numbers given in Eqs (7.3.62a) and (7.3.62b), and (7.3.63) are multiplied by the following correction factors:

$$\left(\frac{T_m}{T_S}\right)^{0.45} \quad \text{for gases, and} \quad \left(\frac{Pr}{Pr_S}\right)^{0.11} \quad \text{for liquids} \tag{7.3.64}$$

The preceding equations, Eqs (7.3.62a) and (7.3.62b), and (7.3.63) yield values for the average heat transfer coefficient over the tube length. To determine the heat transfer coefficient for the fully developed region, the value of D/L should be set to zero. However, the following relation, presented by Petukhov (1970) produces better agreement with experimental data:

$$Nu_D = \frac{(f/8)Re_D\,Pr}{1.07 + 12.7(f/8)^{0.5}(Pr^{2/3} - 1)}\left(\frac{\mu}{\mu_S}\right)^n$$

$$\text{where} \begin{bmatrix} n = 0.11 \text{ for liquids, heating} \\ n = 0.25 \text{ for liquids, cooling} \\ n = 0 \text{ for gases} \\ 0.5 < Pr < 200 \;\; 6\% \text{ error} \\ 200 < Pr < 2000 \;\; 10\% \text{ error} \\ 10^4 < Re_D < 5 \times 10^6 \\ 0.08 < (\mu/\mu_S) < 40 \end{bmatrix} \tag{7.3.65}$$

Example 7.3.2 Comparison of the effect of alternative regressions on the calculated value for the heat transfer rate for turbulent flow in pipes.

Thus far in our study of forced turbulent flow in circular pipes we have presented several alternative regression models for calculating the Nusselt number. In the present example, for the same temperature, fluid, and flow conditions, we compare the results obtained using three of these models, namely Eqs (7.3.62b), (7.3.63), and (7.3.65).

Water at a bulk temperature of 32°C flows through a pipe of diameter 0.05 m at a volume flow rate of $5 \times 10^{-4}\,m^3/s$. If the pipe is 10 m long and has a constant surface temperature of 70°C, calculate the rate of heat transfer using each of the three above mentioned equations. Also, discuss the effect of varying the pipe diameter on the pressure drop within the pipe and on the rate of heat transfer.

Analysis

Each of these equations requires knowledge of fluid properties at both the bulk and surface temperatures. These may be obtained from the appropriate tables for the temperatures specified. For the conditions specified, for each of the three above equations, those parameters that define the validity of their application must be calculated and their values compared with the conditional limits specified. Eqs (7.3.63) and (7.3.65) require knowledge of the friction factor. For both these equations, for smooth pipes, f is calculated as shown

$$f = (0.79 \ln Re_D - 1.64)^{-2}$$

The pressure drop per unit length of the pipe may also be determined

$$\Delta p = \frac{f}{D} \frac{\rho v^2}{2}$$

Solution

Open the workbook Ex07-03-02.xls and select the sheet *Compute* by clicking on its tab. Insert the data given in **cells (B4:B8)**. For the flow rate specified, Excel automatically calculates the fluid velocity, **cell B8**, for the diameter specified in **cell B7**. On inserting the values for the bulk and surface temperatures, Excel returns the fluid properties for both temperatures as shown in **cells (F9:T9)** for the bulk temperature, and **cells (F14:H14)** for the surface temperature.

While the full array of properties is given for the bulk temperature, only those properties actually required for the calculations involved are given for the surface temperature. The user should familiarise themselves with the actions and commands required in Excel to obtain the properties required.

The Reynolds number, **cell B15**, is calculated as 6.6×10^4, indicating that the flow is turbulent. The hydrodynamic entry length, calculated via Eq. (7.3.59), having a value of 1.09 m, **cell B16**, also indicates that fully developed turbulent flow exists for the majority of the pipe length. The friction factor, being a common parameter for Eqs (7.3.63) and (7.3.65) is given, highlighted in green in **cell D19**. It can be seen from comparison of the values given in **cells (B20:D22)**, that the values obtained for the Nu_D, h, and q, are very similar for the three models applied.

Discussion

Considering the limits of application specified for Eq. (7.3.65), as the Prandtl number for the present case lies within the limits $0.5 < Pr < 200$, Petukhov (1970) claims that the error in calculating the Nusselt number is of the order of 6%. This being the case, if the Nusselt number thus calculated is under estimated, the values obtained through the other two models, Eqs (7.3.62b) and (7.3.63), lie within this limit. Comparing the values obtained for the rate of heat transfer per unit length of the pipe, there is little difference between the models applied. Thus, for the present situation, there is little to be gained through application of one model in preference to the others. However, as Eqs (7.3.61b) and (7.3.63) both contain the term $(D/L)^{2/3}$, for shorter pipe lengths, the difference between the Nusselt numbers calculated through these models and that of Eq. (7.3.65) increases. The reader should confirm this for themselves by selecting various lengths of pipe and comparing the results obtained.

In respect of the pressure drop, by halving the pipe diameter, for the given volume flow, fluid velocity necessarily increases by a factor of 4. The pressure loss, being a function of the square of fluid velocity, is observed to increase from 203 to 5607 Pa/m. To maintain the required flow rate, the associated pump hydraulic output would have to be increased. In addition to an increase in the pressure loss, the rate of heat transfer is also seen to increase.

Example 7.3.3 Maximum design length of exhaust gas tubes to avoid condensation.

The exhaust gases from a large diesel engine are used to generate steam in an exhaust gas boiler. The gases, at atmospheric pressure on exit from the engine, enter the 0.025 m diameter boiler tubes with a temperature and velocity of 280°C, and 9 m/s, respectively. If the temperature of the gas drops below 210°C, condensation occurs, and a corrosive acidic mixture is formed. To avoid condensation, the temperature of the gas must therefore be maintained above this minimum. If the tube surface temperature is 150°C, calculate the maximum allowable length of the boiler tubes. Assume that the exhaust gases have the same thermophysical properties as air.

Analysis

On obtaining the relevant properties, the Reynolds number should be calculated to confirm the flow condition. Considering the Prandtl number for air at the gas bulk temperature of $(T_i + T_{o,min})/2$, either Eqs (7.3.62a) or (7.3.62b) may be selected as appropriate. Both these equations require prior knowledge of the tube length. If we consider it likely that the tube length calculated will be much greater than the diameter, implying a negligible influence of the term $(D/L)^{2/3}$ in these equations, we may omit this term and perform a 'one-step' solution. However, if we wish to maintain a higher degree of accuracy, an initial estimate for the tube length should be made, and the Nusselt number calculated through either Eqs (7.3.62a) or (7.3.62b). Thereafter, the surface heat transfer coefficient may be defined.

To define the maximum allowable tube length, first, the gas mass flow rate may be determined via

$$\dot{m} = \dot{V} \rho \pi r^2$$

thus allowing the maximum area for heat transfer to be defined

$$A_S = \frac{\dot{m} c_p}{h} \ln \frac{T_o - T_S}{T_i - T_S}$$

The tube length may thereafter be calculated using simple geometric relations.

As stated previously, an initial estimate of the tube length is required to afford a solution to the Nusselt number. If the difference between the estimated and calculated values is unacceptable, an iterative procedure must be employed until the values converge.

Solution

Open the workbook Ex07-03-03.xls, and enable the 'macro' when prompted. Activate the *Compute* sheet by clicking on its tab. Input the data given above in the **cells** indicated, and put an estimate of the tube length in **cell B21**. Excel performs all necessary calculations. It is unlikely that the estimated and calculated values for the tube length will be the same. The user is therefore required to run the 'macro' *Iterate*, accessed through the 'Tools' menu, to obtain an accurate measure for the tube length. Through running this 'macro' the estimated and calculated tube lengths will be seen to converge to a common value of 0.82 m.

Discussion

As stated in the previous example, all the models presented for the Nusselt number have been obtained from the regression of experimental data. They are therefore subject to error. This error may have a value of up to 6% for Eq. (7.2.65) for the previous example. For the present case, in designing such a system, it would therefore be better to err on the side of caution and use a shorter tube length in the steam boiler.

Considering the above caveat, it may also be noted that, while the *Iterate* 'macro' may produce a more accurate mathematical result, for any reasonable initial estimate for the tube length, the calculated value will not differ greatly from that achieved by iteration. For example, an initial estimate of 2.5 m for the tube length will produce a first calculated value of 0.88 m.

7.3.12 The concentric tube annulus

A common situation encountered in engineering heat transfer is that of the concentric tube annulus, as shown in Fig. 7.3.6. Fluid passes through the annulus, with convection heat transfer occurring either to or from the fluid in the inner tube. As with all previous examples, the heat flux from each surface may be computed individually by Eqs (7.3.66a) and (7.3.66b). Thus,

$$q_i = h_i(T_{s,i} - T_m) \quad \text{(Inner surface)} \tag{7.3.66a}$$

$$q_o = h_o(T_{s,o} - T_m) \quad \text{(Outer surface)} \tag{7.3.66b}$$

with the hydraulic diameter, required for the respective Nusselt numbers, based on Eq. (7.3.58), equal to $D_o - D_i$. For fully developed laminar flow, with one surface insulated and the other at constant temperature, the respective Nusselt numbers may be obtained from Table 7.3.2.

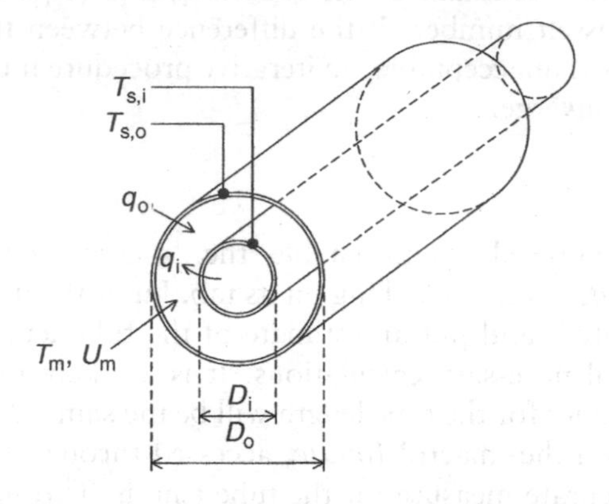

Figure 7.3.6 Concentric tube annulus.

Table 7.3.2 Nusselt numbers for fully developed flow for one surface insulated and the other at constant temperature

D_i/D_o	Nu_i	Nu_o
0	N/A	3.66
0.05	17.46	4.06
0.10	11.56	4.11
0.25	7.37	4.23
0.50	5.74	4.43
1.00	4.86	4.86

If uniform heat flux exists on both surfaces, the respective Nusselt numbers may be found from,

$$Nu_i = \frac{Nu_{ii}}{1 - (q_o/q_i)\theta_i^*} \tag{7.3.67a}$$

$$Nu_o = \frac{Nu_{oo}}{1 - (q_i/q_o'')\theta_o^*} \tag{7.3.67b}$$

where the values of the variables are found from Table 7.3.3.

The parameters (Nu_{ii}, $Nu_{oo}, \theta_i^*, \theta_o^*$) in the above mentioned table are termed the influence coefficients. Care must be taken to ensure that the correct sign, either positive or negative, is assigned to the values of q_i and q_o, depending on whether heat is being transferred to the fluid, or from it.

Table 7.3.3 Influence coefficients for fully developed laminar flow in a circular tube annulus with uniform heat flux maintained on both surfaces

D_i/D_o	Nu_{ii}	Nu_{oo}	θ_i^*	θ_o^*
0	N/A	4.364	∞	0
0.05	17.81	4.792	2.18	0.0294
0.10	11.91	4.834	1.383	0.0562
0.20	8.499	4.833	0.905	0.1041
0.40	6.583	4.979	0.603	0.1823
0.60	5.912	5.099	0.473	0.2455
0.80	5.580	5.240	0.401	0.299
1.00	5.385	5.385	0.346	0.346

7.4 Friction factor regressions

In many engineering applications it is necessary to define the pressure drop associated with the movement of a fluid through a channel or pipe. The pressure drop, as presented in Section 7.3.4 through Eqs (7.3.23) and (7.3.24), is proportional to the friction factor which is dependent on, among other factors, the fluid velocity. Thus far, a number of numerical solutions and a graphical method have been presented that allow an estimation of the friction factor. In Chapter 2 Colebrook's equation (Eq. (2.9.3)) and the Swamee and Jain equation were detailed. In addition, alternative regressions were given through Eqs (7.3.22a), (7.3.22b), and (7.3.63). The Excel workbook Ex07-04-01.xls provides a comparison of the values of the friction factor obtained by the above mentioned equations, and the graphical method.

It is immediately apparent that there is good correlation between both of the more complex Colebrook and Swamee and Jain equations, and the values obtained from the Moody chart. However, as seen in cell **E28**, reading from the Moody chart can produce an error in the predicted value. The limitations of Eqs (7.3.22a), (7.3.26), and (7.3.63), through their narrower range of applicability, are immediately apparent. The latter equations are specified for smooth pipes and do not include a measure of surface roughness. As such, as can be seen through comparison of cells **(C34 and D35)** and cells **(C43 and D43)**, for a given Reynolds number, these formulae return a constant value for f irrespective of the relative roughness. While it may appear that there is little value in their use for general application, these formulae should still be used for calculating the friction factor for the regressed equations with which they are associated. Through the ease of application of Excel's **Goal Seek** function, for more general pressure drop calculations, the Colebrook equation should be used in preference to the other methods presented.

Problems

Problem 7.1

A 15 mm square, surface mounted silicon chip is cooled on its exposed side by air at temperature of 19°C flowing in parallel with a free stream velocity of 8 m/s. The underside of the chip may be assumed to be perfectly insulated. If the chip surface temperature may not exceed 75°C, what is the maximum allowable power that it may dissipate? If the chip is flush mounted in a substrate such that it may be considered as having an unheated starting length of 25 mm, determine the maximum power that may be dissipated.

Problem 7.2

A 0.5 m wide electric air heater comprises five 0.1 m wide, flat, electrical heating elements that are arranged side-by-side presenting a smooth surface perpendicular to an air stream flowing parallel to the surface at 2 m/s. If the surface temperature of each strip is maintained at 150°C, determine the rate of convection heat transfer from each of the individual strips if the initial air temperature is 15°C.

Problem 7.3

An air preheater comprises a square, aligned, array of 100 tubes, each 1.2 m long. The tube outside diameter is 0.012 m with $S_T = S_L = 0.018$ m. If the outer surface of the tubes is maintained at a constant temperature of 98°C calculate the rate of heat transfer to the air if the initial temperature and velocity are 12°C and 0.5 m/s.

Problem 7.4

Exhaust gases in a coal fired power station are used to preheat boiler feed water in an economiser. Each 100 m long water tube in the economiser, while serpentine in nature, may be considered as a straight length of pipe set perpendicular to the exhaust gas flow. Water enters each of the 0.1 m diameter tubes at a flow rate of 10 kg/s with pressure and temperature of 40 bar and 25°C respectively. The exhaust gases, which may be assumed to have the thermophysical properties of air, enter the economiser at a temperature of 180°C and a velocity of 12 m/s. Neglecting the thermal resistance of the tube walls, calculate the rate of heat transfer to the boiler feed water and plot the fluid bulk temperature as a function of tube length.

Problem 7.5

After a period of operation, the inner surface of the tubes in the economiser of Problem 7.4 are found to have a 4 mm thick coating of limescale (k = 0.052 W/mK). Due to the extra resistance to flow caused by these deposits, for the given pressure and temperature, the feed water flow rate is reduced to 9.2 kg/s. Considering the effect of this additional thermal resistance recalculate the rate of heat transfer to the feed water and again plot the fluid bulk temperature as a function of tube length. Comment on your findings.

References

Churchill, S. W. (1976) *AIChEJ*, 22, 264.

Churchill, S. W. and Bernstein, M. (1977) *J. Heat Transfer*, 99, 300.

Churchill, S. W. and Ozoe, H. (1973) *J. Heat Transfer*, 95, 78.

Gnielinski, V. (1976) New equations for heat and mass transfer in turbulent pipe channel flow, *Int. Chem. Eng.* 16, 359.

Gnielinski, V. (1990) Forced convection in ducts, in: Hewitt, G. F. (ed.), *Handbook of Heat Exchanger Design*, Hemisphere Publishing, New York.

Grimison, E. D. (1937) *Trans. ASME*, 59, 583.

Hilpert, R. (1933) *Forsch. Geb. Ingenieurwes*, 4, 215.

Holman, J. P. (1990) *Heat Transfer*, 7th edn, McGraw-Hill, New York.

Incropera, F. P. and DeWitt, P. (2002) *Fundamentals of Heat and Mass Transfer*, 5th edn, John Wiley and Sons, New York.

Kays, W. M. and Crawford, M. E. (1980) *Convective Heat and Mass Transfer*, McGraw-Hill, New York.

Martin, H. (1977) Heat and mass transfer between impinging gas jets and solid surfaces, in: Hartnett, J. P., and Irvine, T. F. (eds), *Advances in Heat Transfer*, vol. 13, Academic press, New York.

Moody, L. F. (1944) *Trans ASME*, 66, 671.

Petukhov, B. S. (1970) in: Hartnett, J. P. and Irvine, T. F. (eds), *Advances in Heat Transfer*, vol. 6, Academic press, New York.

Schlichting, H. (1960) *Boundary Layer Theory*, 4th edn, McGraw-Hill, New York.

Shah, R. K. and Bhatti, M. S. (1987) in: Kakac, S., Shah, R. K., and Aung, W. (eds) *Handbook of Single-Phase Convective Heat Transfer*, Wiley-Interscience, New York, chapter 3.

Sieder, E. N. and Tate, G. E. (1936) *Ind. Eng. Chem.*, 28, 1429.

Suryanarayana, N. V. (1995) *Engineering Heat Transfer*, West Publishing Company, New York.

Whitaker, S. (1972) *AIChEJ*, 18, 361.

Whitaker, S. (1985) in: Malabar, F. E., and Krieger, R. E. (eds), *Fundamental Principles Of Heat Transfer*, R.E. Krieger Publishing Co., US.

Zhi-qing, W. (1982) Study on correction coefficients of laminar and turbulent entrance region effects in round pipes, *Appl. Math. Mech.* 3(3), 433.

Zhukauskas, A. (1972) Heat transfer from tubes in cross flow, in: Hartnett, J. P., and Irvine, T. F. (eds), *Advances in Heat Transfer*, vol. 8, Academic press, New York.

8 Natural convection

8.1 Introduction

Natural convection fluid motion results from buoyancy forces within the fluid. This buoyancy is due to the combined presence of a fluid density gradient and a body force that is proportional to density. The body force may be gravitational, centrifugal, electrical, or magnetic. Density gradients within a fluid may be promoted via a variety of mechanisms. Our discussion will, however, be restricted to the most common form of free convection encountered in engineering practice, that is, convection in which the above mentioned density gradient results from a temperature gradient, and the body force is gravitational.

The term free convection is also used, but its use is restricted to *natural convection* in an unbounded fluid. As for forced convection, natural convection flow can also be categorised by geometry (external or internal), by type (laminar or turbulent), and by the number of phases present in the fluid (single or two phases).

The convective heat transfer rate from or to a solid surface can be determined by Newton's Law of Cooling

$$q = hA(T_S - T_\infty) \tag{8.1.1}$$

where q is the average heat rate from the solid surface to the adjacent fluid (W), h is the average convective heat transfer coefficient (W/m^2 K), A is the surface area (m^2), T_S is the surface temperature (K), and T_∞ is the fluid temperature (K).

As discussed in Section 6.1, velocity is one of the parameters that determines the value of the *convective heat transfer coefficient*. As shown in Table 6.1.1, for the same fluid, *forced convection* heat transfer rates are usually much greater than for natural convection. However, in some systems involving multi-mode heat transfer effects, natural convection is the dominant mode and hence plays an important role in the design or performance of the system.

The Reynolds number, being a measure of the ratio of the fluid inertial to viscous forces within the boundary layer, is used to define the state of fluid motion in forced convection, and is pivotal in determining the convective heat transfer coefficient. However, in the case of *free convection*, as fluid motion results from the interaction of gravity and density gradients, the inertial forces are greatly diminished and an alternative dimensionless grouping, termed the Grashof number, is applied in heat transfer analysis. The Grashof number, as given in Eq. (8.1.2), is a measure of the ratio of the buoyancy to viscous forces within the fluid.

$$\mathrm{Gr_L} = \frac{g\beta(T_S - T_\infty)L^3}{\nu^2} \tag{8.1.2}$$

where g is the gravitational constant (m/s^2), β the thermal expansion coefficient (K^{-1}), T_S and T_∞ the surface and ambient fluid temperatures respectively (K), L the characteristic length (m), and ν is the kinematic viscosity (m^2/s). As discussed in Chapter 6, for ideal gases β is taken as the reciprocal of the absolute temperature. For non ideal gases and liquids appropriate values are obtained from the electronic data files given on the companion CD.

As for forced convection, the equations describing momentum and energy transfer in free convection are also derived via the related conservation principles. Again, should a full description of their derivation be required, the reader is directed to the previously mentioned texts. They are now presented below in a simplified form. It can be seen that, with the exception of Eq. (8.1.4), the momentum equation, they are the same as those for forced convection (Eqs (7.2.1)–(7.2.3)). The additional term $g\beta(T - T_\infty)$ in Eq. (8.1.4) takes account of the effects of gravity and the buoyancy forces that result from temperature differences. Hence, again for a two-dimensional case, for free convection, the governing equations are,

$$\text{Continuity:} \quad \frac{\partial u}{\partial x} + \frac{\partial v}{\partial y} = 0 \tag{8.1.3}$$

$$\text{Momentum:} \quad u\frac{\partial u}{\partial x} + v\frac{\partial u}{\partial y} = g\beta\,(T - T_\infty) + \nu\frac{\partial^2 u}{\partial y^2} \tag{8.1.4}$$

$$\text{Energy:} \quad u\frac{\partial T}{\partial x} + v\frac{\partial T}{\partial y} = \alpha\frac{\partial^2 T}{\partial y^2} \tag{8.1.5}$$

These differential equations must, as for the forced convection scenario, be solved simultaneously to obtain a solution (analytical or numerical) for a given natural convection problem. As previously, they may again be solved through similarity considerations and a *stream function*. A detailed account of this method is presented by Incropera and DeWitt (2002), Burmeister (1983), and by Kays and Crawford (1980). Alternatively, an integral method of analysis, as used by Holman (1990), may be applied.

The regressions used in natural convection calculations are expressed in dimensionless form. They are generally of the form

$$\mathrm{Nu_L} = \mathrm{CRa_L^m} \tag{8.1.6}$$

and

$$\mathrm{Nu_L} = \frac{hL}{k} \tag{8.1.7}$$

where $\mathrm{Nu_L}$ and $\mathrm{Ra_L}$ are, respectively, the Nusselt and Rayleigh numbers based on the characteristic length 'L' of the geometry, k is the thermal conductivity of the fluid, and 'C' and 'm' are numerical constants. The Rayleigh number is the product of the Grashof and Prandtl numbers, as given by

$$\mathrm{Ra_L} = \mathrm{Gr_L Pr} = \frac{g\beta(T_\mathrm{S} - T_\infty)L^3}{\nu\alpha} \tag{8.1.8}$$

where α is the *thermal diffusivity*.

Again, all properties, unless stated otherwise, are evaluated at the film temperature

$$T_\mathrm{f} = \frac{T_\mathrm{S} + T_\infty}{2} \tag{8.1.9}$$

Most natural convection regressions are based on laboratory experiments, where great care is taken to suppress forced motion of the surrounding fluid. In many engineering applications, some forced motion is inevitable. Such forced motion may lead to higher or lower convective heat transfer rates than predicted by the regressions for natural convection. The uncertainties in the value of h predicted by these regressions, as will be seen through comparison of values obtained in some of the forthcoming examples, are generally within ±10%, though in some cases they may be as high as ±20% (Suryanarayana, 1995). Empirical regressions are now presented, first, for the case of *external flow*, then for *internal/enclosed flow*.

8.2 External flow

While, for application in Eq. (8.1.6), Table 8.2.1 presents a summary of the constants 'C' and 'm' that apply to different surface geometries for various Raleigh numbers, many investigators have presented more detailed and accurate (claimed!) regressions for specific scenarios. While reasonable accuracy may be obtained through using the aforementioned constants presented in Table 8.2.1, these alternative regressions are now presented should the reader wish to consider their application.

8.2.1 *Vertical plates*

For the heated vertical plate, we first consider the case of an isothermal, or uniform, surface temperature. Figure 8.2.1 shows the free convection

Table 8.2.1 Constants for use with Eq. (8.2.1)

Geometry	$\mathrm{Ra_L}$	C	*m*
Vertical planes and cylinders (where $D/L \geq 35/\mathrm{Gr}_L^{1/4}$ for the cylinder)	10^{-1} to 10^4	See note I	
	10^4 to 10^9	0.59	
	10^9 to 10^{13}	0.021	0.25
			0.40
Horizontal cylinders	0 to 10^{-5}	0.4	0
	10^{-5} to 10^4	See note II	
	10^4 to 10^9	0.53	0.250
	10^9 to 10^{12}	0.13	0.333
	10^{-10} to 10^2	0.675	0.058
	10^{-2} to 10^2	1.02	0.148
	10^2 to 10^4	0.850	0.188
	10^4 to 10^7	0.480	0.250
	10^7 to 10^{12}	0.125	0.333
Upper surface of heated plates, or lower surface of cooled plates	2×10^4 to 8×10^6	0.54	0.25
	8×10^6 to 10^{11}	0.15	0.333
Lower surface of heated plates, or upper surface of cooled plates	10^5 to 10^{11}	0.27	0.25
Vertical cylinder, height = diameter, or characteristic length = diameter	10^4 to 10^6	0.775	0.21
Irregular solids, characteristic length = distance fluid particle travels in boundary layer	10^4 to 10^9	0.52	0.25

Notes

I From a regression of experimental results presented by McAdams (1954), for the specified range of $\mathrm{Ra_L}$, the Nusselt number may be obtained from the following relationship, $\log \mathrm{Nu_L} = 0.013 \log \mathrm{Ra}_L^2 + 0.091 \log \mathrm{Ra_L} + 0.226$.

II From a regression of experimental results presented by McAdams (1954), for the specified range of $\mathrm{Ra_L}$, the Nusselt number may be obtained from the following relationship, $\log \mathrm{Nu_L} = 0.009 \log \mathrm{Ra}_L^2 + 0.128 \log \mathrm{Ra_L} + 0.060$.

boundary layer on an isothermal vertical plate. Transition from laminar to turbulent flow depends on the relative magnitude of the buoyancy and viscous forces in the fluid. For vertical plates the critical Rayleigh number at which transition occurs is $\mathrm{Ra_{CR}} \approx 10^9$. Examples of natural convection from vertical plates can be seen on the internal surface of walls and windows of buildings, duct surfaces, and domestic space heating systems.

Churchill and Chu (1975a) recommended the following regression for an isothermal vertical plate,

$$\mathrm{Nu_L} = \left\{ 0.825 + \frac{0.387\mathrm{Ra}_L^{1/6}}{\left[1 + (0.492/\mathrm{Pr})^{9/16}\right]^{8/27}} \right\}^2, \quad 10^{-1} < \mathrm{Ra_L} < 10^{12} \tag{8.2.1}$$

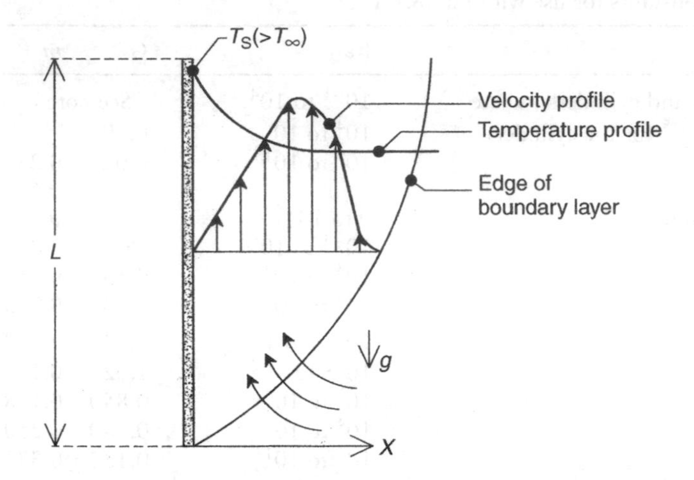

Figure 8.2.1 Free convection boundary layer on an isothermal vertical plate.

They also recommended the following regression, to obtain better accuracy, for laminar flows ($Ra_L \leq 10^9$).

$$Nu_L = 0.68 + \frac{0.670 Ra_L^{1/4}}{\left[1 + (0.492/Pr)^{9/16}\right]^{4/9}}, \quad 10^{-1} < Ra_L < 10^9 \quad (8.2.2)$$

For a uniform heat flux Churchill and Chu recommend that the following regressions be applied. (Note that the constant 0.492 in the denominator of Eqs (8.2.1) and (8.2.2) is replaced by 0.437.)

$$Nu_L = \left\{0.825 + \frac{0.387 Ra_L^{1/6}}{\left[1 + (0.437/Pr)^{9/16}\right]^{8/27}}\right\}^2, \quad 10^{-1} < Ra_L < 10^{12} \quad (8.2.3)$$

$$Nu_L = 0.68 + \frac{0.670 Ra_L^{1/4}}{\left[1 + (0.437/Pr)^{9/16}\right]^{4/9}}, \quad 10^{-1} < Ra_L < 10^9 \quad (8.2.4)$$

For the present case of a uniform heat flux, the plate temperature, and hence the film temperature, will not be uniform. The temperature at the mid-height of the plate $T_{L/2}$ should be used in determining both fluid properties and the Grashof number. However, as $T_{L/2}$ is unknown, its value must be determined by iteration as follows:

- Assume a value for $T_{L/2}$.
- Calculate T_f from Eq. (8.1.9) by replacing T_S with the assumed value of $T_{L/2}$ and determine fluid properties at T_f.

- Calculate $Ra_{L/2}$ from Eq. (8.1.8) by again replacing T_S with $T_{L/2}$.
- Calculate $Nu_{L/2}$ from Eq. (8.2.3) or (8.2.4), and calculate $h_{L/2}$ from Eq. (8.1.6).
- Finally, find the new value of $T_{L/2}$ from the relation, $q_S = h_{L/2}(T_{L/2} - T_\infty)$.
- If this value is outwith a specified tolerance (e.g. 5% of the difference in ambient and plate temperatures based on the assumed $T_{L/2}$), from the originally assumed value of $T_{L/2}$, the latest available value of $T_{L/2}$ is used for the next iteration. The process is repeated until two consecutive values of $T_{L/2}$ are sufficiently close to each other.

This method is now demonstrated through Example 8.2.1, using the software presented in Chapter 2.

Example 8.2.1 Natural convection heat loss from a heated vertical plate.

A vertical plate 0.4 m high by 0.6 m wide is maintained at a constant surface temperature of 60°C. In the first instance, if the fluid surrounding the plate is air at a temperature of 10°C, calculate the rate of heat loss from the plate. For a plate of similar dimensions with a uniform heat flux 200 W/m^2, calculate the plate surface temperature at the vertical midpoint. Again, the surrounding quiescent fluid is air at a temperature of 10°C.

Analysis

In the first instance we simply require to define the fluid properties for the film temperature, calculate the Rayleigh number, and select either Eq. (8.2.1) or (8.2.2) as appropriate. From the Nusselt number calculated, the *surface heat transfer coefficient* may be defined, and thence q. For the second case, as outlined above, an iterative method of solution is required. For the present example, this is detailed in the solution now presented.

Solution

Open the workbook Ex08-02-01.xls. For the initial case of uniform surface temperature, select the worksheet *Uniform surface temp*. Input the temperature and dimension data given in cells **(B4:B7)**.The workbook performs all the necessary calculations, namely:

Calculation of the mean fluid temperature, T_f, and β cells **(B13 and B14)** with the relevant property data imported from the *Properties* worksheet to cells **(G8:N8)**.

The Rayleigh number is calculated, cell **B15**, then the Nusselt number. Two Nusselt numbers are given, namely, those obtained via Eqs (8.2.1) and (8.2.2). For the present case, as the value for the Rayleigh number is less than 10^9, the value given in cell **B18** for the Nusselt number should be used for all subsequent calculations. For other cases, where the Rayleigh number exceeds this limit, the alternative regression, providing the value in cell **E18** should be used.

A value for the surface heat transfer coefficient is reported in **cell B19**, and a final value of 53.1 W returned for q in **cell B20**.

It is useful to note at this point that, for the present case of the Rayleigh number being close to the upper limit of application of Eq. (8.2.2), there is a considerable variance in the values obtained for the heat flux q, **cells (B20 and E20)** via the two respective relations. As discussed previously, all relations are 'best' approximations to trends observed in experimental data. Variance in the above mentioned values will therefore tend to be evident near limits of application.

For the second case of uniform heat flux, again, select worksheet *Uniform heat flux*. For this case, as discussed in the main body of the text, an iterative procedure is required to determine the plate midpoint temperature. This is an extremely laborious manual procedure as, for reasonable accuracy, all fluid property values need to be recalculated at each iterative step. The **Solver** function, as presented in Chapter 2, is an ideal tool for this application.

Having selected the *Uniform heat flux* worksheet, input the given data in **cells (B4:B7)**. The user must then insert an estimated value for the mid point temperature in **cell B22**. As before, the workbook performs the necessary calculations and provides a result for the calculated heat flux per unit area in **cell B20**. If this value is sufficiently close to that in **cell B4** (200 W), then the initial estimate for the mid point temperature may be considered acceptable. However, if the difference between the above mentioned values, **cells (B4** and **B20)** is unacceptable, then the user should employ the **Solver** function.

Pressing the '**Alt**' and '**t**' keys simultaneously, followed by the '**v**' key accesses the **Solver** function input form. Select the 'Set target cell' box by initially placing and then clicking the cursor over the box, then clicking on **cell B20**. The 'Set target cell' box should then read 'B20'. On the next 'Equal to' line, for the present case, the user manually inserts the numerical value displayed in **cell B4**. Insert the cell number B22 in the 'By changing cells' box in the same manner as described for the 'Set target cell' box.

The degree of tolerance, convergence, and iterations required can be selected via the 'Options' button. By selecting the 'Solve' button on the top right hand corner, the software performs the necessary number of iterations to obtain the defined accuracy. For the present case, a value of 55.6°C is returned in **cell B22** for the midpoint temperature, and the values in **cells (B4:B20)** are seen to have converged.

Discussion

For the first case of a uniform surface temperature of 60°C the value of 53.1 W obtained for q equates to a heat flux of 221 W/m^2 for the plate in question. For the second case, the heat flux is defined as 200 W/m^2, equating to a midpoint temperature of 55.6°C. (If the problem is redefined such that the plate has a uniform heat flux of 221 W/m^2, the midpoint temperature is evaluated as 59.4°C.) For the plate characteristic length and air temperature, the heat flux specified infers a Rayleigh number close to the upper limit of application of the regression for laminar flow (Eq. (8.2.4)). For increased heat flux the user will therefore necessarily include the turbulent flow regression in the *Uniform heat flux* worksheet in a similar manner to that for the *Uniform surface temp* worksheet.

8.2.2 *Horizontal plates*

Figure 8.2.2a–d show the four possible arrangements for natural convection heat transfer from heated and cooled horizontal plates. Based on the experimental results of McAdams (1954), Goldstein *et al.* (1973), and Lloyd and Moran (1974), the following regressions for horizontal plates with uniform surface temperature are recommended.

For upward-facing heated plates and downward-facing cooled plates:

$$\mathrm{Nu_L} = 0.96\mathrm{Ra_L^{1/6}}, \quad 1 < \mathrm{Ra_L} < 200 \tag{8.2.5}$$

$$\mathrm{Nu_L} = 0.59\mathrm{Ra_L^{1/4}}, \quad 200 < \mathrm{Ra_L} < 10^4 \tag{8.2.6}$$

$$\mathrm{Nu_L} = 0.54\mathrm{Ra_L^{1/4}}, \quad 2.2 \times 10^4 < \mathrm{Ra_L} < 8 \times 10^6 \tag{8.2.7}$$

$$\mathrm{Nu_L} = 0.15\mathrm{Ra_L^{1/3}}, \quad 8 \times 10^6 < \mathrm{Ra_L} < 1.5 \times 10^9 \tag{8.2.8}$$

For upward-facing cooled plates and downward-facing heated plates:

$$\mathrm{Nu_L} = 0.27\mathrm{Ra_L^{1/4}}, \quad 10^5 < \mathrm{Ra_L} < 10^{10} \tag{8.2.9}$$

For Eqs (8.2.5)–(8.2.9), the fluid properties are determined at T_f (Eq. (8.1.7)). The characteristic length $L = A/P$, where A and P are, respectively, the area and perimeter of the plate.

The following regressions developed by Fujii and Imura (1972) can also be used for both isothermal and constant heat flux horizontal plates.

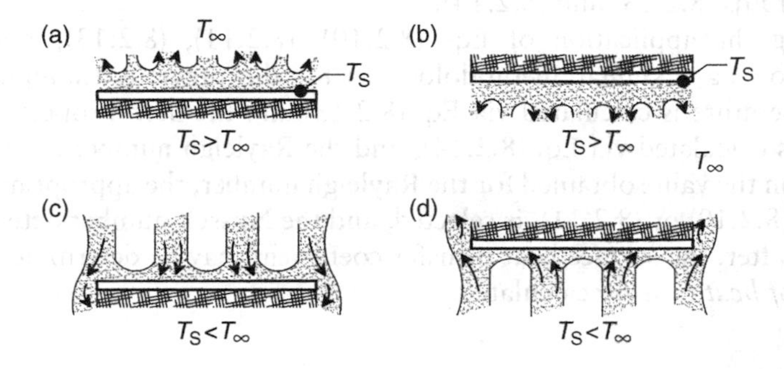

Figure 8.2.2 Form of buoyancy induced flows on horizontal plates: (a) heated top surface, (b) cooled bottom surface, (c) cooled top surface, (d) heated bottom surface.

Upward-facing heated plates and downward-facing cooled plates and

$$\mathrm{Nu_L} = 0.16\mathrm{Ra_L^{1/3}}, \quad 6\times10^6 < \mathrm{Ra_L} < 2\times10^8 \qquad (8.2.10)$$

$$\mathrm{Nu_L} = 0.13\mathrm{Ra_L^{1/3}}, \quad 5\times10^8 < \mathrm{Ra_L} < 10^{11} \qquad (8.2.11)$$

Upward-facing cooled plates and downward-facing heated plates

$$\mathrm{Nu_L} = 0.58\mathrm{Ra_L^{1/5}}, \quad 10^6 < \mathrm{Ra_L} < 10^{11} \qquad (8.2.12)$$

For Eqs (8.2.10)–(8.2.12), the fluid properties are determined at

$$T_r = T_S - 0.25\,(T_S - T_\infty) \qquad (8.2.13)$$

and

$$\beta = \frac{1}{[T_\infty + 0.25\,(T_S - T_\infty)] + 273.15} \qquad (8.2.14)$$

Example 8.2.2 Convective heat loss from a heated horizontal plate.

For a specific setting of the draft plate, the upper surface of a back-flue wood burning stove remains at an even and constant temperature of 110°C in a room at 25°C. If the surface dimensions are 0.6 m by 0.4 m, what is the rate of convective heat loss from the top surface of the stove.

Analysis

For the present case of an upward facing heated plate, as given above, there are two possible sets of relations that apply, namely Eqs (8.2.5)–(8.2.8), (8.2.10), and (8.2.11). Equations (8.2.10) and (8.2.11) require the application of Eqs (8.2.13) and (8.2.14).

Considering the application of Eqs (8.2.10), (8.2.11), (8.2.13), and (8.2.14), following a similar methodology as for the previous example, the film temperature is calculated via Eq. (8.2.13) and the fluid properties obtained. β is calculated via Eq. (8.2.14), and the Rayleigh number calculated. Based on the value obtained for the Rayleigh number, the appropriate relation, Eq. (8.2.10) or (8.2.11), is selected, and the Nusselt number determined. Thereafter, the surface heat transfer coefficient may be determined, and the *rate of heat transfer* calculated.

Solution

Open the workbook Ex08-02-02.xls and insert the given data in the appropriate cells in the worksheet *Compute*. As outlined in the analysis above, the various calculations are performed by Excel as follows:

- The fluid properties are obtained based on the film temperature **cell B13**, as calculated via Eq. (8.2.13).
- The coefficient of thermal expansion β is calculated via Eq. (8.2.14), **cell B14**, the characteristic length $L(= A/P)$ in **cell B15**, and thereafter the Rayleigh number in **cell B16**.
- Embedded in **cell B17**, as shown below, is a simple *string*

  [=IF(B16<2*10^8,0.16*B16^0.333,0.13*B16^0.333)]

  The Excel function **IF** allows the selection of the appropriate relation (Eq. (8.2.10) or (8.2.11)), based on the Rayleigh number, and the subsequent calculation of the Nusselt number.
- Thereafter, the heat transfer coefficient h is determined, **cell B18**, and the total rate of convection heat transfer from the stove top to the surrounding air is calculated as 155.0 W, **cell B19**.

Discussion

It is assumed that the flows established on the stove sides, bottom, and flue, have a negligible effect on that from the stove top.

While Eqs (8.2.10) and (8.2.11) have been used for the present example, with due consideration of their respective ranges of applicability, as stated previously, Eqs (8.2.5)–(8.2.8) could equally have been applied. For this latter set of equations, T_f, and β, are evaluated at the mean fluid temperature, and not through Eqs (8.2.13) and (8.2.14). Hence, the Rayleigh number applicable to each set of equations will necessarily be different. Using Eqs (8.2.5)–(8.2.8), we would obtain a value of 140.4 W for q. However, the Raleigh number approaches the upper limit for the equation applied (Eq. (8.2.7)). Using the alternative equation, Eq. (8.2.8), a value of 145.1 W would have been returned for q.

8.2.3 *Inclined plates*

We now consider the case of free convection heat transfer from inclined plates. Figures 8.2.3a,b, and 8.2.4a,b, show the representative forms of the boundary layers developed for the different scenarios presented. It is clear that there will necessarily be a different interaction between the buoyancy and gravitational forces than for either purely horizontal or vertical plates.

A *Uniform surface temperature*

Upward-facing cooled and downward-facing heated inclined plate. Where θ is the angle between the plate and the vertical axis, as shown in Fig. 8.2.3a and b, the form of the Rayleigh number, as given in Eq. (8.2.15), may

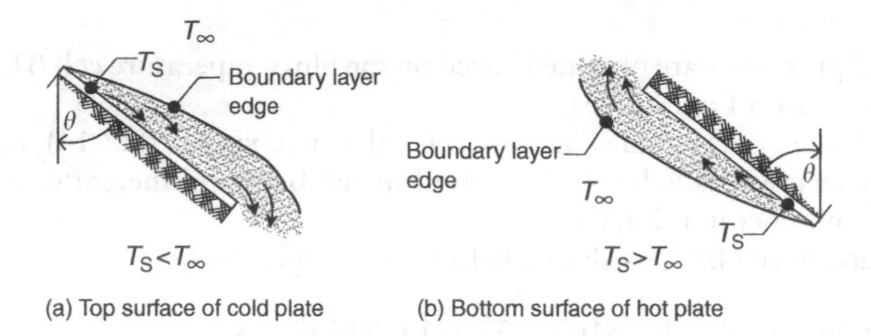

Figure 8.2.3 Buoyant flows over inclined isothermal plates: (a) cooled surface facing up, (b) heated surface facing down.

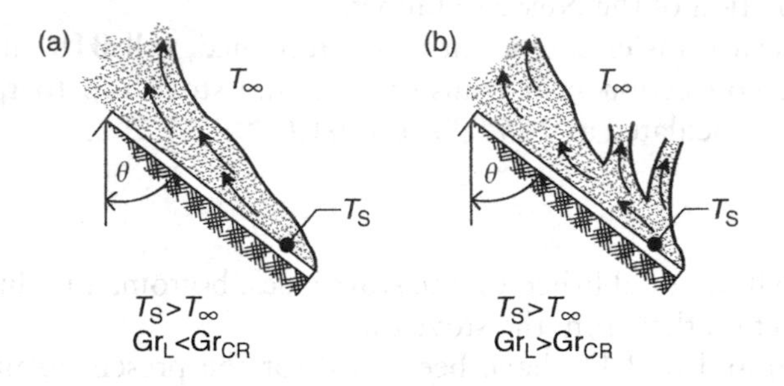

Figure 8.2.4 Form of the buoyant flows developed on an upward facing inclined plate: (a) $Gr_L < Gr_{CR}$, (b) $Gr_L > Gr_{CR}$.

be employed in Eqs (8.2.1) and (8.2.2) (the vertical plate). In this form, the Rayleigh number is evaluated using the component of the gravitational force parallel to the plate

$$Ra_L = \frac{g(\cos\theta)\beta(T_S - T_\infty)L^3}{\nu\alpha} \tag{8.2.15}$$

where L is the length of the plate in the flow direction. If $\theta > 88°$, the regressions for horizontal plates can be used.

Upward-facing heated and downward-facing cooled plates. As shown in Fig. 8.2.4a, where the characteristic Grashof number Gr_L, again calculated using $g(\cos\theta)$, is less than the critical value Gr_{CR} (Eqs (8.2.16) and (8.2.17)), fluid motion remains parallel to the plate. As the flow pattern for $Gr_L < Gr_{CR}$ is similar to that for the downward-facing heated plate (Fig. 8.2.3b), Eqs (8.2.1) and (8.2.2) may again be used. However, when Gr_L exceeds Gr_{CR} (Fig. 8.2.4b), the component of the buoyancy force perpendicular to the plate causes the boundary layer to detach from the plate surface, and

Table 8.2.2 Selected values of $Gr_{CR}Pr$ and θ

θ(°)	15	30	60	70
$Gr_{CR}Pr$	5×10^9	2×10^9	10^8	10^6

hence the above mentioned equations no longer apply. For both the above scenarios, the critical value of the Grashof number depends on the angle θ.

Table 8.2.2 provides a summary of the effect of the plate angle on the critical Grashof number. The values presented, in the form of the combined Grashof and Prandtl numbers, are derived from the following regressions that were developed by Fujii and Imura (1972)

$$\ln(Gr_{CR}Pr) = 14.223 + 8.3945\cos\theta, \quad 15° < \theta < 60° \qquad (8.2.16)$$

$$\ln(Gr_{CR}Pr) = 3.848 + 29.146\cos\theta, \quad 60° < \theta < 70° \qquad (8.2.17)$$

For $Gr_L Pr > Gr_{CR}Pr$, Fujii and Imura (1972) developed the following regression for Nu_L, for $10^6 < Gr_L Pr < 10^{11}$,

$$Nu_L = 0.14\left[(Gr_L\,Pr)^{1/3} - (Gr_{CR}\,Pr)^{1/3}\right] + 0.56(Gr_{CR}\,Pr\cos\theta)^{1/4} \qquad (8.2.18)$$

where $Gr_L Pr (= Ra_L)$ is calculated from Eq. (8.2.15) and the properties are again determined at

$$T_r = T_S - 0.25(T_S - T_\infty) \qquad (8.2.13)$$

and

$$\beta = \frac{1}{[T_\infty + 0.25(T_S - T_\infty)] + 273.15} \qquad (8.2.14)$$

Example 8.2.3 Heat loss from an isothermal inclined plate.

A metal plate of dimensions 0.2 m high by 0.6 m wide is maintained at a constant surface temperature of 40°C. If the plate is immersed, at an angle of 58° to the vertical, in water at 10°C, calculate the rate of heat loss from both the upper and lower surfaces.

Analysis

The two surfaces of the plate must be considered separately. For the upward facing surface, for the conditions specified, we require to define the critical Rayleigh number via Eqs (8.2.16) and (8.2.17). If the plate Rayleigh number, Eq. (8.2.15), is greater than Ra_{CR}, Eq. (8.2.18) should be used in conjunction with Eqs (8.2.13) and (8.2.14). For the present case of water the tabulated value for the *coefficient of thermal expansion* should be used in preference to

Eq. (8.2.14). On obtaining a value for the Nusselt number, the *surface heat transfer coefficient* and heat transfer rate are calculated in the usual manner.

For the downward facing surface, the Rayleigh number is again calculated via Eq. (8.2.15). Unlike the upward facing surface, for higher values of Ra_L, the boundary layer will not detach from the surface. Hence, as for Example 8.2.1, Eqs (8.2.1) and (8.2.2) may be used as appropriate.

Solution

Open the workbook Ex08-02-03.xls. This workbook has two worksheets, namely, *Compute* and *Properties*. The thermophysical properties of water are given in the *Properties* worksheet. Insert the given data in **cells (B4:B8)** of worksheet *Compute*. Equation (8.2.13) is used to define the film temperature **cell B14**. Excel then returns the fluid properties to **(cells G7:U7)**. For the present case the coefficient of thermal expansion β is provided via the fluid properties table, where it is not provided, it should be calculated via Eq. (8.2.14).

The critical Rayleigh number is calculated via Eqs (8.2.16) and (8.2.17) **cells (D15 and E15)** for the respective limits defined **cells (D14 and E14)**. As the thermal diffusivity α is not provided directly from the properties it is calculated ($\alpha = k/\rho c_p$) in **cell B16**. Similarly, the kinematic viscosity $\nu(= \mathrm{Pr}\alpha)$ is calculated in **cell B17**.

Using this data, the Rayleigh number is then calculated via Eq. (8.2.15), **cell B18**. As the value obtained is greater than that defined as critical **cells (D15:E15)** Eq. (8.2.18) is used to determine the Nusselt number, **cells (D19 and E19)**. Thereafter, Excel computes the heat transfer coefficient, h, **cells (D20 and E20)**, and, finally, the heat loss q. For the present case, as the plate angle lies within the limits of $15° < \theta < 60°$, the value for q given in **cell D21** should be chosen. Hence, heat is lost from the upper surface at a rate of 2155 W.

For the lower surface, following the steps outlined in the above *Analysis*, values for q are returned in **cells (H21 and J21)**. As the plate Rayleigh number is greater than 10^9, the value for q obtained using Eq. (8.2.1) should be used. Hence, for the lower surface, the rate of heat transfer is 2488 W.

Discussion

For the present case, as the plate angle is close to the limit of applicability for the angle range specified in **cell D14**, the respective values calculated for q via Eq. (8.2.18) for each range are relatively close. The user may wish to explore the effect of altering the plate angle on the relative values obtained.

Also, in considering the lower surface, while the plate Rayleigh number is only slightly greater than the threshold value of 10^9, the values obtained for the Nusselt number from Eqs (8.2.1) and (8.2.2), and subsequently q, are markedly different. While this in itself is significant, highlighting the inaccuracies that may be incurred through using the present relations, the higher of these values implies a greater heat loss from the lower surface than from the upper, while the lower value implies the converse. In such situations where the Rayleigh number is close to a conditional limit, one may be well advised to take the average of the two values obtained.

Through varying the plate angle, the reader may wish to investigate the relative rates of heat transfer from the plate upper and lower surfaces, and also consider the critical Rayleigh numbers obtained for the different angles.

B Uniform heat flux

Equation (8.2.18) was derived by the above mentioned authors from data obtained from experimental measurements with uniform heat flux. Suryanarayana (1995) states that, while Eq. (8.2.18) may be used for the above case of uniform surface temperature, there will be a greater uncertainty in the predicted value of h than for the case of uniform heat flux.

For an inclined plate with the heated surface facing down, for uniform heat flux, Fussey and Warneford (1978) recommend the use of Eqs (8.2.19) and (8.2.20).

$$\mathrm{Nu}_x = 0.592(\mathrm{Ra}_x^* \cos\theta)^{0.2}, \quad 0 < \theta < 86.5, \quad \mathrm{Ra}_x^* < \mathrm{Ra}_{x,\mathrm{CR}} \qquad (8.2.19)$$

$$\mathrm{Nu}_x = 0.889(\mathrm{Ra}_x^* \cos\theta)^{0.205}, \quad \theta < 31, \quad \mathrm{Ra}_x^* > \mathrm{Ra}_{x,\mathrm{CR}} \qquad (8.2.20)$$

where $\mathrm{Ra}_x^* = g\beta q x^4/k\nu\alpha$.

Example 8.2.4 Heat loss from an inclined plate with uniform heat flux. A plate of similar dimensions to that of the previous example (0.2 m high by 0.6 m wide) has an evenly distributed embedded electric heater such that, when held at an angle of 35° in air at 18°C, the heater draws 200 W. The plate may therefore be considered as having a uniform heat flux. If the heated surface is upward-facing, calculate the mid point temperature of the plate.

Analysis

As demonstrated for the case of uniform heat flux in Example 8.2.1, an iterative method of solution is again required. An initial estimate of the mid point temperature is required to obtain the fluid properties, and initiate the solution. The present example demonstrates the use of the **Goal Seek** function. As the plate is inclined, Eq. (8.2.15) is again used to calculate the Rayleigh number, which must be compared with the critical value (Eqs (8.2.16) and (8.2.17)) to define the relevant Nusselt number relation. The iterative method is detailed in the *Solution* now presented.

Solution

Open the workbook Ex08-02-04.xls. This workbook has two worksheets, namely *Compute* and *Properties*. The thermophysical properties of air are given in the *Properties* worksheet. Insert the given data in cells **(B4:B8)**, and an estimate for the midpoint temperature in **cell D23**. Equation (8.2.13) is used to define the film temperature, **cell B14**. Excel then returns the fluid properties to **cells (G8:U8)**. The coefficient of thermal expansion β calculated via Eq. (8.2.14) **cell B15**, with the critical Rayleigh numbers, Eqs (8.2.16) and (8.2.17), given in **cells (D16 and E16)**, respectively.

The Rayleigh number is then calculated via Eq. (8.2.15), **cell B17**. As the calculated Rayleigh number is less than that defined as critical, Eqs (8.2.3) and (8.2.4) may be used to determine the Nusselt number, **cell D20**. Thereafter, Excel computes the heat transfer coefficient, h, **cell D21**, and then the heat loss q. On running **Goal Seek**, by setting the value of q, **cell D22**, to that given in **cell B8** (200), a value of 79.0°C is obtained for the plate midpoint temperature.

Discussion

For a heat flux of 200 W, we find the midpoint temperature to be 79.0°C. For the previous example, where the fluid considered was water, for a uniform surface temperature of 40°C, and a plate angle of 50°, the rate of heat flux from the upper surface was found to be 2056 W. This represents approximately a ten-fold increase in heat flux for similar conditions.

8.2.4 *Vertical and inclined cylinders*

The regressions used for isothermal vertical plates (Eqs (8.2.1) and (8.2.2)) and uniform heat flux vertical plates (Eqs (8.2.3) and (8.2.4)) can be employed for vertical cylinders if the following criterion is satisfied

$$\frac{D}{L} > \frac{35}{\mathrm{Gr}_{\mathrm{L}}^{1/4}} \tag{8.2.21}$$

where D and L are, respectively, the diameter and the axial length of the cylinder.

Churchill (1990) developed the following regression for vertical and inclined cylinders with insulated end surfaces

$$\mathrm{Nu}_{\mathrm{L}} = \frac{0.518(\mathrm{Ra}_{\mathrm{L}}\cos\theta)^{1/4}(1+[2.8(D/L)\tan\theta]^{3/2})^{1/6}}{[1+(0.559/\mathrm{Pr})^{9/16}]^{4/9}} \tag{8.2.22}$$

where $10^5 < \mathrm{Ra}_{\mathrm{L}} < 10^9$ (Ra_{L} is calculated from Eq. (8.1.6)), θ is the angle of inclination of the cylinder axis from the vertical, and the fluid properties are determined at T_{f} (Eq. (8.1.7)).

For narrow vertical and inclined cylinders, the transverse curvature having a greater influence on boundary layer development enhances the rate of heat transfer. Minkowycz and Sparrow (1974) and Cebeci (1974) present results for slender cylinders that do not meet the criterion specified in Eq. (8.2.20).

8.2.5 Horizontal cylinders

Figure 8.2.5 shows the boundary layer development on a heated horizontal cylinder. For an isothermal cylinder, the average Nusselt number over the entire circumference can be calculated from the following relations (Churchill and Chu, 1975b)

$$\mathrm{Nu_D} = \left\{ 0.60 + \frac{0.387\mathrm{Ra_D^{1/6}}}{[1+(0.559/\mathrm{Pr})^{9/16}]^{8/27}} \right\}^2, \quad 10^3 < \mathrm{Ra_D} < 10^{13} \tag{8.2.23}$$

$$\mathrm{Nu_D} = 0.36 + \frac{0.518\mathrm{Ra_D^{1/4}}}{[1+(0.559/\mathrm{Pr})^{9/16}]^{4/9}}, \quad 10^6 < \mathrm{Ra_D} < 10^9 \tag{8.2.24}$$

Equation (8.2.24) is recommended for better accuracy in the lower ranges of Ra. The diameter of the cylinder is the characteristic length for both the Nusselt and Rayleigh numbers. The fluid properties are determined at T_f (see Eq. (8.1.7)), and the convective heat transfer coefficient, h, is calculated by replacing L with D, as given in Eq. (8.2.25).

$$\mathrm{Nu_D} = \frac{hD}{k} \tag{8.2.25}$$

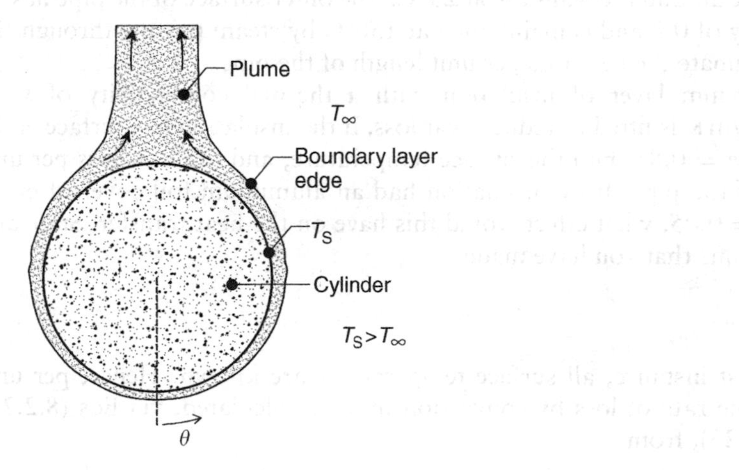

Figure 8.2.5 Boundary layer development on a heated horizontal cylinder.

Table 8.2.3 Constants for use with Eq. (8.2.26)

Ra_D	C	m
10^{-10} to 10^{-2}	0.675	0.058
10^{-2} to 10^{2}	1.020	0.148
10^{2} to 10^{4}	0.850	0.188
10^{4} to 10^{7}	0.480	0.250
10^{7} to 10^{12}	0.125	0.333

Morgan (1975), considering the standard form for natural convection regressions, Eq. (8.2.26), developed the following regressions for horizontal cylinders

$$Nu_D = C\,Ra_D^m \tag{8.2.26}$$

As previously, the numerical constants 'C' and exponent 'm', as given in Table 8.2.3 are functions of Ra_D.

Example 8.2.5 Effect of insulation and surface emissivity on the heat loss from a long horizontal cylinder.

Example 8.2.5 comprises two parts. In the first instance we consider both the convective and radiative heat losses from an uninsulated long horizontal copper pipe. We then consider the effect of adding insulation, with a lower external surface emissivity, on the overall heat loss from the same pipe.

(a) A horizontal pipe of 70 mm outside diameter passes through a room in which the air and the walls are at 25°C. The outer surface of the pipe has an emissivity of 0.8 and is maintained at 150°C by steam passing through the pipe. Estimate the heat loss per unit length of the pipe.
(b) A 25 mm layer of insulation with a thermal conductivity of k = 0.026 W/m K is fitted to reduce heat loss. If the insulation has surface emissivity of $\varepsilon = 0.85$, find the surface temperature, and the heat loss per unit length of the pipe. If the insulation had an aluminium foil cover of emissivity $\varepsilon = 0.05$, what effect would this have on the heat transfer. State any assumptions that you have made.

Analysis

In the first instance, all surface temperatures are known. Hence, per unit length, the rate of loss by convection may be calculated, via Eqs (8.2.22) and (8.2.23), from

$$q_{conv} = h\pi D(T_S - T_\infty)$$

and the rate of loss by thermal radiation, as will be presented in Chapter 9, by

$$q_{\text{rad}} = \varepsilon \pi D \sigma (T_S^4 - T_{\text{sur}}^4)$$

The total heat loss from the pipe is the sum of the above convective and radiative components, thus,

$$q = q_{\text{conv}} + q_{\text{rad}}$$

The relations for external free convection flow from a horizontal pipe, selected on the basis of the Rayleigh number, should be used in this example (Eqs (8.2.22) and (8.2.23)).

For the second case, natural convection heat transfer to the ambient air q_{conv} and net radiation transfer to the surroundings q_{rad} depend on the surface temperature, T_{ins}, of the insulation. This temperature is unknown and may be obtained by performing an energy balance at the outer surface, such that,

$$q_{\text{cond}} - (q_{\text{conv}} + q_{\text{rad}}) = 0$$

where

$$q_{\text{conv}} = h \pi D_{\text{ins}} (T_{\text{ins}} - T_{\infty})$$

$$q_{\text{rad}} = \varepsilon_{\text{ins}} \pi D_{\text{ins}} \sigma (T_{\text{ins}}^4 - T_{\text{sur}}^4)$$

and

$$q_{\text{cond}} = \frac{2\pi k_{\text{ins}} (T_S - T_{\text{ins}})}{\ln (D_{\text{ins}}/D)} \quad \text{(see Chapter 3)}$$

(Note that T_{ins} and T_{sur} in the above equations are absolute temperatures.)

Solution

Open the workbook Ex08-02-05.xls. This workbook consists of three worksheets, namely *Compute* (a), *Compute* (b), and *Properties*.

Part A

Activate the worksheet *Compute* (a) by clicking on its tab. This worksheet includes the computations performed by Excel for the first case considered. Read the example text carefully and insert the given data in cells **(B6:B10)**.

Excel performs the necessary calculations as follows,

- Using Eq. (8.1.7), $T_f = 87.5°\text{C}$ **cell B16**, with the thermophysical properties of air again obtained by using the **VLOOKUP** function, as discussed earlier in Chapter 2. $\beta = 2.77 \times 10^{-3}\ \text{K}^{-1}$ **cell B17**.

- Using Eq. (8.1.6), $Ra_D = 1.7 \times 10^6$ (**cell B18**). Since $10^6 < Ra_D < 10^9$, as stated in the above *Analysis*, Eq. (8.2.25) is used to calculate $Nu_D(= 14.41$ **cell B19**).
- Using Eq. (8.1.5), $h = 6.3\,W/m^2K$ (**cell B20**), and the heat loss by convection per unit length, q_{conv}, as given in the above equation is found to be 174 W/m, **cell B21**.
- The heat loss by radiation per unit length, calculated via the appropriate above equation, is found to be 241 W/m **cell B22**. The total heat loss per unit length is therefore 415 W/m, **cell B23**.

Part B

Using the same workbook as above, click on the tab of the worksheet *Compute* (b). Insert the additional data in **cells (E4:E6)**. (The value for diameter (**cell E4**) is the original diameter plus two times the insulation thickness.)

The above convection, radiation, and conduction equations are inserted in **cells (B20:B22)**, respectively.

The energy balance equation, $q_{cond} - (q_{conv} + q_{rad}) = 0$, is inserted in **cell B23**. To fulfil the energy balance the value of this cell should be zero. We must therefore find the value of T_{ins}, **cell B14**, at which $q_{cond} - (q_{conv} + q_{rad}) = 0$, **cell B23**. An iterative method of solution is therefore again required. The user must first assume a value for T_{ins} and insert it in **cell B14**, (e.g. $T_{ins} = 30°C$). Use the **Goal Seek** function as explained previously in Chapter 2. **Goal Seek** yields $T_{ins} = 34.9°C$ and $q = q_{cond} = q_{conv} + q_{rad} = 34.9\,W/m$, **cell B24**.

Changing the value for the emissivity (**cell E6**) to 0.05 and re-running the **Goal Seek** function as above, yields a value for the total heat loss per unit length **cell B24** of 32.6 W/m with $T_{ins} = 42.4°C$.

Discussion

It is assumed that the pipe is sufficiently long as to disregard end effects. Also, the additional resistance to conductive heat flow attributed to the foil layer is negligible.

By adding the insulation, the heat loss from the pipe has been dramatically reduced from 415 to 34.9 W/m. By adding the low emissivity aluminium foil to the insulation, the overall heat loss is only minimally reduced. However, the relative proportions of the convective and radiative losses are dramatically altered (where $\varepsilon = 0.85, q_{conv} = 14.8$, and $q_{rad} = 20.1\,W/m$, and, for $\varepsilon = 0.05, q_{conv} = 30.5$, and $q_{rad} = 2.2\,W/m$). Lowering the surface emissivity, while reducing radiative losses, has the effect of increasing the surface temperature and, hence, the convective losses. While radiative losses are proportional to the surface temperature to the fourth power, the effect of an increase in surface temperature is more than compensated for by the reduction in emissivity.

8.2.6 Spheres

Churchill (1983) developed the following regression for spheres, in fluids of $Pr > 0.7$, and $Ra_D \leq 10^{11}$.

$$Nu_D = 2 + \frac{0.589 Ra_D^{1/4}}{[1 + (0.469/Pr)^{9/16}]^{4/9}}, \quad Ra_D < 10^{11}, \quad Pr \geq 0.7 \quad (8.2.27)$$

The diameter of the sphere is the characteristic length for both the Nusselt and Rayleigh numbers. The fluid properties are again determined at T_f Eq. (8.1.7), and h via Eqs (8.2.23) and (8.2.24) as appropriate.

Example 8.2.6 Heat loss from a hot sphere suspended in oil.

A 0.1 m diameter sphere with a uniform surface temperature of 80°C is suspended in oil at a temperature of 10°C. Determine the rate of heat loss from the sphere. Further develop this workbook to determine what the rate of heat loss would be if the sphere were immersed in water.

Analysis

For the present case, the relation as given in Eq. (8.2.26) should be used. As stated in the above text, this relation is only applicable for $Pr > 0.7$, and $Ra_D < 10^{11}$. We must therefore check that, for the conditions specified in the present problem, Eq. (8.2.27) is applicable.

Solution

Open the workbook Ex08-02-06.xls and activate the worksheet *Compute*. Insert the given data in the cells indicated. As for all previous examples in this section, Excel returns the fluid properties for the fluid film temperature. As the Prandtl and Rayleigh numbers are seen to be within the above mentioned limits, Eq. (8.2.26) may be used.

The rate of heat loss is from the sphere is calculated as 2987 W.

Discussion

It is intended that the user, through developing the workbook, will further familiarise themselves with the simple manipulation of the appropriate fluid property data and use of the **LOOKUP** function. A comparison of the relative rates of heat transfer for different sphere and fluid temperatures should be made. Appropriate plots may also be generated.

In Section 8.2 the recommended regressions for some of the more common surface geometries encountered in engineering practice have been presented. These are summarised in Table 8.2.4. A rigorous review of results and regressions for other geometries and special conditions has been presented by both Churchill (1983) and Raithby and Hollands (1975).

We now consider the case of natural convection within channels and cavities.

Table 8.2.4 Summary of free convection correlations for common immersed geometries

Geometry	*Recommended correlation*	*Restrictions*
Vertical plates	Isothermal – Eqs (8.2.1) and (8.2.2) Constant heat flux – Eqs (8.2.3) and (8.2.4)	None
Horizontal plates		
Hot surface up, cold surface down	Eq. (8.2.5)	$1 < Ra_L < 200$
	Eq. (8.2.6)	$200 < Ra_L < 10^4$
	Eq. (8.2.7)	$2.2 \times 10^4 < Ra_L < 8 \times 10^6$
	Eq. (8.2.8) or	$8 \times 10^6, Ra_L < 10^9$
	Eq. (8.2.10)	$6 \times 10^6 < Ra_L < 2 \times 10^8$
	Eq. (8.2.11)	$5 \times 10^8 < Ra_L < 10^{11}$
Cold surface up, hot down	Eq. (8.2.9) or Eq. (8.2.12)	$10^5 < Ra_L < 10^{10}$ $10^6 < Ra_L < 10^{11}$ also $T_r = T_S - 0.25(T_S - T_\infty)$, and $\beta = \dfrac{1}{[T_\infty + 0.25(T_S - T_\infty)] + 273.15}$
Inclined plates		
Cold surface facing up, hot surface down	$g \rightarrow g\cos\theta$ (Eq. (8.2.15)) use Eqs (8.2.1) and (8.2.2.)	$\theta < 88°$
Hot surface facing up, cold surface down	Eq. (8.2.18) with $\left\{\begin{array}{l}\text{Eq. (8.2.16) } 15 < \theta < 60 \\ \text{Eq. (8.2.17) } 60 < \theta < 70\end{array}\right\}$	$10^6 < Gr_L Pr < 10^{11}$
Vertical cylinder	Eqs (8.2.1) and (8.2.2)	$\dfrac{D}{L} > \dfrac{35}{Gr_L^{1/4}}$
Inclined cylinder, insulated ends	Eq. (8.2.21)	
Horizontal cylinders		
Isothermal	Eq. (8.2.22)	$10^{-3} < Ra_D < 10^{13}$
	Eq. (8.2.23)	$10^{-6} < Ra_D < 10^9$
Constant heat flux	Eq. (8.2.24)	
	Eq. (8.2.25)	
Spheres	Eq. (8.2.26)	$Ra_D < 10^{11}$, $Pr \geq 0.7$

8.3 Enclosed natural convection

8.3.1 Vertical channels

Parallel plate channels (vertical or inclined) are used in many engineering applications. For example, interacting convection currents will be established between plates in the fin array attached to a base surface of an electronic component or an internal combustion engine for cooling purposes. Surface thermal conditions, as shown in Fig. 8.3.1, may be idealised as being isothermal or having constant heat flux, and symmetrical ($T_{S,1} = T_{S,2}; q_{S,1} = q_{S,2}$) or asymmetrical ($T_{S,1} \neq T_{S,2}; q_{S,1} \neq q_{S,2}$).

For vertical channels ($\theta = 0°$) buoyancy forces act exclusively to induce motion in the vertical plane. For short channels and/or large spacings (small L/S), independent boundary layer development occurs at each surface and

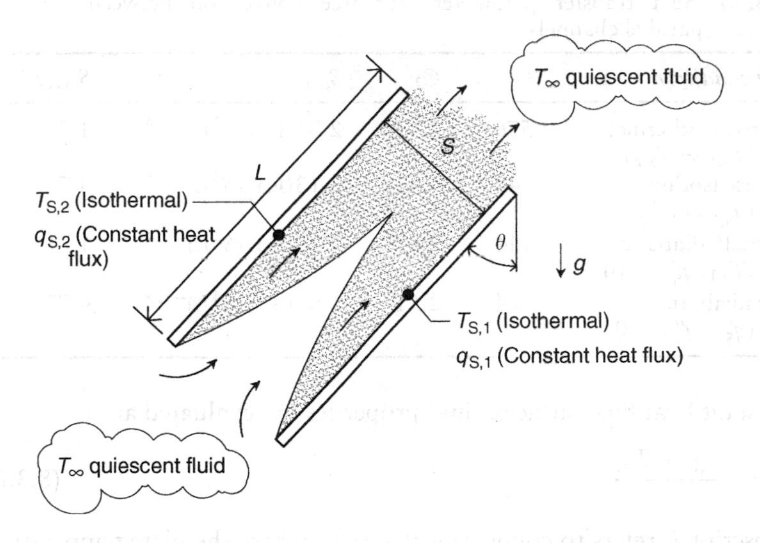

Figure 8.3.1 Free convection flow between heated parallel plates with opposite ends open to a quiescent fluid.

conditions correspond to those for an isolated plate in a finite, quiescent medium. For large (L/S), however, boundary layers developing on opposing surfaces merge to yield a fully developed condition.

Bar-Cohen and Rohsenow (1984) recommended the following relations for vertical channels with isothermal and isoflux conditions respectively. For any value of S/L

$$\mathrm{Nu_S} = \left[\frac{C_1}{(\mathrm{Ra_S} S/L)^2} + \frac{C_2}{(\mathrm{Ra_S} S/L)^{1/2}}\right]^{-1/2} \tag{8.3.1}$$

$$\mathrm{Nu_{S,L}} = \left[\frac{C_1}{\mathrm{Ra_S^*} S/L} + \frac{C_2}{(\mathrm{Ra_S^*} S/L)^{2/5}}\right]^{-1/2} \tag{8.3.2}$$

where

$$\mathrm{Ra_S^*} = \frac{g\beta q_S S^4}{k\alpha\upsilon} \tag{8.3.3}$$

and

$$\mathrm{Ra_S} = \frac{g\beta(T_S - T_\infty)S^3}{\nu\alpha} \tag{8.3.4}$$

Fluid properties for isothermal surfaces are, as before, evaluated at

$$T_f = \frac{T_S + T_\infty}{2} \tag{8.3.5}$$

Table 8.3.1 Heat transfer parameters for free convection between vertical parallel channels

Surface condition	C_1	C_2	S_{opt}	S_{max}/S_{opt}
Symmetric isothermal plates ($T_{S,1} = T_{S,2}$)	576	2.87	$2.71(\mathrm{Ra}_S/S^3L)^{-1/4}$	1.71
Symmetric isoflux plates ($q''_{S,1} = q''_{S,2}$)	48	2.51	$2.13(\mathrm{Ra}_S^*/S^4L)^{-1/5}$	4.77
Isothermal/adiabatic plates ($T_{S,1}, q''_{S,2} = 0$)	144	2.87	$2.15(\mathrm{Ra}_S/S^3L)^{-1/4}$	1.71
Isoflux/adiabatic plates ($q''_{S,1}, q''_{S,2} = 0$)	24	2.51	$1.69(\mathrm{Ra}_S^*/S^4L)^{-1/5}$	4.77

For constant heat flux surfaces, fluid properties are evaluated at

$$T_f = \frac{T_{S,L} + T_\infty}{2} \tag{8.3.6}$$

The subscript L refers to conditions at $x = L$, where the plate temperature is maximum.

The numerical constants 'C_1' and 'C_2' for application in Eqs (8.3.1) and (8.3.2) are given in Table 8.3.1.

In each case the fully developed and isolated plate limits correspond to Ra_S (or Ra_S^*)$S/L \leq 10$ and Ra_S(or Ra_S^*)$S/L \geq 100$, respectively.

The optimum plate spacing S_{opt} is the space required to maximise the heat transfer from an array of plates, and the maximum plate spacing S_{max} is the space required to maximise the heat transfer from each plate in the array. Values of S_{opt} and S_{max}/S_{opt} given in Table 8.3.1 are for plates of negligible thickness. Although heat transfer from each plate decreases with decreasing S, the number of plates that may be placed in a prescribed volume increases. Hence S_{opt} maximises heat transfer from the array by yielding a maximum for the product of h and the total plate surface area. In contrast, to maximise heat transfer from each plate, S_{max} must be large enough to preclude overlap of adjoining boundary layers (Incropera and DeWitt, 2002).

Example 8.3.1 Convective heat loss in a vertical channel.

Consider the case of two parallel vertical plates separated by a 3 cm air gap. One of the plates is maintained at a constant surface temperature of 75°C, and the other is perfectly insulated. If both plates measure 1.0 m by 0.5 m wide, calculate the rate of heat loss if the ambient air temperature is constant at 16°C.

Analysis

As the temperature of the heated plates is constant, the relationship given in Eq. (8.3.1) should be used, with the constants 'C_1' and 'C_2' obtained from

Table 8.3.1. A method of solution similar to that applied in many of the previous examples in this chapter is used. The fluid properties are obtained for the film temperature, followed by calculation of the Rayleigh number, Eq. (8.3.4). The Nusselt number, h, and then q, are calculated.

Solution

Open the workbook Ex08-03-01.xls. This workbook has two worksheets, namely, *Compute* and *Properties*. The thermophysical properties of air are given in the *Properties* worksheet. Insert the given data in cells (B4:B8). As one surface is isothermal and one insulated, the constants 'C_1' and 'C_2', from Table 8.3.1, take the respective values of 144 and 2.87. The rate of convective heat transfer from the heated plate is calculated as 121.6 W.

Example 8.3.2 Optimum spacing of the plates in an air heater.

A natural convection air heater comprises an array of 2 mm wide plates that are each 0.35 m high and 0.3 m wide. The plates surface temperature is maintained constant at a temperature of 65°C by embedded electrical trace heater wires. If the width of the array is fixed at 0.2 m, determine the optimum plate spacing, the number of plates required, and the rate of convective heat transfer if the ambient air temperature is 18°C.

Analysis

For the present case of the air heater plates being symmetrical and isothermal, from Table 8.3.1, the constants 'C_1' and 'C_2', for application in Eq. (8.3.1), take the values of 516 and 2.87, respectively. The optimum plate spacing is found from the equations detailed below.

$$S_{\text{opt}} = 2.71/(\text{Ra}_S S^3 L)^{0.25}$$

where

$$\text{Ra}_S/S^3 L = \frac{g\beta(T_S - T_a)}{L\nu\alpha}$$

From the optimum plate spacing, the optimum number of plates may be defined,

$$N_{\text{p}} = 1 + (W_{\text{max}} - \delta)/(S_{\text{opt}} + \delta)$$

where W_{max} is the maximum width of the array, and δ is the plate thickness.

It is unlikely that N_{p} will be an integer, and hence should be rounded down to the nearest whole number, $N_{\text{p,act}}$. Rounding up will reduce spacing, implying an overlapping of boundary layers, and hence a reduction in the heat transfer rate. The optimum plate spacing may then be redefined by rearranging the third of the above equations. As can be seen in the second of the above equations, while the left-hand side contains the term S^3, the right-hand side does not. The value of $\mathrm{Ra}_S S^3 L$ is therefore unaffected by the aforementioned redefinition of S_{opt}.

Having obtained a value for S_{opt}, the Nusselt number may be calculated via Eq. (8.3.1), with the appropriate values for the constants 'C_1' and 'C_2' applied. Thereafter, the surface heat transfer coefficient may be calculated, and thence, via the equation below, also the *heat transfer rate*:

$$q = N_{p,act}(2LW)h(T_S - T_a)$$

where L and W are the plate length and width respectively.

Solution

Open the workbook Ex08-03-02.xls, and activate the worksheet *Compute* by clicking on its tab. Insert the data given above in **cells (B4:B9)**. Excel performs all necessary calculations. The Excel **INT** function, applied to the third of the above equations, returns the rounded down integer value, 19, as given in **cell B19**. The actual value for N_p as calculated is 19.65. Through the method outlined in the above analysis, the rate of heat transfer from the array is found to be 820.3 W.

Discussion

As from many of the examples thus far presented in Chapter 8, the accuracy of the solution depends, among other factors, on the correct choice of equation and associated values for the constants required from the appropriate table.

In the present example, additional calculations are provided in the workbook. The user may input alternative integer values for the optimum number of plates in **cell C19**, and compare the values obtained for q with that obtained above. For example, for 20 plates, q is found to be 813.3 W, whereas, for 18 plates q is 813.9 W, verifying that the optimum number of plates is, as defined, 19.

8.3.2 *Inclined channels*

Azevedo and Sparrow (1985) recommended the following regression for inclined channels in water with symmetric isothermal plates ($T_{S,1} = T_{S,2}$) or isothermal-insulated plates ($T_{S,1}, q_{S,2} = 0$)

$$\mathrm{Nu}_S = 0.645[\mathrm{Ra}_S(S/L)]^{1/4}, \quad \left\{\begin{matrix} 0 \le \theta \le 45 \\ \mathrm{Ra}_S(S/L) > 200 \end{matrix}\right\} \tag{8.3.7}$$

Fluid properties are evaluated at

$$T_f = \frac{T_S + T_\infty}{2} \tag{8.3.8}$$

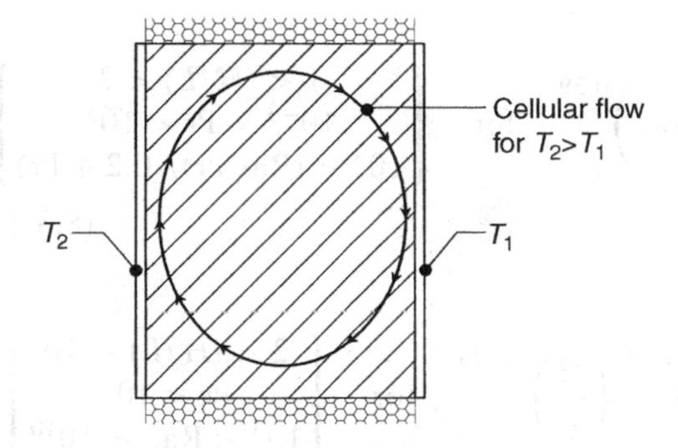

Figure 8.3.2 Cellular flow in a rectangular cavity with different wall temperatures.

8.3.3 *Rectangular cavities*

Figure 8.3.2 shows a rectangular cavity with a height H and a width L. Two of the opposing walls are maintained at different temperatures ($T_1 > T_2$), while the remaining walls are insulated from the surroundings. When a fluid is enclosed within the cavity, it starts to circulate within the cavity and the heat transfers by natural convection from the hot wall to the cold one. The average heat flux is expressed as

$$q = h(T_1 - T_2) \tag{8.3.9}$$

Regressions for determining the average heat transfer coefficient, h, for different cavity configurations are given below. In all the regressions, L is the characteristic length in $\mathrm{Ra_L}$ and $\mathrm{Gr_L}$ and the properties of the infill fluid are evaluated at the arithmetic mean temperature

$$T_\mathrm{f} = \frac{T_1 + T_2}{2} \tag{8.3.10}$$

Rectangular cavities may be horizontal ($\theta = 0°$ or $180°$), vertical ($\theta = 90°$) or inclined to the horizontal plane at an angle θ. The ratio between the height and width of the cavity is called the aspect ratio of the cavity (H/L). Engineering applications such as solar collectors and double-glazed windows are examples of rectangular cavities. The above cases are now considered in detail.

A *Vertical cavities*

For $\mathrm{Ra_L} < 1000$, the viscous forces in the fluid are greater than the buoyancy forces and the heat transfer is primarily by conduction across the fluid ($\mathrm{Nu_L} = 1$). With increasing $\mathrm{Ra_L}$, the fluid starts to circulate within the cavity inducing heat transfer by natural convection.

Catton (1978) recommended the following regressions

$$\mathrm{Nu_L} = 0.18\left(\frac{\mathrm{Pr}}{0.2+\mathrm{Pr}}\mathrm{Ra_L}\right)^{0.29} \quad \text{for} \quad \left\{\begin{array}{c} 1 < (H/L) < 2 \\ 10^{-3} < \mathrm{Pr} < 10^{5} \\ 10^{3} < (\mathrm{Ra_L\,Pr})/(0.2+\mathrm{Pr}) \end{array}\right\} \tag{8.3.11}$$

$$\mathrm{Nu_L} = 0.22\left(\frac{\mathrm{Pr}}{0.2+\mathrm{Pr}}\mathrm{Ra_L}\right)^{0.28}\left(\frac{H}{L}\right)^{-0.25} \quad \text{for} \quad \left\{\begin{array}{c} 2 < (H/L) < 10 \\ \mathrm{Pr} < 10^{5} \\ 10^{3} < \mathrm{Ra_L} < 10^{10} \end{array}\right\} \tag{8.3.12}$$

For larger aspect ratios, MacGregor and Emery (1969) recommended the following regressions,

$$\mathrm{Nu_L} = 0.42\mathrm{Ra_L^{0.25}}\mathrm{Pr}^{0.012}\left(\frac{H}{L}\right)^{-0.3} \quad \text{for} \quad \left\{\begin{array}{c} 10 < (H/L) < 40 \\ 1 < \mathrm{Pr} < 2\times 10^{4} \\ 10^{4} < \mathrm{Ra_L} < 10^{7} \end{array}\right\} \tag{8.3.13}$$

$$\mathrm{Nu_L} = 0.046\mathrm{Ra_L^{1/3}} \quad \text{for} \quad \left\{\begin{array}{c} 1 < (H/L) < 40 \\ 1 < \mathrm{Pr} < 20 \\ 10^{6} < \mathrm{Ra_L} < 10^{9} \end{array}\right\} \tag{8.3.14}$$

In keeping apace with technological developments, much work has been undertaken recently in relation to heat transfer within, and heat loss from, double glazed windows. Convection in a sealed glazing unit is complicated as it is affected by many parameters, such as the width of the cavity, the height of the cavity, the temperatures of the hot and cold panes, the temperature differential and, most importantly, the properties of the infill gases. Traditionally, heat transfer characteristics of multiple glazed windows were researched using experimental methods. Owing to an economic and efficiency advantage, computer modelling has however now gained favour.

From experimental measurements and computational fluid dynamics (CFD) software modelling of double glazed windows, Muneer and Han (1999) provided the following regression for the aspect ratio constraints of $12 < H/L < 150$. The window aspect ratio, A_r, is the ratio of window height (H) to cavity width (L)

$$\mathrm{Nu_L} = \mathrm{Max}\left[1, 0.36(\mathrm{Gr_L\,Pr})^{0.245}\left(\frac{H}{L}\right)^{-0.28}\right] \tag{8.3.15}$$

Table 8.3.2 Critical angles for inclined rectangular cavities

H/L	1	3	6	12	>12
θ_{CR}	25°	53°	60°	67°	70°

B Horizontal cavities

For $Ra_L < 1080$, the heat transfer in a horizontal cavity heated from below ($\theta = 0°$) is essentially by conduction from the bottom to the top surface. Conditions correspond to one-dimensional conduction through a plane fluid layer with $h = k/L$ and $Nu_L = 1$. However, for $Ra_L > 1080$, convection currents are established and heat is transferred by natural convection.

For a horizontal cavity heated from below and filled with air, Jacob (1949) recommended,

$$Nu_L = 0.195Gr_L^{1/4}, \quad 10^4 < Gr_L < 4 \times 10^5 \tag{8.3.16}$$

$$Nu_L = 0.068Gr_L^{1/3}, \quad 4 \times 10^5 < Gr_L < 10^7 \tag{8.3.17}$$

For a horizontal cavity heated from below and filled with a liquid, Globe and Dropkin (1959) recommended,

$$Nu_L = 0.069Ra_L^{1/3}Pr^{0.074}, \quad 3 \times 10^5 < Ra_L < 7 \times 10^9 \tag{8.3.18}$$

C Inclined cavities

For $H/L \geq 12$ and $0° < \theta \leq \theta_{CR}$, where θ_{CR} is the critical tilt angle (see Table 8.3.2), Hollands *et al.* (1976) recommend the following relation

$$Nu_L = 1 + 1.44\left[1 - \frac{1708}{Ra_L \cos\theta}\right]^*\left[1 - \frac{1708(\sin 1.8\theta)^{1.6}}{Ra_L \cos\theta}\right] + \left[\left(\frac{Ra_L \cos\theta}{5830}\right)^{1/3} - 1\right]^* \tag{8.3.19}$$

The notation $*$ implies that, if the quantity in brackets is negative, it must be set equal to zero.

For $H/L \leq 12$ and $0° < \theta \leq \theta_{CR}$, Catton (1978) suggested the following relation

$$Nu_L = Nu_L(\theta = 0°)\left[\frac{Nu_L(\theta = 90°)}{Nu_L(\theta = 0°)}\right]^{\theta/\theta_{CR}}(\sin\theta_{CR})^{(\theta/4\theta_{CR})} \tag{8.3.20}$$

For all values of H/L and $\theta_{CR} < \theta < 90°$, Ayyaswamy and Catton (1973) suggested the following relation

$$Nu_L = Nu_L(\theta = 90°)(\sin\theta)^{1/4} \tag{8.3.21}$$

For all values of H/L and $90° < \theta < 180°$, Arnold *et al.* (1975) suggested the following relation

$$\mathrm{Nu_L} = 1 + [\mathrm{Nu_L}(\theta = 90) - 1] \sin\theta \tag{8.3.22}$$

Example 8.3.3 Convective heat loss from the absorber plate of a solar collector.

A solar collector with a height of 1.4 m and width equal to 2.0 m is mounted at an angle of 45° on a southerly facing roof. For a specific set of operating and environmental conditions the absorber and cover plates have temperatures of 73°C and 27°C, respectively. Determine the rate of convective heat transfer between the absorber and the cover plate for a spacing of 0.02 m. The heat lost from the absorber by natural convection depends on the plate spacing. Plot the rate of heat loss as a function of plate spacing for $0.005\ \mathrm{m} < L < 0.05\ \mathrm{m}$ and comment on your findings.

Analysis

From the data given, $H/L > 12$ for all values within the specified plate spacing range. From Table 8.3.2, the critical angle θ_{CR} is 70°. Equation (8.3.19) may therefore be applied. For a range of spacing values, the rate of convective heat loss may therefore be determined.

Due to the complexity of Eq. (8.3.19), to ease calculation, terms 1 and 2, which are required to assume a value of zero if their respective values are negative, may be separated out of the main equation and their appropriate values established:

$$\left[1 - \frac{1708}{\mathrm{Ra_L}\cos\theta}\right] \qquad \text{(Term 1)}$$

$$\left[\left(\frac{\mathrm{Ra_L}\cos\theta}{5830}\right)^{1/3} - 1\right] \qquad \text{(Term 2)}$$

Solution

Open the workbook Ex08-03-03.xls and activate the worksheet *Compute*. Input the given data in the **cells** indicated. Excel performs all necessary calculations. A range of cover to absorber spacings have been included in **cells (B8:L8)**. The Rayleigh number, Terms 1 and 2, followed by the Nusselt number, and thence the surface heat transfer coefficient and rate of convective heat loss are displayed for each spacing given in the above mentioned cells. While the spacings and associated heat losses are displayed in rows in the *Compute* worksheet. They are copied into adjacent columns in worksheet *Chart* to enable a plot of heat loss versus spacing width to be generated. The plot obtained can be viewed by accessing the *Chart* worksheet in the usual manner.

Discussion

As can be seen from the plot obtained, there is a local minimum of 380 W at a spacing of 0.01 m. To further reduce the convective heat losses the absorber/cover gap must be increased to beyond 0.05 m (359 W). The use of either of these spacings is preferable to any intermediate sized gap as this will incur a greater convective loss.

8.3.4 *Concentric cylinders*

Free convection heat transfer in the annular space between long, horizontal concentric cylinders (Fig. 8.3.3) is characterised by two cells that are symmetric about the vertical mid-plane. If $T_i > T_o$, fluid ascends and descends along the inner and outer cylinders, respectively. If $T_i < T_o$, the cellular flows are reversed. The heat transfer rate per unit length of the inner cylinder may be expressed (Raithby and Hollands, 1975) as

$$q = \frac{2\pi k_{eff}(T_i - T_o)}{\ln(D_o/D_i)} \tag{8.3.23}$$

where k_{eff} is the thermal conductivity of a stationary medium with identical inner and outer surface temperatures for the same heat transfer rate. The value of k_{eff} can be determined from the relation

$$\frac{k_{eff}}{k} = 0.386\left(\frac{\Pr}{0.861 + \Pr}\right)^{1/4} \mathrm{Ra}_C^{*1/4}, \quad 10^2 < \mathrm{Ra}_C^* < 10^7 \tag{8.3.24}$$

where

$$\mathrm{Ra}_C^* = \frac{[\ln(D_o/D_i)]^4}{L^3\left(D_i^{-0.6} + D_o^{-0.6}\right)^5}\mathrm{Ra}_L \tag{8.3.25}$$

$$\mathrm{Ra}_L = \frac{g\beta(\Delta T)L^3}{\nu\alpha} \quad \text{and} \quad \Delta T = |T_i - T_o| \tag{8.3.26}$$

$$L = (D_o - D_i)/2 \tag{8.3.27}$$

Fluid properties are evaluated at

$$T_f = \frac{T_i + T_o}{2} \tag{8.3.28}$$

As stated above, Eq. (8.3.24) may be used within the limits specified ($10^2 < \mathrm{Ra}_C^* < 10^7$). For $\mathrm{Ra}_C^* < 100$, as the influence of buoyancy induced flows is minimal, k_{eff} is effectively equal to k. As such k_{eff} may be replaced by k in Eq. (8.2.21).

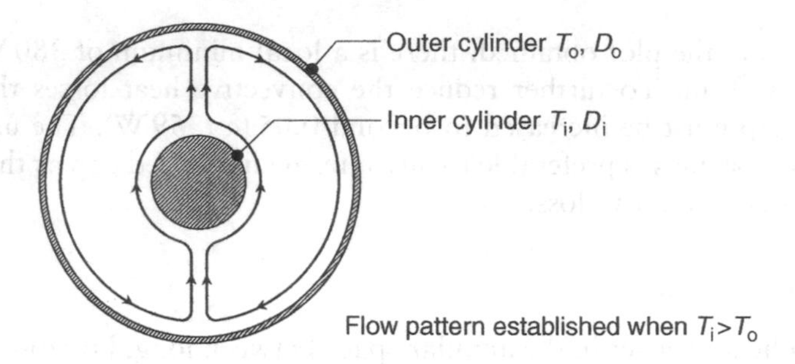

Figure 8.3.3 Pattern of flow established in the annular space between two long concentric cylinders.

Example 8.3.4 Convective heat transfer in the annular space between two concentric cylinders.

Two concentric cylinders have diameters of 0.2 and 0.14 m. If the inner surface of the outer cylinder is maintained at a constant temperature of 65°C, and the outer surface of the inner at 24°C, calculate the rate of convective heat transfer between the two cylinders per unit length if the fluid in the annulus is air.

Analysis

Thus far, all solutions presented for the examples detailed in this chapter have required calculation of the Nusselt number in order to define the heat transfer coefficient. In the present case however, as detailed in the above text, through application of a modified Rayleigh number, we consider the effective thermal conductivity of the fluid medium between the two surfaces. We therefore do not require to calculate the Nusselt number. The method of solution is as follows.

For the fluid mean temperature, the thermophysical properties are obtained. The Rayleigh number is calculated based on the characteristic length (the difference between the two radii). The value thus obtained is thereafter used to define the modified Rayleigh number, Eq. (8.3.25). The effective conductivity k_{eff} is determined via Eq. (8.3.24), with q defined via Eq. (8.3.23).

Solution

Open the workbook Ex08-03-04.xls and activate the worksheet *Compute*. Input the data given above in the cells indicated. Excel performs all necessary calculations as detailed in the *Analysis*. A value of 56.8 W/m is returned for q in cell B20. To accommodate cases where the inner cylinder wall temperature is greater than that of the outer, ΔT, cell **B13**, is calculated as the absolute difference using Excels **ABS** function. Irrespective of which surface is hotter, a positive value will always be returned for ΔT.

Discussion

Through defining an effective thermal conductivity k_{eff}, for the present case of two concentric cylinders, we redefine our convection problem in terms of those parameters and equations more commonly associated with conduction in a radial system. However, k_{eff}, having a value of 0.0786 W/m K is considerably larger than k (0.0276 W/m K) as it accounts for the transfer of heat via the convective flows established.

8.3.5 *Concentric spheres*

Raithby and Hollands (1975) suggested the following relation for the heat transfer rate between two concentric spheres

$$q = k_{eff}\pi \frac{D_i D_o}{L}(T_i - T_o) \tag{8.3.29}$$

For the present case, the effective thermal conductivity is

$$\frac{k_{eff}}{k} = 0.74\left(\frac{Pr}{0.861 + Pr}\right)^{1/4} Ra_S^{*1/4}, \quad 10^2 < Ra_S^* < 10^4 \tag{8.3.30}$$

where

$$Ra_S^* = \frac{L}{(D_o D_i)^4} \frac{Ra_L}{(D_i^{-1.4} + D_o^{-1.4})^{.5}} \tag{8.3.31}$$

$$Ra_L = \frac{g\beta(\Delta T)L^3}{\nu\alpha} \quad \text{and} \quad \Delta T = |T_i - T_o| \tag{8.3.32}$$

$$L = (D_o - D_i)/2 \tag{8.3.33}$$

As for the case of concentric cylinders, if $Ra_S^* < 100, k_{eff}$ should be set equal to k.

8.4 Combined forced and natural convection

As we have already seen in this chapter, natural convection occurs whenever density gradients are present. If an externally forced flow is imposed, as the fluid free stream velocity increases, forced convection effects play an increasing role. The range of values given for the surface convection heat transfer coefficient in Table 6.1.1 show that, while there is a degree of overlap between the values associated with forced and natural convection, forced convection effects are generally dominant. In considering the combined effects of forced and natural convection Burmeister (1983) suggests that one rationale for the design procedure is to calculate the respective forced and natural convection coefficients separately, and to then adopt the larger value.

This simple method however can obviously produce a large error in the estimated heat transfer for the given situation. An alternative method is to use the dimensionless group $\mathrm{Gr_L/Re_L^2}$, being a measure of the ratio of buoyancy to inertial forces, to provide an indication of the relative effects of the two convection heat transfer mechanisms considered. Where $\mathrm{Gr_L/Re_L^2}$ is large (i.e. $\gg 1$), the effects of natural convection will be dominant, whereas, if the above ratio is small ($\ll 1$), then forced convection will prevail. Holman (1990) states that when $\mathrm{Gr_L/Re_L^2} > 10$, natural convection will be of primary importance.

In presenting the relevant relations, it is necessary to consider the direction of the forced flow in relation to that established through buoyancy effects. We shall therefore first consider the case of aiding flow.

8.4.1 *Aiding flow*

Where the direction of forced flow aids that established through buoyancy effects, Churchill (1977) suggests that the relationship defined in Eq. (8.4.1) is applied:

$$\mathrm{Nu^3_{L,combined}} = \mathrm{Nu^3_{L,forced}} + \mathrm{Nu^3_{L,free}} \tag{8.4.1}$$

where the *forced* and *free* regressions applicable for the specific geometry are used to determine the respective Nusselt numbers.

For $\mathrm{Gr_L/Re_L^2} < 0.1$, for the present case of *aiding* flow, Lloyd and Sparrow (1970) found that, for $\mathrm{Pr} \approx 1$, the influence of natural convection was less than 5%. Conversely, Acrivos (1958) had earlier shown that, for laminar flow over a vertical plate, the effect of forced convection was less than 10% for both $\mathrm{Gr_L/Re_L^2} > 4$ and $0.73 < \mathrm{Pr} < 10$ and also for $\mathrm{Gr_L/Re_L^2} > 60$ and $\mathrm{Pr} \approx 100$.

For turbulent natural convection from an isothermal vertical plate, Hall and Price (1970) found that, for $\mathrm{Gr_L/Re_L^2} > 10$, forced flow influenced natural convection by less than 10%.

8.4.2 *Opposing and cross flow*

Where the forced flow is in opposition to that caused by natural convection Tsuruno and Iguchi (1979) found that, for laminar flow over a vertical plate, boundary layer separation occurred at $\mathrm{Gr_L/Re_L^2} = 0.2$. Moreover, the effects of the buoyancy forces were observed to decrease as the plate was inclined to the horizontal. For many cases of opposing and cross flow, an adapted form of Eq. (8.4.1) may be employed, as given in

$$\mathrm{Nu}^n_{\mathrm{L,combined}} = \mathrm{Nu}^n_{\mathrm{L,forced}} - \mathrm{Nu}^n_{\mathrm{L,free}} \tag{8.4.2}$$

where $n = 3$ for vertical plates (as for aiding flow), 3.5 for transverse flow over horizontal plates, and 4 for opposing flow over cylinders and spheres.

For airflow normal to long horizontal cylinders Oosthuizen and Madan (1970) recommend the following relation

$$\frac{\overline{\mathrm{Nu}_\mathrm{D}}}{\overline{\mathrm{Nu}_{\mathrm{D,forced}}}} = 1 + 0.18\frac{\mathrm{Gr_D}}{\mathrm{Re_D^2}} - 0.011\left(\frac{\mathrm{Gr_D}}{\mathrm{Re_D^2}}\right)^2, \quad \left\{\begin{matrix} 100 < \mathrm{Re_D} < 300 \\ 2.5 \times 10^4 < \mathrm{Gr_D} < 3 \times 10^5 \end{matrix}\right\} \tag{8.4.3}$$

where $\mathrm{Nu_{D,forced}} = 0.464\,\mathrm{Re_D^{1/2}} + 0.0004\,\mathrm{Re_D}$.

In addition, Oosthuizen and Madan (1970) found that natural convection effects were negligible if $\mathrm{Gr_D}/\mathrm{Re_D^2} < 0.28$. Holman (1990), in citing the work of Metais and Eckert (1964), presents the results of their review of heat transfer relations for flow through horizontal tubes in graphical form.

Example 8.4.1 Mixed convection heat transfer from a long horizontal pipe.

A long horizontal copper pipe has internal and external diameters of 0.195 and 0.200 m, respectively. Water at a bulk temperature of 70°C flows through the pipe at a rate of 2 kg/s. If the fluid surrounding the pipe is air with an ambient temperature of 12°C, calculate the pipe surface temperature, the rate of heat loss form the pipe, and the rate of change in the water bulk temperature with respect to the direction of flow. Discuss your findings.

Analysis

The rate of change of temperature of the water per unit length of the pipe, for a loss of heat, is given by

$$\dot{m}c_\mathrm{p}\frac{\mathrm{d}T_\mathrm{b}}{\mathrm{d}x} = -q$$

From the thermal circuit

$$q = \frac{T_\mathrm{b} - T_\mathrm{a}}{(1/h_\mathrm{i}\pi d_\mathrm{i}) + (1/2\pi k)\ln(d_\mathrm{o}/d_\mathrm{i}) + (1/h_\mathrm{o}\pi d_\mathrm{o})}$$

As the thermal conductivity of the pipe walls is very high, $k = 401\,\mathrm{W/m\,K}$, and the walls are thin, (with $d_\mathrm{o} \approx d_\mathrm{i}$, $\ln(d_\mathrm{o}/d_\mathrm{i}) \approx 0$) the conductive resistance offered by the walls will be negligible compared to that of the inner and outer surface boundary layers. The above equation may therefore be reduced to

$$q = \frac{T_b - T_a}{(1/h_i \pi d_i) + (1/h_o \pi d_o)}$$

To afford a solution to q however, we require to make an initial estimate of the wall surface temperature in order to define the respective film temperatures, fluid properties and surface heat transfer coefficients. The inner surface Nusselt number may be determined via the appropriate regressions for flow in circular pipes, (Eqs (7.3.62a) and (7.3.62b)) while the outer Nusselt number may similarly be calculated via Eqs (8.2.22) and (8.2.23).

On assuming a value for T_S the computed values of h_i and h_o may be used, via the energy balance equation given below, to compute a value for T_S

$$h_i(T_b - T_S) = h_o(T_S - T_\infty)$$

As for all other previous examples that have required an estimated input value, to obtain an accurate solution, an iterative method is necessarily employed.

Solution and discussion

Open the workbook Ex08-04-01.xls. This workbook comprises three worksheets, namely *Compute, PropWater*, and *PropAir*. Select the worksheet *Compute* and input the given data in the appropriate cells. Input an estimate for the surface temperature in **cell B15**. (Obviously, your estimate would lie between the air and water bulk temperatures.) Excel performs all necessary calculations. Fluid properties are returned for the estimated film temperatures for both the air and water flows, and also for water at the estimated surface temperature. Equations (7.3.62a) and (7.3.62b) require knowledge of the Prandtl number at this temperature.

For the inner fluid, the Reynolds number is calculated as 2.6×10^4. As the Prandtl number is greater than 1.5, Eq. (7.3.62b) is used. For the present case of a long cylinder, this may be reduced to

$$\mathrm{Nu_D} = 0.012(\mathrm{Re_D^{0.87}} - 280)\,\mathrm{Pr}^{2/5}\left(\frac{\mathrm{Pr}}{\mathrm{Pr_S}}\right)^{0.11}$$

Thereafter, based on an initial estimate of 40°C for the surface temperature, the inner surface heat transfer coefficient is calculated as 411.2 W/m^{2}K.

For the outer surface, the Rayleigh number is found to be 2.1×10^7, and, via Eq. (8.2.24), the outer surface heat transfer coefficient is determined as 3.5 W/m^{2}K. For an initial estimate of 40°C for the surface temperature, a value of 69.5°C is returned for that calculated. As stated in the *Analysis* above, under such circumstances one would normally adopt an iterative method to afford an accurate solution for the surface temperature. However, for the present case, due to the difference in the respective *heat transfer coefficients*, the calculated surface temperature is highly insensitive to the initial estimate. For the above mentioned estimated and calculated values, if the user inserts a second try value of 69.5 the calculated value remains unchanged. However, the respective inner and outer heat transfer coefficients change significantly, with $h_i = 482.9\,\text{W/m}^2\text{K}$, and $h_o = 4.1\,\text{W/m}^2\,\text{K}$.

Thereafter q is calculated as 149.2 W/m, and the rate of change of the bulk fluid temperature as $-0.02°\text{C/m}$.

Due to the high degree of insensitivity of the calculated surface temperature on the initial estimate, the present workbook contains neither an *Iterate* 'macro' nor instruction on employing Excels **Goal Seek** or **Solver** functions. Should the user wish to further develop this workbook, the *Iterate* 'macro', often referred to in Chapter 7, may easily be incorporated.

Problems

Problem 8.1

A 0.4 m long electric immersion heater with diameter 0.015 m has a rating of 1 kW. If the heater is placed horizontally in a large tank containing water at 20°C, what is the heater surface temperature. How would the rate of heat transfer compare if the heater were vertical. What would be the surface temperature if the heater was held horizontally in air.

Problem 8.2

A thin walled cylindrical vessel containing oil at a uniform temperature of 50°C is placed in a water bath at a temperature of 15°C. Neglecting the thermal resistance of the vessel wall, calculate the overall rate heat transfer from the hot to the cool fluid if the vessel is 0.1 m in diameter, and walls are vertical with a height of 0.2 m.

Problem 8.3

The external surface of a square section duct of side 0.2 m is maintained at a constant temperature of 64°C. For a mean ambient air temperature of 15°C, calculate the rate of heat loss per unit length of the duct. If the duct were rotated by 45° about its horizontal axis such that all sides were at 45°, would this increase or decrease the rate of hat transfer. Discuss your findings.

Problem 8.4

A double glazed window has the following dimensions: height 1.5 m, width 1.8 m, cavity width 0.012 m. If the inner and outer panes are at temperatures of 24°C and 15°C, respectively, calculate the rate of convective heat transfer between the panes. Also, generate a plot of heat transfer rate versus cavity width. Comment on your findings.

References

Acrivos, A. (1958) Combined laminar free- and forced-convection heat transfer in external flows, *JAIChE* 4, 285–289.

Arnold, J. N., Catton, I., and Edwards, D. K. (1975) Experimental investigation of natural convection in inclined rectangular regions of differing aspect ratios. *ASME* Paper 75-HT-62.

Ayyaswamy, P. S. and Catton, I. (1973) The boundary layer regime for natural convection in a differentially heated, tilted rectangular cavity, *J. Heat Transfer* 95, 543.

Azevedo, L. F. A. and Sparrow, E. M. (1985) Natural convection in open-ended inclined channel, *J. Heat Transfer* 107, 893.

Bar-Cohen, A. and Rohsenow, W. M. (1984) Thermally optimum spacing of vertical natural convection cooled, parallel plates, *J. Heat Transfer* 106, 116.

Burmeister, L. C. (1983) *Convective Heat Transfer*, John Wiley and Sons, New York.

Catton, I. (1978) Natural convection in enclosures, in: *Proceedings of the Sixth International Heat Transfer Conference*, Vol. 6, Toronto.

Cebeci, T. (1974) Laminar free convective heat transfer from the outer surface of a vertical slender circular cylinder, in: *Proceedings of the Fifth International Heat Transfer Conference*, Paper number NC1.4, 15–19.

Churchill, S. W. (1983) Free convection around immersed bodies, in: Schlunder, E. U. (ed.), *Heat Exchange Design Handbook*, Hemisphere Publishing, New York.

Churchill, S. W. (1990) Free convection around immersed bodies, in: Hewitt, G. F. (ed.), *Handbook of Heat Exchanger Design*, Hemisphere Publishing, New York.

Churchill, S. W. and Chu, H. H. S. (1975a) Correlating equations for laminar and turbulent free convection from a vertical plate *Int. J. Heat Mass Transfer* 18, 1323.

Churchill, S. W. and Chu, H. H. S. (1975b) Correlating equations for laminar and turbulent free convection from a horizontal cylinder, *Int. J. Heat Mass Transfer* 18, 1049.

Churchill, S. W. (1977) A comprehensive correlating equation for laminar, assisting, forced and free convection, *JAIChE* 16, 10–16.

Fussey, D. E. and Warneford, I. P. (1978) Free convection from a downward facing inclined flat plate, *Int. J. Heat Mass Transfer* 21, 119.

Fujii, T. and Imura, H. (1972) Natural convection heat transfer from a plate with arbitrary inclination, *Int. J. Heat Mass Transfer* 15, 755.

Globe, S. and Dropkin, D. (1959) Natural convection heat transfer in liquids confined between two horizontal plates, *Trans. ASME* 81C, 24–28.

Goldstein , R. J., Sparrow, E. M., and Jones, D. C. (1973) Natural convection mass transfer adjacent to horizontal plates, *Int. J. Heat Mass Transfer* 16, 1025.

Hollands, K. G. T., Unny, S. E., Raithby, G. D., and Konicek, L. (1976) Free convective heat transfer across inclined air layers, *Trans. ASME* 98C, 189.

Holman, J. P. (1990) *Heat Transfer*, 7th edn, McGraw-Hill.

Incropera, F. P. and DeWitt, D. P. (2002) *Fundamentals of Heat and Mass Transfer*, 5th edn, John Wiley and Sons, New York.

Jacob, M. (1949) *Heat Transfer*, Wiley, New York.

Kays, W. M. and Crawford, M. E. (1980) *Convective Heat and Mass Transfer*, McGraw-Hill, New York.

Lloyd, J. R. and Moran, W. R. (1974) Natural convection adjacent to horizontal surfaces of various plan forms, *J. Heat Transfer* 96, 443.

Lloyd, J. R. and Sparrow, E. M. (1970) Combined forced and free convection flow on vertical surfaces. *Int. J. Heat Mass Transfer* 13, 434–438.

McAdams, W. H. (1954) *Heat Transmission*, 3rd edn, McGraw-Hill, New York.

MacGregor, R. K. and Emery, A. P. (1969) Free convection through vertical plane layers: moderate and high Prandtl number fluids, *J. Heat Transfer* 91.

Metais, B. and Eckert, E. R. G. (1964) Forced, mixed and free convection regimes, *Trans. ASME J Heat Transfer* 86, 295–296.

Minkowycz, W. J. and Sparrow, E. M. (1974) Local non similar solutions for natural convection on a vertical cylinder, *J. Heat Transfer* 96, 178.

Morgan, V. T. (1975) The overall convective heat transfer from smooth circular cylinders, in: Irvine, T. F. and Hartnett, J. P. (eds), *Advances in Heat Transfer*, Vol. 11, Academic Press, New York.

Muneer, T. and Han, B. (1999) Simplified analysis for free convection in enclosures-application to an industrial problem, *Energy Conversion & Management* 37, 1463.

Oosthuizen, P. H. and Madan, S. (1970) Combined convective heat transfer from horizontal cylinders in air, *J. Heat Transfer* 92, 194–196.

Raithby, G. D. and Hollands, K. G. T. (1975) A general method of obtaining approximate solutions to laminar and turbulent free convection problems, in: Irvine, T. F. and Hartnett, J. P. (eds), *Advances in Heat Transfer*, Vol. 11, Academic Press, New York.

Suryanarayana, N. V. (1995) *Engineering Heat Transfer*, West Publishing Company, New York.

Tsuruno, S. and Iguchi, I. (1979) Mechanism of heat and momentum transfer of combined free and forced convection with opposing flow, *J. Heat Transfer* 101, 422–426.

9 Thermal radiation

9.1 Introduction

In contrast to convection and conduction, heat transfer by *thermal radiation* does not require the presence of matter. However, as we shall see in the following sections, its interaction with matter is of great importance in many natural and industrial processes. In introducing *radiation*, we will first present the fundamental concepts necessary for its understanding.

9.1.1 Fundamental concepts

Through the vibration and transition of electrons within atoms, all matter emits radiation. Radiation may be considered as the propagation of either *electromagnetic waves*, or as *photons* or *quanta*. The internal energy, or temperature, of a medium is a measure of the degree of vibration and excitation of the electrons within that medium. For the present case, as $T_S > T_{sur}$, while both the surroundings and the solid emit radiation, the net radiant energy exchange from the solid to the surroundings, $q_{rad,\,net}$, will be positive. The solid will therefore cool as it's internal energy decreases until *thermal equilibrium* is reached.

For most solids and liquids, radiation emitted by interior molecules is absorbed by those adjacent, and hence does not reach the medium's surface. Consequently, radiation may be considered to be a surface phenomenon, originating from molecules that are within a distance of approximately 1 μm of the surface. As radiation may be considered to have the form of electromagnetic waves, it will therefore have standard wave properties of frequency, f, and wavelength, λ. The relationship between these two properties is as given in

$$\lambda = \frac{c}{f} \tag{9.1.1}$$

where c is the speed of light in the medium (in a vacuum $c_o = 2.988 \times 10^8$ m/s).

Within the spectrum of electromagnetic radiation, thermal radiation is emitted between approximately 0.1 and 1000 μm. This includes a portion of the UV range (0.01–0.4 μm), and all of the visible and infrared. All surfaces emit radiation. Both the magnitude of the radiation at a specific wavelength, and its spectral distribution, are dependent on both the temperature, and the properties of the emitting surface. Hence, the rate of *thermal radiation heat* transfer will also be dependent on these factors. In addition, again depending on the nature of the radiating surface, the surface may emit preferentially in specific directions.

9.1.2 *Emissivity*

Applying the principles of quantum statistical thermodynamics, it is possible to derive an expression for the radiation energy density per unit volume, and per unit wavelength, as,

$$u_\lambda = \frac{8\pi hc\lambda^{-5}}{e^{hc/\lambda kT} - 1} \tag{9.1.2}$$

where h is Planck's constant (6.625×10^{-24} J s), and k is Boltzmann's constant (1.38066×10^{-23} J/molecule K).

If we integrate over all wavelengths, for all directions, that is, considering a hemisphere as viewed by the surface, the total emitted energy, E_b, is found, as given in the Stefan–Boltzmann law below, to be proportional to the absolute temperature to the fourth power

$$E_b = \sigma T^4 \tag{9.1.3}$$

where σ, the Stefan–Boltzmann constant, is equal to 5.67×10^{-8} W/m^2 K^4, E_b is in W/m^2, and T is expressed in Kelvin.

The subscript b in Eq. (9.1.3) denotes that this is the radiation from a *blackbody*. It is referred to as a blackbody as it absorbs all radiation that falls on it, and reflects none. E_b is therefore called the *emissive power* of a blackbody, and the blackbody is therefore an ideal emitter.

While the term black is used to define the emissive power of a surface, while a surface may appear black to the eye, it may not necessarily absorb all incident radiation. The converse is also true. For example, while snow and ice appear bright to the eye, they, and many white paints, are effectively black for long wavelength thermal radiation.

As the blackbody is an ideal emitter, the *radiant energy* emitted by any real surface will be less than that emitted by a blackbody at the same temperature. For a given wavelength, the ratio of the monochromatic emissive power, E_λ, of a given surface, to the monochromatic emissive power of a blackbody, $E_{b,\lambda}$, at the same temperature, is the spectral hemispherical emissivity of the surface,

$$\varepsilon_\lambda = \frac{E_\gamma}{E_{b,\lambda}} \tag{9.1.4}$$

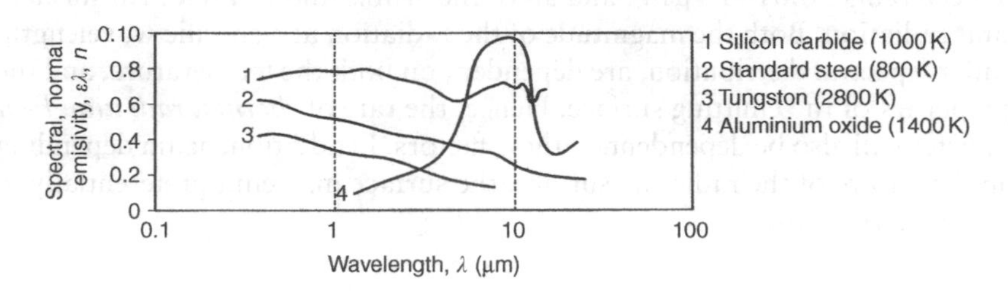

Figure 9.1.1 Spectral dependence of emissivity for selected materials.

and

$$E = \int_0^\infty E_\lambda \, d\lambda = \int_0^\infty \varepsilon_\lambda E_{b,\lambda} \, d\lambda \tag{9.1.5}$$

For many surfaces ε varies with λ, therefore, to quantify E, we need to know the functional relationship between ε and λ. Figure 9.1.1 presents plots for the spectral dependence of the spectral emissivity normal to the surface for a selection of materials at specified temperatures. It is clear that, to obtain a value for the total or average emissivity, we must integrate the above mentioned relationship over all wavelengths. The total *hemispherical emissivity*, ε, is therefore defined as

$$\varepsilon = \frac{E}{E_b} = \frac{1}{E_b} \int_0^\infty \varepsilon_\lambda E_{b,\lambda} \, d\lambda \tag{9.1.6}$$

The above relationship implies that ε may be considered to be either the *emissivity* of a surface which is wavelength independent, or, alternatively, the average emissivity of the surface at the given temperature.

A surface whose radiation properties are independent of wavelength is termed a *gray surface*. The emissive power of a gray surface is given by,

$$E = \varepsilon E_b \tag{9.1.7}$$

A *black* surface is therefore a *gray* surface with $\varepsilon = 1$ for all wavelengths. While, as stated earlier, ε_λ and α_λ may be dependent on both temperature and wavelength, for many cases there is only a weak relation. However, as will be seen in Section 9.7, temperature can have a marked effect on the emissivity of some gases.

9.2 Radiation properties

9.2.1 *Reflectivity, absorptivity, and transmissivity*

As shown in Fig. 9.2.1, considering a semi-transparent surface, when radiation strikes the surface (known as *incident radiant flux* or *irradiation*, G)

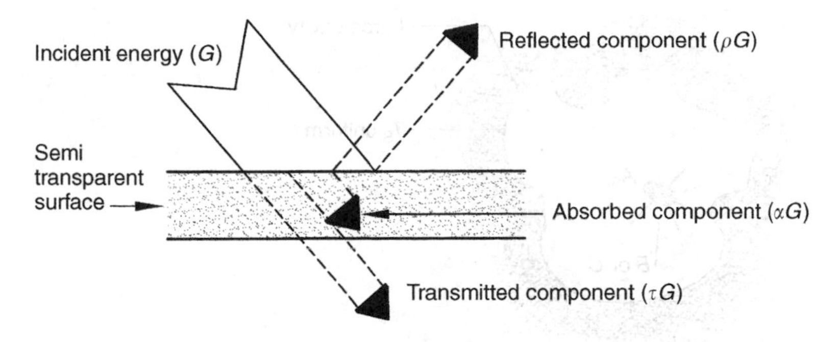

Figure 9.2.1 A radiant beam incident on a semi-transparent surface.

part is reflected, ρ (reflectivity), part is absorbed, α (absorptivity), and the remainder is transmitted, τ (transmissivity)

$$\rho + \alpha + \tau = 1 \tag{9.2.1}$$

A surface with zero transmissivity ($\tau = 0$), is termed *opaque*, and $\alpha + \rho = 1$. For a given surface, while each of the above perameters may vary with wavelength, their summation will always be equal to 1.

It is more difficult to specify reflectivity than the other radiative properties such as absorptivity. As indicated by Siegel and Howell (1992), no less than eight types of reflectivity are in current use: bi-directional spectral; directional-hemispherical spectral; hemispherical-directional spectral; hemispherical spectral; bi-directional total; directional-hemispherical total; hemispherical-directional total and hemispherical total. Of these, hemispherical total reflectivities are sufficient for applications such as meteorology and solar heating design, while spectral reflectivities are required for other applications such as electricity generation by photovoltaic cells, photobiology, and solar controlled glazing.

9.2.2 *Radiosity*

Radiosity accounts for all the radiant energy leaving a surface. It therefore includes the reflected portion of the incident radiant heat flux, and the direct emission. The spectral radiosity represents the rate at which radiation of a specific wavelength leaves a surface. This is considered in more detail in Section 9.3.

9.2.3 *Blackbody radiation*

In considering the above definitions of α, ρ, and τ, and that of the blackbody as an ideal or perfect absorber, for all wavelengths it is therefore clear that $\alpha_\lambda = 1$. For a black opaque surface, both ρ and τ are equal to zero.

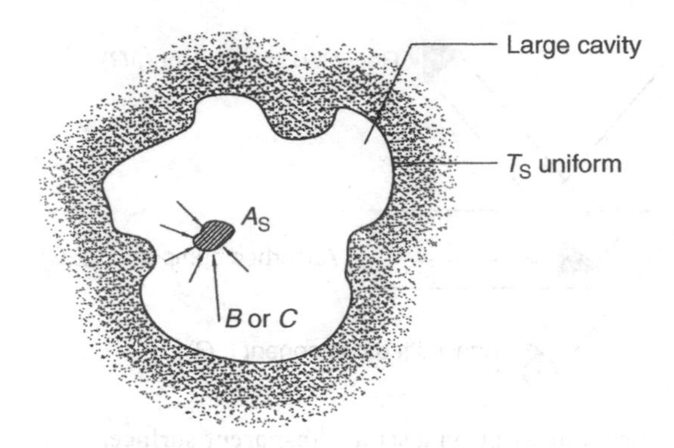

Figure 9.2.2 A large cavity with uniform surface temperature.

In addition, the blackbody is a perfect and diffuse emitter, with $\varepsilon = 1$, and its radiant emissions independent of direction.

In considering the case of a number of small bodies enclosed within, and in thermal equilibrium with, a large cavity at uniform surface temperature, (Fig. 9.2.2), if α_λ and ε_λ are wavelength independent for each of the surfaces, as for *gray surfaces*, Kirchoff's law, as presented by Suryanarayana (1995), shows that $\alpha_\lambda = \varepsilon_\lambda$, and that $\alpha_n = \varepsilon_n$. However, if, independently, they are functions of wavelength, then the above equality does not hold. Only for the specific case of the cavity and body surfaces being identical does the equality hold.

9.2.4 *Radiation intensity*

Radiation from a surface propagates in all directions. Considering Fig. 9.2.3, the area of an element of the hemisphere as viewed from the surface is given by

$$\mathrm{d}A_n = r^2 \sin\theta \, \mathrm{d}\theta \, \mathrm{d}\phi \tag{9.2.2}$$

where $\mathrm{d}A_n$ is normal to the (θ, ϕ) direction.

We define the spectral intensity, $I_{\lambda,\mathrm{e}}$, as the rate at which radiant energy of wavelength λ passes through the differential element $\mathrm{d}A_n$, per unit area of the emitting surface normal to this direction ($\mathrm{d}A_1 \cos\theta$), per unit area of the solid angle about this direction, per unit wavelength interval. Thus

$$I_{\lambda,\mathrm{e}}(\lambda, \theta, \phi) = \frac{\mathrm{d}q}{\mathrm{d}A_i \cos\theta \, \mathrm{d}\varpi \, \mathrm{d}\lambda} \tag{9.2.3}$$

If we define $\mathrm{d}q/\mathrm{d}\lambda = \mathrm{d}q_\lambda(\mathrm{W}/\mu\mathrm{m})$ as the rate at which radiation of wavelength λ leaves $\mathrm{d}A_1$ and passes through $\mathrm{d}A_n$, Eq. (9.2.3) can be rearranged

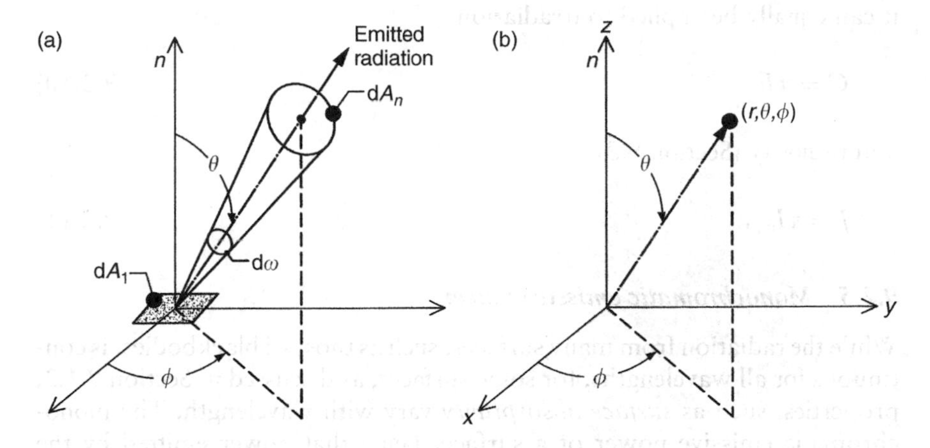

Figure 9.2.3 The view of a planar surface from an element in the surrounding hemisphere.

as follows

$$dq_\lambda = I_{\lambda,e}(\lambda, \theta, \phi)\, dA_i \cos\theta\, d\varpi \qquad (9.2.4)$$

Where the *spectral hemispherical emissive power* $E_\lambda(\text{W/m}^2\ \mu\text{m})$ is defined as the rate at which radiation of wavelength λ is emitted in all directions, it can be shown (Siegel and Howell, 1992) that

$$E_\lambda(\lambda) = \int_0^{2\pi} \int_0^{\pi/2} I_{\lambda,e}(\lambda, \theta, \phi) \cos\theta \sin\theta\, d\theta\, d\phi \qquad (9.2.5)$$

and that the total hemispherical emissive power $E(\text{W/m}^2)$ is given by

$$E = \int_0^\infty E_\lambda(\lambda)\, d\lambda \qquad (9.2.6)$$

While the directional distribution of surface emission varies with the nature of the surface, for a *diffuse emitter*, such as a gray surface (Section 9.3), the intensity of the emitted radiation is independent of direction. Therefore

$$I_{\lambda,e}(\lambda, \theta, \phi) = I_{\lambda,e}(\lambda) \qquad (9.2.7)$$

If we remove $I_{\lambda,e}$ from the integrand of Eq. (9.2.5), and perform the integration we get

$$E_\lambda(\lambda) = \pi I_{\lambda,e}(\lambda) \qquad (9.2.8)$$

and, hence,

$$E = \pi I_e \qquad (9.2.9)$$

While the above analysis has been presented for the case of surface emission, it can equally be applied to irradiation

$$G = \pi I_{\mathrm{i}} \qquad (9.2.10)$$

and radiosity (Section 9.3)

$$J = \pi I_{\mathrm{e}+r} \qquad (9.2.11)$$

9.2.5 *Monochromatic emissive power*

While the radiation from many surfaces, such as those of blackbodies, is continuous for all wavelengths, for some surfaces, as discussed in Section 9.1.2, properties, such as *surface absorptivity* vary with wavelength. The monochromatic emissive power of a surface, being that power emitted by the surface over an infinitesimal wavelength band is defined in

$$\mathrm{d}E = E_{\lambda}\,\mathrm{d}\lambda \qquad (9.2.12)$$

where $\mathrm{d}E$ is the emissive power in the infinitesimal wavelength band $\lambda + \mathrm{d}\lambda$ ($\mathrm{W/m^2}$), and E_{λ} is the monochromatic emissive power ($\mathrm{W/m^2\,\mu m}$). As a black surface is a perfect emitter, the monochromatic emissive power for a black surface is therefore the highest of any surface at any wavelength, for any temperature. This maximum power is given by *Planck's distribution.*

$$E_{\mathrm{b}\lambda} = \frac{c_1}{\lambda^5[\exp(c_2/\lambda T) - 1]} \qquad (9.2.13)$$

where $c_1 = 3.74047\,\mathrm{\mu m^4/m^2}$, and $c_2 = 1.43868\,\mathrm{\mu m\,K}$.

A plot of the monochromatic emissive power of a blackbody, as a function of wavelength, is given in Fig. 9.2.4, for a selection of temperatures. It is immediately apparent that, for any wavelength, the monochromatic emissive power increases with temperature, and the wavelength λ_{max} at which $E_{\mathrm{b}\lambda}$ is maximum decreases with temperature. The logarithmic scale belies the relative magnitude of the monochromatic emissive power for different temperatures. Through differentiation of Eq. (9.2.13) with respect to wavelength, and solving for the resultant being set to zero, we get

$$\lambda_{\mathrm{max}}T = 2897.6\,\mathrm{\mu m\,K} \qquad (9.2.14)$$

This relationship is known as *Wien's displacement law.*

Equation (9.1.13) gives the maximum energy that a surface can emit for a given temperature over all wavelengths. Where knowledge of the energy emitted over a specific waveband is required, for a blackbody, we need to

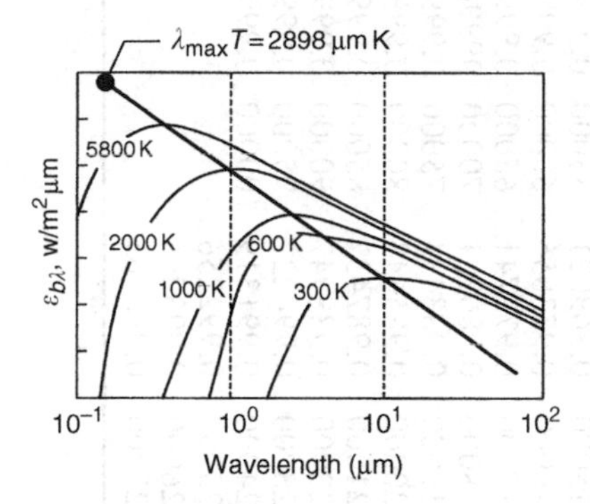

Figure 9.2.4 Blackbody monochromatic emissive power as a function of wavelength.

integrate as shown below:

$$E_{b,\lambda_1\to\lambda_2} = \int_{\lambda_1}^{\lambda_2} E_{b\lambda}\,d\lambda \tag{9.2.15}$$

For a given temperature and specified interval of 0–λ, the fraction of the total emission from a blackbody emitted over this waveband may be expressed as

$$f_{0\to\lambda} = \frac{\int_0^{\lambda} E_{b\lambda}\,d\lambda}{\int_0^{\infty} E_{b\lambda}\,d\lambda} = \frac{\int_0^{\lambda} E_{b\lambda}\,d\lambda}{\sigma T^4} \tag{9.2.16}$$

which may be rewritten as follows

$$f_{0\to\lambda} = \int_0^{\lambda T} \frac{E_{\lambda,b}}{\sigma T^5}\,d(\lambda T) \tag{9.2.17}$$

Equation (9.2.17) can be evaluated to obtain values of $f_{0\to\lambda}$ as a function of λT only. These values are presented in Table 9.2.1. In addition, these values may also be used to obtain the fraction of the total *blackbody radiation* between any two specific wavelengths. Thus

$$E_{b,\lambda_1\to\lambda_2} = \frac{\int_0^{\lambda_2} E_{b\lambda}\,d\lambda - \int_0^{\lambda_1} E_{b\lambda}\,d\lambda}{\sigma T^4} = f_{0\to\lambda_2} - f_{0\to\lambda_1} \tag{9.2.18}$$

Example 9.2.1 Blackbody radiation.

In this example we consider the effect of blackbody temperature on the percentage of radiant energy emitted that lies within the visible spectrum. If we consider both the sun and the tungsten filament of an electric light bulb

Table 9.2.1 Blackbody radiation functions

λT (μm K)	$f_{0-\lambda}$	λT (μm K)	$f_{0-\lambda}$	λT (μm K)	$f_{0-\lambda}$	λT (μm K)	$f_{0-\lambda}$	λT (μm K)	$f_{0-\lambda}$	λT (μm K)	$f_{0-\lambda}$	λT (μm K)	$f_{0-\lambda}$
100	0.000000	2000	0.066733	3900	0.462446	5800	0.720192	7800	0.848042	9800	0.910002	28000	0.994376
200	0.000000	2100	0.083058	4000	0.480902	5900	0.729196	7900	0.852258	9900	0.912153	29000	0.994911
300	0.000000	2200	0.100895	4100	0.498767	6000	0.737852	8000	0.856325	10000	0.914238	30000	0.995381
400	0.000000	2300	0.120037	4200	0.516040	6100	0.746173	8100	0.860251	11000	0.931929	35000	0.997044
500	0.000000	2400	0.140266	4300	0.532723	6200	0.754174	8200	0.864040	12000	0.945138	40000	0.998008
600	0.000000	2500	0.161366	4400	0.548823	6300	0.716869	8300	0.867699	13000	0.955179	45000	0.998605
700	0.000002	2600	0.183132	4500	0.564348	6400	0.769268	8400	0.871232	14000	0.962938	50000	0.998994
800	0.000016	2700	0.205370	4600	0.579309	6500	0.776386	8500	0.874645	15000	0.969021	55000	0.999259
900	0.000087	2800	0.227904	4700	0.593718	6600	0.783234	8600	0.877943	16000	0.973855	60000	0.999444
1000	0.000321	2900	0.250577	4800	0.607589	6700	0.789823	8700	0.881129	17000	0.977741	65000	0.999579
1100	0.000911	3000	0.273248	4900	0.620937	6800	0.796164	8800	0884210	18000	0.980901	70000	0999678
1200	0.002134	3100	0.295797	5000	0.633777	6900	0.802268	8900	0.887188	19000	0.983494	75000	0.999754
1300	0.004317	3200	0.318120	5100	0.646127	7000	0.808144	9000	0.890068	20000	0.985643	80000	0.999812
1400	0.007791	3300	0.340130	5200	0.658001	7100	0.813803	9100	0.892853	21000	0.987437	85000	0.999857
1500	0.012850	3400	0.361755	5300	0.669417	7200	0.819253	9200	0.895548	22000	0.988947	90000	0.999893
1600	0.019720	3500	0.382937	5400	0.680392	7300	0.824504	9300	0.898156	23000	0.990227	95000	0.999922
1700	0.028535	3600	0.403628	5500	0.690940	7400	0.829563	9400	0.900680	24000	0.991319	100000	0.999946
1800	0.039344	3700	0.423794	5600	0.701079	7500	0.834439	9500	0.903124	25000	0.992256		
1900	0.052111	3800	0443406	5700	0.710824	7600	0.839139	9600	0.905490	26000	0.993064		
						7700	0.843671	9700	0.907782	27000	0.993765		

as black bodies, with effective surface temperatures of 5760 and 2800 K, respectively, calculate the percentage of the energy emitted by both surfaces that lies within the visible spectrum.

Analysis

If we define the visible spectrum wave band as lying between 0.38 and 0.78 μm, we can calculate the $T\lambda$ product for each of the above mentioned bodies for each limit of the wave band. From the blackbody radiation functions presented in Table 9.2.1 we then calculate $f_{0-\lambda}$, and thereafter are able to determine the percentage of energy emitted by each body in the visible spectrum wave band.

Solution

Open the workbook Ex09-02-01.xls, and 'Enable' the macro. (The function of the embedded macro is explained in the *Discussion* section.) The workbook contains two worksheets, namely, *Compute* and *Blckbdy rdn fns*. Click on the *Compute* worksheet tab and enter the data given above in the appropriate cells, namely **cells (C4:C7)**. Excel performs all necessary calculations, as displayed in **cells (B16:B21)** for body (1), and **cells (D16:D21)** for body (2). As with many of the worked examples presented in the preceding chapters on Convection, the present example also utilises the Excel **Lookup** function. In this case this function is use to obtain the respective values of $f_{0-\lambda}$ from the 'Lookup' table in the *Blckbdy rdn fns* worksheet. Interpolation is also carried out in this worksheet. The values obtained are returned to the *Compute* worksheet, where the respective percentages are calculated as simply the difference between the f_1 and f_2 values. For body (1), the tungsten filament, a value of 9.7% **cell B21** is obtained, while, for body (2) it can be seen that the percentage of total energy emitted that lies within the visible spectrum is of the order of 46.4% **cell D21**.

Discussion

It is immediately apparent that the Sun is a much more efficient light source than a common filament type light. The electric light, dissipating a greater proportion of its radiant energy at higher wavelengths (90% in the near infrared, between 0.78 and 25 μm) may be considered more effective as a heater. You may wish to further explore the effect of blackbody temperature on thermal radiation distribution by considering the operating temperature of alternative filament materials.

Additionally, the Sun may be considered as emitting most of its radiant energy within the approximate waveband 0.2–2.5 μm. As the upper limit for $T\lambda$ given in Table 9.2.1 is 100,000, the present workbook via the small program written in the workbook 'macro' will automatically return a value of 1 for $f_{0-\lambda}$ if the user inputs values T and λ that combine to give a value above the numerical value of 100,000.

Example 9.2.2 The effect of surface properties on plate temperature.

An unglazed flat plate solar water heater is orientated perpendicular to the sun such that the incident radiation reaching the plate is $1100\,\text{W/m}^2$. The plate has a semi selective coating such that the plate absorptivity and emissivity is equal to unity for all wavelengths up to $2\,\mu\text{m}$, and 0.13 for all $\lambda > 2\,\mu\text{m}$. Coolant flowing in the collector tubes removes heat at such a rate as to maintain the plate temperature at 80°C. Considering the sun to be a blackbody at 5760 K, and the sky to be at 0 K (i.e. non participating), calculate the rate of heat removal by the coolant. If the system pump fails, what would be the plate stagnation temperature. Furthermore, calculate the rate of heat removal required to maintain the plate at its initial operational temperature if the surface were black, that is, $\alpha = \varepsilon$ for all λ. What would be the plate stagnation temperature if the pump were again to fail.

Analysis

To determine the rate at which energy is absorbed, if the coolant maintains the surface at a temperature of 100°C, we require to perform an energy balance at the surface such that, heat transfer from the surface is equal to the rate of energy reaching the surface less that leaving the surface. Thus, the rate at which energy is transferred to the fluid, q, is equal to the difference in the rate of absorbed and emitted energy where the emissive power is given by

$$E = \int_0^\infty \varepsilon_\lambda E_{b\lambda}\, d\lambda$$

and the absorbed energy flux by

$$A = \int_0^\infty \alpha_\lambda G_\lambda\, d\lambda$$

However, as the surface is selectively coated, both the above equations need to be considered in terms of the variance in surface properties with wavelength. Therefore

$$q = \int_0^{\lambda_1} \varepsilon_1 E_{b\lambda}\, d\lambda + \int_{\lambda_1}^\infty \varepsilon_\lambda E_{b\lambda} d\lambda - \int_0^{\lambda_1} \alpha_1 G_\lambda\, d\lambda + \int_{\lambda_1}^\infty \alpha_2 G_\lambda\, d\lambda$$

Which, as λ_2 is defined as being ∞, may be reduced to,

$$q = \lfloor \varepsilon_1 f_{0\to\lambda_1} + \varepsilon_2(1 - f_{0\to\lambda_1}) \rfloor E_b - \lfloor \alpha_1 f_{0\to\lambda_1,s} + \alpha_2(1 - f_{0\to\lambda_1,s}) \rfloor G$$

allowing q to be solved for the conditions specified.

If the coolant stops flowing, under steady state conditions, the net rate of heat transfer from the plate will be zero and, hence, the emissive power will equal the rate of energy absorbed. Thus,

$$\lfloor \varepsilon_1 f_{0\to\lambda_1} + \varepsilon_2(1 - f_{0\to\lambda_1})\rfloor E_b = \lfloor \alpha_1 f_{0\to\lambda_1,s} + \alpha_2(1 - f_{0\to\lambda_1,s})\rfloor G$$

The plate surface temperature may then be found through E_b. However, as both $f_{0-\lambda_1}$ and E_b depend on the plate temperature, we need to employ an iterative method. We therefore require to make an initial estimate of temperature and thereafter compare this with the value calculated via the above method. If the difference between the estimated and calculated values is unacceptably large, a second estimate, being the average of the original estimate and calculated values is made. The process is repeated until the estimated and calculated values converge.

Solution

Open the workbook Ex09-02-02.xls, enable the macro, and open the worksheet *Compute*. Input the data given in **cells (B4:B7)**, **cells (B10:C11)** for the surface properties, and **cells (B13:B15)** for the surface, sky, and sun surface temperatures respectively. As with Example 9.2.1, the present example also uses the Excel **Lookup** function and iterative method to determine the respective values for the blackbody radiation functions.

For the first part of the problem, Excel computes the rate of heat transfer as $-926.8\,\mathrm{W/m^2}$ via the above method, and posts the result in **cell B30**. The radiant energy emitted by the plate and that absorbed, obtained via the above equations, are given in **cells (B27 and B28)**, respectively. To afford a solution to the part (b), the user is required to input an initial estimate for the plate temperature in the 'Try' cell, **cell B33**, then run the macro *Iterate*. For the conditions specified, the 'Try' and 'Computed' values will be seen to converge at 610.5 K.

For a black surface, with $\alpha = 1$ for all λ, the energy transferred to the coolant is equal to that emitted less that absorbed, hence $q = -218.3\,\mathrm{W/m^2}$, and the plate temperature would be 373.2 K if the coolant ceased to flow, **cells (B34 and B35)**. This represents a rise of only 20 K above the normal plate operating temperature as opposed to that of approximately 250 K for the case with the selective low emissivity coating.

Discussion

From the above analysis it can be seen that, for the conditions specified the special coating yields a four- to five-fold increase in the energy transferred to the coolant. The selective surface is therefore much more effective as a solar collector. While the majority of incident energy from the sun is in the first waveband, the majority of the blackbody radiation from the surface at the specified temperature is in the second waveband. As the emissivity of the selective coating is only 0.13 over this range, the collector only loses 13% of the energy that would otherwise be lost for $\varepsilon = 1$.

9.3 Radiosity and view factor

9.3.1 *Radiosity*

Let us consider a surface '*A*', surrounded by a number of other surfaces, including those marked as '*B*' and '*C*' in Fig. 9.3.1. The net rate of radiant heat transfer from surface '*A*' will be the sum of that leaving the surface, less that which it receives. The total radiant energy leaving a surface, per unit area and per unit time is defined as the radiosity, *J*, of that surface. It therefore comprises both the emitted energy, and the reflected part of the incident energy, as shown in Fig. 9.3.2, and is therefore defined as

$$J = \varepsilon E_b + \rho G \tag{9.3.1}$$

where, ε is the surface emissivity, E_b the blackbody emissive power at the temperature of the surface, ρ the surface reflectivity, and G the incident

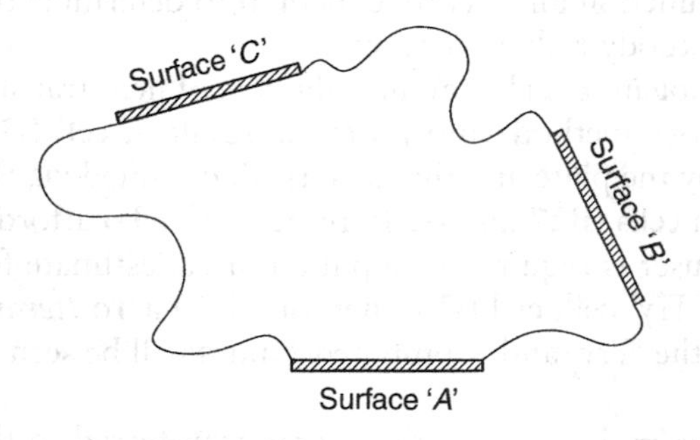

Figure 9.3.1 Surface '*A*' in an enclosure comprising surfaces '*B*' and '*C*'.

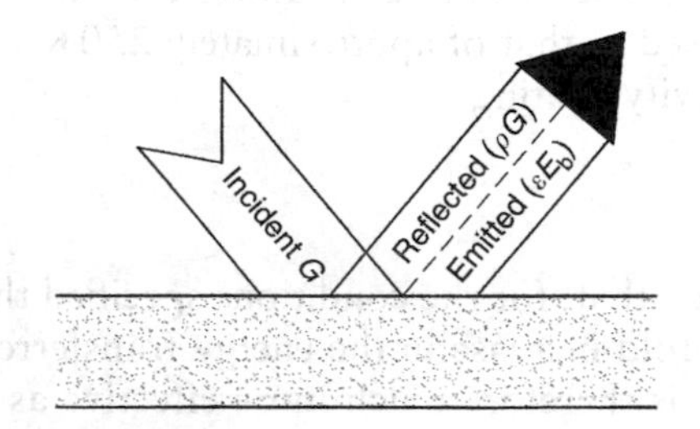

Figure 9.3.2 Radiosity of surface.

radiant heat flux. For a black surface, where $\varepsilon = 1$, and $\rho = 0$, Eq. (9.3.1) therefore reduces to

$$J = E_b \tag{9.3.2}$$

For a grey, diffuse surface, the *emissive power, radiosity* and heat transfer are interrelated. For a surface of given area 'A', the rate of energy leaving the surface is equal to AJ, and that arriving, AG. The net rate of radiant energy leaving the surface, q, is therefore given by

$$q = AJ - AG \tag{9.3.3}$$

For an opaque grey surface, $\tau = 0, \rho = 1 - \alpha$, and $\alpha = \varepsilon$. Hence, from Eq. (9.3.1), we get

$$G = \frac{J - \varepsilon E_b}{1 - \rho} \tag{9.3.4}$$

Substituting Eq. (9.3.4) into Eq. (9.3.3) and rearranging, we get

$$q = \frac{A\varepsilon}{1 - \varepsilon}(E_b - J) \tag{9.3.5}$$

For a black surface, for which $J = E_b$, Eq. (9.3.5) is not applicable. However, Eq. (9.3.3) may be rearranged to give

$$q = A(E_b - G) \tag{9.3.6}$$

Example 9.3.1 Radiant heat transfer from a flat diffuse, gray plate.

A flat, gray diffuse, opaque surface with an absorptivity, α, of 0.75, receives radiant energy at a rate of 8.4 kW/m^2. If the surface is maintained at a temperature of 400°C and has an area of 2 m^2, find the rate of energy absorbed, the rate of radiant energy emitted, the total rate of energy leaving the surface, and the net radiative heat transfer from the surface.

Solution

(a) Rate of energy absorbed $= \alpha AG = 0.75 \times 2 \times 8400 = 12.6$ kW.
(b) Rate of radiant energy emitted for a gray surface $\varepsilon = \alpha = 0.75$

$$\begin{aligned} A\varepsilon\sigma T^4 &= 2 \times 0.75 \times 5.67 \times 10^{-8} \times (400 + 273.15)^4 \\ &= 17.46\,\text{kW} \end{aligned}$$

(c) Total rate of energy leaving the surface = radiant energy emitted + refl-ected component. For an opaque surface $\tau = 0$, therefore $\rho = 1 - \alpha = 0.25$

$$
\begin{aligned}
\text{Total energy flux from the surface} &= (\varepsilon E_b + \rho G)A \\
&= 17.46 \times 10^3 + (0.25 \times 8400 \times 2) \\
&= 17.46 \times 10^3 + 4.2 \times 10^3 \\
&= 21.66\,\text{kW}
\end{aligned}
$$

(d) The net radiative heat transfer rate = total energy leaving the surface – total energy reaching the surface

$$
\begin{aligned}
&= 21.66 \times 10^3 - (2 \times 8400) \\
&= 4.86\,\text{kW}
\end{aligned}
$$

9.3.2 *View factor*

We have thus far considered radiation from a surface in relation to its total hemispherical emissivity. In defining the radiant exchange of heat between two surfaces we need to consider the relevant view factor, or how much of their respective hemispherical views the surfaces occupy. The *view factor*, among other names, is also sometimes referred to as the *shape factor*, *geometry factor*, and *angle factor*. The view factor is simply a function of the geometry of the two surfaces in question, and, as such, considers the orientation of one surface with respect to the other, and the spacing between them. For two surfaces, i and j, as shown in Fig. 9.3.3, we define the view

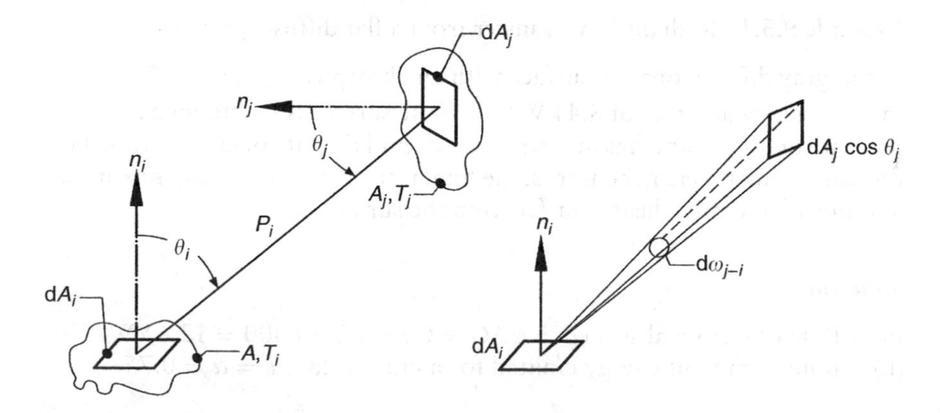

Figure 9.3.3 View factor associated with the radiation heat exchange between two elemental surfaces of area dA_i and dA_j.

factor as the fraction of the radiation leaving i that is intercepted by surface j, F_{ij}. We now present the development of a general expression for F_{ij} that will allow the calculation of the view factor for all situations.

Let us consider an elemental area on each surface, dA_i, and dA_j, connected by a straight line P, which forms the polar angles θ_i and θ_j with the respective surface normals, n_i and n_j. Obviously, the values of P, θ_i, and θ_j, vary with the position of the differential elements dA_i and dA_j. The rate at which radiation leaves dA_i, and is intercepted by dA_j is given by,

$$dq_{i\to j} = I_i \cos\theta_i \, dA_i \, d\varpi_{j-i} \tag{9.3.7}$$

where I_i is the intensity of the radiation leaving surface i, and $d\varpi_{j-i}$ is the solid angle subtended by dA_j when viewed from dA_j, thus

$$d\varpi_{j-i} = (\cos\theta_j \, dA_j)/P^2 \tag{9.3.8}$$

and Eq. (9.3.7) may therefore be re-expressed as,

$$dq_{i-j} = I_i \frac{\cos\theta_i \cos\theta_j}{P^2} \, dA_i \, dA_j \tag{9.3.9}$$

If we assume that surface i *emits* and *reflects* diffusely, then

$$J_i = \pi I_{i(e+r)} \tag{9.3.10}$$

and Eq. (9.3.9) can be re-written as

$$dq_{i-j} = J_i \frac{\cos\theta_i \cos\theta_j}{\pi P^2} \, dA_i \, dA_j \tag{9.3.11}$$

Assuming that the radiosity J_i is uniform over the surface A_i, the rate at which radiation leaves surface i and is intercepted by surface j can be found by integrating over the two surfaces, thus

$$dq_{i-j} = \int_{A_i} \int_{A_j} J_i \frac{\cos\theta_i \cos\theta_j}{\pi P^2} \, dA_i \, dA_j \tag{9.3.12}$$

From our earlier definition of the view factor, F_{ij} may be expresses as follows,

$$F_{ij} = \frac{q_{i-j}}{A_i J_i} \tag{9.3.13}$$

Therefore,

$$F_{ij} = \frac{1}{A_i} \int_{A_i} \int_{A_j} \frac{\cos\theta_i \cos\theta_j}{\pi P^2} \, dA_i \, dA_j \tag{9.3.14}$$

Similarly, the view factor F_{ji} may be obtained from

$$F_{ji} = \frac{1}{A_j} \int_{A_i} \int_{A_j} \frac{\cos\theta_i \cos\theta_j}{\pi P^2} \, dA_i \, dA_j \tag{9.3.15}$$

Equations (9.3.14) and (9.3.15) may be used to determine the view factor where the two surfaces in question are both diffuse emitters and reflectors, and have uniform radiosity. It can also be seen that

$$F_{ij}A_i = F_{ji}A_j \tag{9.3.16}$$

Equation (9.3.16) is termed the *reciprocity relation*.

If one considers radiation from one surface to all those that comprise its hemispherical view, such as in an enclosure, the sum of all the individual view factors, the *summation rule*, equals one

$$\sum_{j=1}^{N} F_{ij} = 1 \tag{9.3.17}$$

If a surface is *concave*, it will *see* itself and F_{ii} will be *nonzero*. However, if the surface is plane or *convex*, $F_{ii} = 0$. To calculate radiation exchange in an enclosure of N surfaces, clearly we require a total of N^2 view factors. However, considering the *reciprocity relation* (Eq. (9.3.16)) and the *summation rule* (Eq. (9.3.17)), it can easily be shown that only $N(N-1)/2$ view factors need to be determined directly.

To further develop our understanding of view factors let us consider the simple case of the long duct with isosceles cross-section represented in Fig. 9.3.4. Let us first assume that, as the duct is long, we can ignore the effects of any radiant energy entering or leaving the duct.

From Eq. (9.3.17),

$$F_{a-a} + F_{a-b} + F_{a-c} = 1$$

As surface a is planar ($F_{a-a} = 0$), and, by symmetry, $F_{a-b} = F_{a-c}$, then, $F_{a-b} = F_{a-c} = 0.5$.

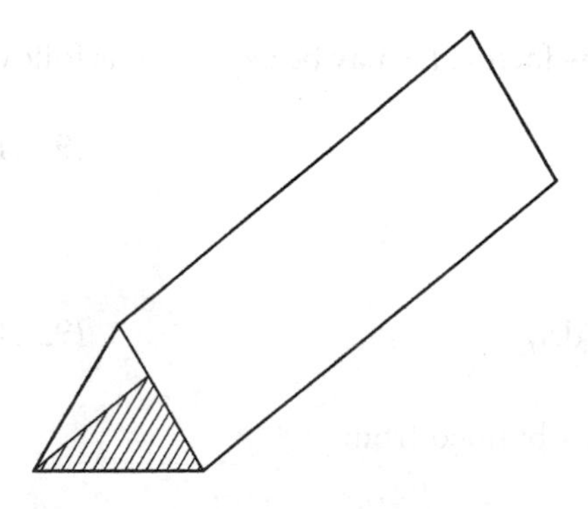

Figure 9.3.4 A long duct with isosceles cross-section.

To find F_{b-c}, again, as b is planar and $F_{b-b} = 0, F_{b-c} + F_{b-a} = 1$. From the reciprocity relation, Eq. (9.3.16), $A_i F_{i-j} = A_j F_{j-i}$ therefore,

$$F_{b-a} = \frac{F_{a-b}A_a}{A_b} = 0.5\frac{A_a}{A_b} \quad \text{and} \quad F_{b-a} = 1 - 0.5\frac{A_a}{A_b}$$

For more complicated geometries, the view factor may be calculated from the double integral equation, Eq. (9.3.14), or, more commonly, obtained from tables or graphical interpretation. Table 9.3.1, adapted from Howell (1982), presents the view factor relation equations for some common two and three dimensional geometries.

To provide an idea of the level of involvement of the view factor formulae required for the analysis of even relatively simple cases, the concerned relationship for radiation exchange between a spherical cup and a planar surface is presented in Box 9.3.1. Siegel and Howell (1992) have catalogued an encyclopaedic collection of view factor formulae.

We have thus far considered, through application of the view factor analysis, that the radiant energy intercepted by a surface is evenly distributed over that surface. Let us presently consider the case of the room shown in Fig. 9.3.5a. Surface a is maintained at a uniform temperature of 40°C while surface b is at 20°C. The analysis thus far presented affords a measure of the total radiative heat transfer between the respective surfaces, and surface c, but not its distribution over these surfaces. Incident radiant heat flux from surfaces a and b decreases with distance. It is therefore appropriate in many situations to consider a given surface as comprising smaller sections, as shown in Fig. 9.3.5b, and to treat each of these sections as a separate surface. Rather than considering each of the walls of the room separately, by symmetry, as the view factor for surfaces a and b will be the same for each horizontal plane (c_1, c_2, etc.) on each of the four walls, one may consider the surfaces of all the walls in one plane, for example, plane 'c_4', as one radiation surface.

In selecting an appropriate number of sections for each surface in question, a balance between accuracy of solution, and ease of computation must be sought.

Example 9.3.2 Radiant heat transfer in an enclosure with three gray surfaces.

In the present example we consider the radiant exchange of heat between the inner surfaces that comprise a small hollow cylinder. The respective surfaces may be assigned specific values of temperature, heat flux, or emissivity, allowing the effects of changing these parameters on those dependent to be investigated.

Table 9.3.1 View factor relation equations for selected geometries

1. Identical, aligned rectangles directly opposite each other

$$F_{12} = \frac{2}{\pi xy}\left\{\ln\left[\frac{(1+x^2)(1+y^2)}{1+x^2+y^2}\right] + x(1+y^2)^{1/2}\tan^{-1}\left[\frac{x}{(1+y^2)^{1/2}}\right]\right.$$

$$\left. + y(1+x^2)^{1/2}\tan^{-1}\left[\frac{y}{(1+x^2)^{1/2}}\right] - x\tan^{-1}x - y\tan^{-1}y\right\}$$

where $x = L/D$ and $y = W/D$ (L and W are the plate dimensions, and D the separation distance).

2. Two coaxial circular discs

$$F_{12} = \frac{1}{2}\left[x - \left(x^2 - 4\frac{R_2^2}{R_1^2}\right)^{1/2}\right]$$

where $R_1 = r_1/L$, $R_2 = r_2/L$, $x = 1 + \left[\frac{(1+R_2^2)}{R_1^2}\right]$.

3. Two perpendicular rectangles sharing a common side

$$F_{12} = \frac{1}{\pi y}\left[y\tan^{-1}\left(\frac{1}{y}\right) + x\tan^{-1}\left(\frac{1}{x}\right) - (a+b)^{1/2}\tan^{-1}\left(\frac{1}{(a+b)^{1/2}}\right)\right.$$

$$\left. + \frac{1}{4}\ln\left\{\frac{(1+a)(1+b)}{1+a+b}\left(\frac{b(1+a+b)}{(1+b)(b+a)}\right)^b\left(\frac{a(1+a+b)}{(1+a)(b+a)}\right)^a\right\}\right]$$

where $x = W/D$, $y = L/D$, $a = x^2$, $b = y^2$.

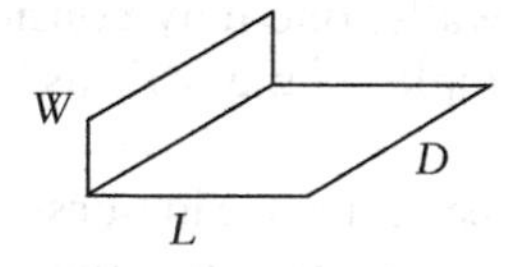

4. Two coaxial cylinders (length L)(outer cylinder radius r_2, inner cylinder, r_1)

$R = r_2/r_1$ $M = L/r_1$ $A = M^2 + R^2 - 1$ $B = M^2 - R^2 + 1$

$$x = \frac{(R^2-1) + (M^2/R^2)(R^2-2)}{M^2 + 4(R^2-1)}$$

$$F_{21} = \frac{1}{R} - \frac{1}{\pi R}\left[\cos^{-1}\left(\frac{B}{A}\right) - \frac{1}{2M}\left[(A+2)^2 - 4R\right]^{1/2}\cos^{-1}\left(\frac{B}{RA}\right)\right.$$

$$\left. + B\sin^{-1}\left(\frac{1}{R}\right) - \frac{\pi A}{2}\right]$$

Table 9.3.1 (Continued)

$$F_{22} = 1 - \frac{1}{R} + \frac{2}{\pi R}\tan^{-1}\left[\frac{2(R^2-1)^{1/2}}{M}\right]$$
$$-\frac{M}{2\pi R}\left\{\frac{(4R^2+M^2)^{1/2}}{M}\sin^{-1}(x) - \sin^{-1}\left(\frac{(R^2-2)}{R^2}\right)\right.$$
$$\left.+\frac{\pi}{2}\left[\frac{(4R^2+M^2)^{1/2}}{M} - 1\right]\right\}$$

$$F_{23} = \frac{1}{2}(1 - F_{21} + F_{22})$$

for all $-\pi/2 \leq \sin^{-1} y \leq \pi/2$ and $0 \leq \cos^{-1} y \leq \pi$.

Box 9.3.1 View factor between a plane element and a spherical sector

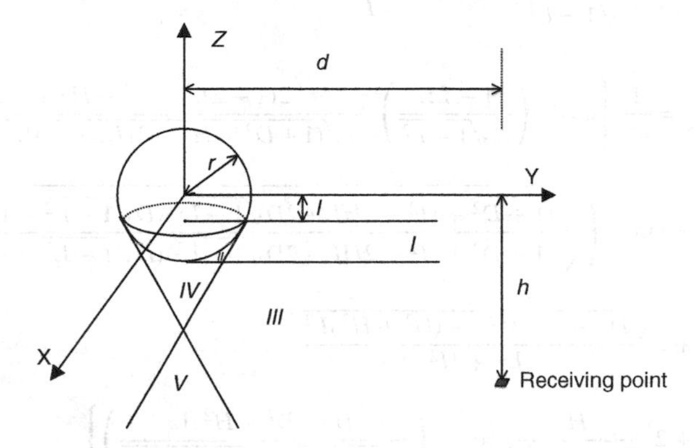

I: $D > \dfrac{1-LH}{\sqrt{1-L^2}}$ and $-1 < H < L$

$\left(\text{where } D = \dfrac{d}{r}, H = \dfrac{h}{r}, L = \dfrac{l}{r}\right)$:

$$F_{dA_1-A_2} = \frac{1}{2\pi}\left\{\cos^{-1}\left(\frac{1-LH}{D\sqrt{1-L^2}}\right) + \frac{2(1-2L^2-D^2-H^2+2HL)}{\sqrt{(1+D^2+H^2-2HL)^2-4D^2(1-L^2)}}\right.$$
$$\times\tan^{-1}\left(\sqrt{\frac{(1+D^2+H^2-2HL+2D\sqrt{1-L^2})(D\sqrt{1-L^2}-1+LH)}{(1+D^2+H^2-2HL-2D\sqrt{1-L^2})(D\sqrt{1-L^2}+1-LH)}}\right)$$
$$+2\frac{\sqrt{D^2-1+2HL-(D^2+H^2)L^2}-\sqrt{D^2-1+2H^2-(D^2+H^2)H^2}}{D^2+H^2}$$
$$+2\frac{H}{(D^2+H^2)^{\frac{3}{2}}}\left[\sin^{-1}\left(\frac{H-(D^2+H^2)L}{\sqrt{H^2+(D^2-1)(D^2+H^2)}}\right)\right.$$
$$\left.\left.-\sin^{-1}\left(\frac{H(1-D^2-H^2)}{\sqrt{H^2+(D^2-1)(D^2+H^2)}}\right)\right] + 2\sin^{-1}\left(\frac{\sqrt{1-H^2}}{D}\right)\right\}$$

II: $D < \dfrac{1-LH}{\sqrt{1-L^2}}$ and $-1 < H < L$:

$$F_{dA_1-A_2} = \frac{1}{\pi}\left\{\tan^{-1}\sqrt{\frac{1-H^2}{D^2+H^2-1}} - \frac{\sqrt{(D^2+H^2-1)(1-H^2)}}{D^2+H^2} - \frac{H}{(D^2+H^2)^{3/2}}\cos^{-1}\left(\frac{H\sqrt{D^2+H^2-1}}{D}\right)\right\}$$

III: when $L > 0$ $D > \dfrac{1-LH}{\sqrt{1-L^2}}$ and $H < -1$

when $L < 0$ $D > \dfrac{1-LH}{\sqrt{1-L^2}}$ and $\dfrac{1}{L} < H < -1$

when $D > \dfrac{LH-1}{\sqrt{1-L^2}}$ and $H < \dfrac{1}{L}$:

$$F_{dA_1-A_2} = \frac{1}{2\pi}\left\{\cos^{-1}\left(\frac{1-LH}{D\sqrt{1-L^2}}\right) + \frac{2(1-2L^2-D^2-H^2+2HL)}{\sqrt{(1+D^2+H^2-2HL)^2-4D^2(1-L^2)}}\right.$$

$$\times\tan^{-1}\left(\sqrt{\frac{(1+D^2+H^2-2HL+2D\sqrt{1-L^2})(D\sqrt{1-L^2}-1+LH)}{(1+D^2+H^2-2HL-2D\sqrt{1-L^2})(D\sqrt{1-L^2}+1-LH)}}\right)$$

$$+2\frac{\sqrt{D^2-1+2HL-(D^2+H^2)L^2}}{D^2+H^2}$$

$$\left. +2\frac{H}{(D^2+H^2)^{3/2}}\cos^{-1}\left(\frac{H-(D^2+H^2)L}{\sqrt{H^2+(D^2-1)(D^2+H^2)}}\right)\right\}$$

IV: when $L > 0$ $D < \dfrac{1-LH}{\sqrt{1-L^2}}$ and $H < -1$

when $L < 0$ $D < \dfrac{1-LH}{\sqrt{1-L^2}}$ and $\dfrac{1}{L} < H < -1$

$$F_{dA_1-A_2} = -\frac{H}{D^2+H^2)^{3/2}}$$

V: when $D < \dfrac{LH-1}{\sqrt{1-L^2}}$ and $H < \frac{1}{L}$

$$F_{dA_1-A_2} = \frac{1}{2\pi}\left\{1 + \frac{1-2L^2-D^2-H^2+2HL}{\sqrt{(1+D^2+H^2-2HL)^2-4D^2(1-L^2)}}\right\}$$

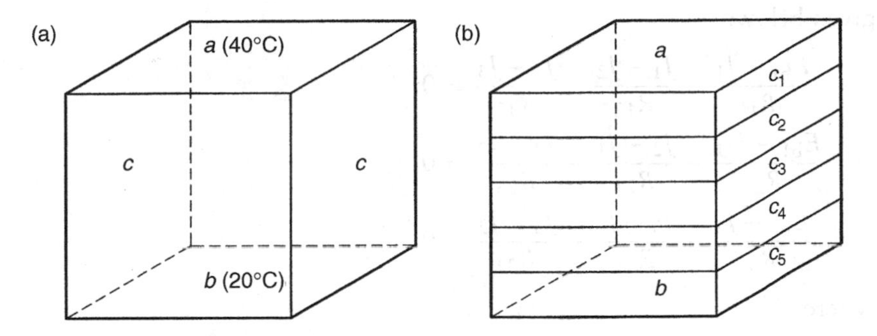

Figure 9.3.5 (a) A room with heated floor, cooled ceiling, and insulated walls; (b) suitable view factor interpretation of the room in (a).

The dimensions of the cylinder are 12 cm long and 8 cm diameter. For the purpose of this example, the cylinder ends are termed surfaces 1 and 3, while the curved surface will be known as surface 2. In the first instance, where the temperature of surface 1 is 280°C, with emissivity $\varepsilon = 0.8$, and the temperature of surface 3 is 100°C, with $\varepsilon = 0.7$, determine the respective surface heat fluxes, q_1, q_2, and q_3, if the temperature and emissivity of surface 2 are 500°C and 0.8, respectively.

In the second instance, for the same initial conditions for surfaces 1 and 3 as outlined above, redefine the respective surface heat fluxes if the temperature of surface 3 remains at 500°C, but the emissivity increases to unity.

Finally, again for the same initial conditions for surfaces 1 and 3, determine q_1, q_3, and T_2 if $q_3 = 30$ W, and $\varepsilon = 1$.

Analysis

A similar method of solution may be employed for each of the three scenarios outlined above. In the first instance we require to determine the view factors and geometrical resistances for the cylinder with dimensions given. The cylinder ends may be considered as coaxial discs, allowing the view factor relation given in Table 9.3.1 to be applied.

For each case we require to perform an energy balance at each surface, such that, for example, for surface 1

$$q_1 = q_{12} + q_{13}$$

or, alternatively,

$$q_1 - q_{12} - q_{13} = 0$$

For the three surfaces, the respective radiation energy balance equations are given below:

$$\frac{E_{b1} - J_1}{R_1} - \frac{J_1 - J_2}{R_{12}} - \frac{J_1 - J_3}{R_{13}} = 0$$

$$\frac{E_{b1} - J_{21}}{R_2} - \frac{J_2 - J_1}{R_{21}} - \frac{J_2 - J_3}{R_{23}} = 0$$

$$\frac{E_{b3} - J_3}{R_3} - \frac{J_3 - J_1}{R_{31}} - \frac{J_3 - J_2}{R_{32}} = 0$$

where

$$R_1 = \frac{1 - \varepsilon_1}{\varepsilon_1 A_1}, \quad R_2 = \frac{1 - \varepsilon_2}{\varepsilon_2 A_2}, \quad R_3 = \frac{1 - \varepsilon_3}{\varepsilon_3 A_3}$$

and

$$R_{12} = \frac{1}{A_1 F_{12}} = \frac{1}{A_2 F_{21}} = R_{21}$$

$$R_{13} = \frac{1}{A_1 F_{13}} = \frac{1}{A_3 F_{31}} = R_{31}$$

$$R_{23} = \frac{1}{A_2 F_{23}} = \frac{1}{A_3 F_{32}} = R_{32}$$

For each of the three scenarios, the first three equations given above, via the respective data given, may be resolved to give a set of three linear equations with three variables. In the first instance, as we are given the temperatures and emissivities of the three surfaces, we may solve for the surface radiosities, J_1, J_2, and J_3, and thence, via the first term on the left-hand side of each of the aforementioned equations, the respective surface heat fluxes. The matrix inversion method of solution, as detailed in Chapter 2, may be used to afford a solution for each condition considered.

In the second instance the temperature and emissivity of surface 2 are again specified. However, as the emissivity of surface 2 is unity, we may again solve our linear equations for J_1 and J_3, but also obtain q_2 directly from the solution. The heat fluxes for surfaces 1 and 3 may be determined in the same manner as before. For scenario three, the heat flux and emissivity of surface 2 are specified. In this instance the set of linear equations are used to solve for J_1, E_{b2}, and J_3.

Solution

Open the workbook Ex09-03-02.xls. The work book comprises four worksheets, namely, *Main, Case (a)*, *Case (b)*, and *Case (c)*. Select worksheet *Main* by clicking on its tab and input the data given above in the cells indicated.

The coaxial disc view factor formula is embedded in this worksheet, and Excel returns the required values for the view factors and surface resistances as shown. The solution for Case (a) is presented in worksheet Case (a), and so on. The respective *Coefficient, Inverse, Constant*, and *Solution* matrices are presented in each of the three case worksheets. Users should familiarise themselves with the use of the solution technique applied, and verify the authenticity of the values presented in the respective coefficient and constant matrices.

Discussion

Comparing the results of Cases (a) and (b), where q_1, q_2, and q_3 have values of −51.6, 112.0, −60.5, −55.1, 118.5 and −63.5 W, respectively, it can be seen that the increase in emissivity of surface 2 increases the heat flux at all surfaces. Radiant heat flux from surface 2 is increased, and consequently that absorbed by surfaces 1 and 3 also increases. Where the heat flux of a given surface is known, and the temperature of the remaining surfaces can be measured, the method of solution presented in Case (c) will allow the cooling load, or heating requirement, for the remaining surfaces to be defined for a given situation.

9.4 Blackbody radiation exchange

If all surfaces considered may be approximated as blackbodies, the study of radiative heat transfer between them is simplified as there is no reflection. If, as previously, we define q_{a-b} as the rate at which radiation leaves surface a and is intercepted by surface b, it therefore follows that

$$q_{a\to b} = (A_a J_a)F_{ab} = A_a E_{b,a} F_{ab} \tag{9.4.1}$$

and the *net radiative exchange* between the surfaces may be expressed as

$$q_{ab} = q_{a\to b} - q_{b\to a} \tag{9.4.2}$$

Considering the Stefan–Boltzmann law, Eq. (9.3.3) and the reciprocity relation, Eq. (9.3.16) and (9.4.2) may be expressed as

$$q_{ab} = A_a F_{ab} \sigma (T_a^4 - T_b^4) \tag{9.4.3}$$

This provides a measure of the net rate at which radiation leaves surface a through its interaction with surface b. As such it does not provide a measure of the total radiative loss or gain from b to a. For any surface in an enclosure of N black surfaces, maintained at different surface temperatures, the above mentioned net transfer of radiation to all N surfaces from surface a can be found through summation, such that,

$$q_a = \sum_{b=1}^{N} A_a F_{ab} \sigma (T_a^4 - T_b^4) \tag{9.4.4}$$

Sparrow and Minkowycz (1962) undertook an analytical study of the effect of radiator fin configuration on the overall heat transfer rate from radiators. In the first stage of their analysis they assumed the fins to behave as black bodies. In addition to defining the conduction heat transfer paths within the tube and fin material, they considered the respective component view factors and radiant energy exchange for various tube and fin thicknesses for each configuration. They also considered the effect of tube spacing.

Through this analysis they were able to define the fin temperature distribution, and to show that the overall rate of radiant heat transfer was not significantly different for either a closed sandwich or open configuration, or for a centre-line fin arrangement. In the second stage of their analysis, they assumed all surfaces to be gray and diffuse, with identical properties. For the three above mentioned configurations they considered the effect of equal fin weight on radiant performance. Obviously, with this constraint, the fins of the closed sandwich configuration will be thinner than the single fin of either the open or centre-line arrangement. Again they found that there was not a marked difference in the heat transfer rate for the configurations studied.

Example 9.4.1 Radiant heat exchange between two rectangular black surfaces.

In the present example we consider the exchange of heat, via radiation, between two opposing black rectangular surfaces. This situation is often encountered in, for example, the walls of a furnace where the wall properties may be approximated as those of a black surface. The simple problem posed, and solved, may be further developed to allow more complex analyses to be tackled. Also, the ease with which the necessary calculations are made allows a ready analysis of the effect of deign parameters on the predicted performance of the system considered.

Two parallel plates with surface emissivity $\varepsilon = 1$, have dimensions of 0.6 m long and 0.3 m wide. If the plates are separated by a distance of 0.1 m, determine the rate of heat loss or gain to each plate if plate 1 is at 150°C, and plate 2 is at 100°C. Assume that the plate enclosure is the sky at −273.15°C. Also, determine the rate of heat loss to the sky.

Analysis

For the given dimensions, the view factor can be calculated for each plate via the relationships given in Table 9.3.1. As the emissivity of each plate is equal unity, the individual plate surface resistance is therefore zero. We need therefore only consider the inter plate resistance, where, for parallel plates with the same dimensions

$$R_{12} = \frac{1}{(A_1 F_{12})} = R_{21}$$

If we assume that the enclosure surrounding the plates is the sky at −273.15°C, the enclosure will therefore only absorb thermal radiation, and will not participate, other than via its view factor relation with the plates, with the exchange of heat between them.

In respect of the plate view factors, as the plates are identical, each will have the same view of the other, and of the sky. We need only calculate two view factors for one of the plates (e.g. for plate 1, we may find R_{12} and R_{13} via the above mentioned relationships of Table 9.3.1) and apply those to the other. The rate of radiant heat flux from each plate is then calculated via the equations given by

$$q_1 = (E_{b1} - E_{b2}/R_{12}) + (E_{b1}/R_{13})$$

$$q_2 = (E_{b2} - E_{b1}/R_{12}) + (E_{b2}/R_{13})$$

Solution

Open the workbook Ex09-04-01.xls. The workbook contains only one worksheet, namely *Compute*. Input the data given above in **cells (B4:B11)**. The parallel rectangular plate view factor relations are embedded in **cells (E18:E21)**. Using the values calculated in these **cells**, Excel, via calculation of the appropriate network resistances and surface emissive powers, computes the respective plate heat fluxes, **cells (B26 and B27)**. The heat loss to the sky is the sum of the aforementioned cell values.

Discussion

For the plate dimensions, separation distance, and temperatures specified the heat loss from plate 1 is calculated as 202.5 W, while the negative sign associated with the heat flux from plate 2 implies that it gains heat at a rate of 8.4 W from plate 1. If the temperature of plate 2 were raised to 103.9°C, q_2 would effectively become zero. As discussed in the introduction to this example, through varying the input parameters, the user may gain a better understanding of their influence on the solution outputs.

Example 9.4.2 Effect of insulation on radiation heat transfer between surfaces.

In the present example we consider the effect of having an insulated backing on one of the participating surfaces in an enclosure with surfaces at uniform but different temperatures. If a surface is insulated, under steady state conditions, the net radiant heat flux from that surface will be zero.

A circular plate of 600 mm diameter is maintained at a temperature of 650 K by means of an internal electrical heating element embedded in its base. The plate is positioned below, and coaxial to, a well insulated conical surface of the same radius. The plate and cone are separated by a distance of 650 mm. If the plate and cone are enclosed within surroundings with a surface temperature of 300 K, determine the surface temperature of the cone and the electrical heat input required to maintain the base plate at the temperature specified if all surfaces are black.

Analysis

Assuming that the cone behaves as an insulated re-radiating surface, and that the surroundings form a three surface enclosure, with all surfaces black, we perform a radiation balance on the cone, such that,

$$q_2 = 0 = q_{21} + q_{23} = A_2F_{21}\sigma(T_2^4 - T_1^4) + A_2F_{23}\sigma(T_2^4 - T_3^4)$$

As the plate and cone are parallel and coaxial, we may apply the coaxial parallel disc relationship given in Table 9.3.1 in order to define the respective view factors. Having obtained the necessary view factors, we can then solve the above equation for T_2. With knowledge of T_2 we can then solve the radiation balance for the heated plate in the same manner as above, thus,

$$q_1 = q_{12} + q_{13} = A_1F_{12}\sigma(T_1^4 - T_2^4) + A_1F_{13}\sigma(T_1^4 - T_3^4)$$

Solution

Open the workbook Ex09-04-02.xls. This workbook contains only one worksheet, namely, *Main*. Input the data given in the above problem in cells **(B4:B8)**. Excel performs all necessary calculations of view factors, and presents the results for the cone surface temperature, T_2, and the electrical power input q_1, in cells **(B21 and B22)** respectively. For the present case, the cone surface temperature is found to be 429.9 K, while the heat input is 2.7 kW.

Discussion

By varying the plate and cone diameters, and surface temperatures, the user can explore the effects of altering these parameters on the heat input required to maintain the conditions specified.

We now consider radiative heat transfer between gray, diffuse, and opaque surfaces.

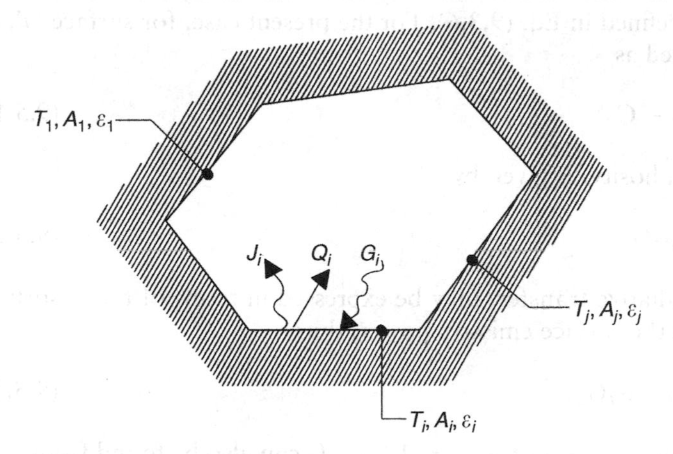

Figure 9.5.1 Radiant heat exchange in an enclosure comprising surfaces with different properties.

9.5 Radiative heat transfer between gray, diffuse, and opaque surfaces

When surfaces exchange energy by radiation, we generally know either the temperature at which the surface is maintained, or the net radiant heat transfer rate. If we know the former, we may wish to determine the latter, and vice versa. For example, in considering the design of a voltage regulator, we may wish to design for a specified radiant heat transfer to ensure a safe working temperature for the device. We therefore require a thorough knowledge of all adjacent surface emissivities and view factors.

While this is relatively simple for the case of blackbodies, for non black surfaces, such as the enclosure shown in Fig. 9.5.1, radiation may experience multiple reflections off many surfaces, with partial absorption at each. The analysis of radiant heat exchange may be somewhat simplified by assuming that each surface of the enclosure is isothermal, opaque, diffuse, and gray. Assuming that each surface may be characterised by uniform radiosity and irradiation, and that the medium is non-participating simplifies the analysis. However, as discussed in Section 9.6, it is not always appropriate to assume that the medium will be non-participating.

In many engineering situations, the temperature of each of the surfaces of the enclosure is known. In this instance, the aim is to quantify the net radiative heat flux from each surface.

9.5.1 *Net radiative heat exchange at a surface*

The net radiative heat exchange is the rate at which energy must be transferred to a surface to keep its temperature constant. Being equal to the difference between the surface radiosity and irradiation, for a black surface, this was defined in Eq. (9.3.6). For the present case, for surface 'a', it may be expressed as

$$q_a = A_a(J_a - G_a) \tag{9.5.1}$$

where J_a, the radiosity, is given by

$$J_a = E_a + \rho_a G_a \tag{9.5.2}$$

Also the net radiative transfer, may be expressed in terms of the absorbed irradiation and the surface emissive power, thus

$$q_a = A_a(E_a - \alpha_a G_a) \tag{9.5.3}$$

As, for a gray surface, $\rho_a = 1 - \alpha_a = 1 - \varepsilon_a$, J_a can also be found from

$$J_a = \varepsilon_a E_{ba} + (1 - \varepsilon_a)G_a \tag{9.5.4}$$

Solving for G_a and substituting into Eq. (9.5.1) we get

$$q_a = A_a\left(J_a - \frac{J_a - \varepsilon_a E_a}{1 - \varepsilon_a}\right) \tag{9.5.5}$$

which rearranges to

$$q_a = \frac{E_{ba} - J_a}{(1 - \varepsilon_a)/\varepsilon_a A_a} \tag{9.5.6}$$

The numerator in Eq. (9.5.6) may be likened to a motive power for radiative heat transfer, while the denominator is equivalent to a surface radiative resistance. The greater the emissivity, the lower the resistance, and hence the greater the radiative heat transfer.

To use Eq. (9.5.6) we must have knowledge of J_a. We therefore need to know the radiation exchange between the surfaces of the enclosure. In considering the radiosities of all other surfaces in the enclosure, the total radiation incident on surface a, including itself if it is non-planar, is

$$A_a G_a = \sum_{b=1}^{N} F_{ba} A_b J_b \tag{9.5.7}$$

which, from *reciprocity*, means that

$$A_a G_a = \sum_{b=1}^{N} F_{ab} A_a J_a \tag{9.5.8}$$

Cancelling out the term A_a, and substituting into Eq. (9.5.1) for G_a, we get

$$q_a = A_a \left(J_a - \sum_{b=1}^{N} F_{ab} J_b \right) \tag{9.5.9}$$

From the summation rule

$$q_a = A_a \left(\sum_{b=1}^{N} F_{ab} J_a - \sum_{b=1}^{N} F_{ab} J_b \right) \tag{9.5.10}$$

which reduces to

$$q_a = \sum_{b=1}^{N} A_a F_{ab} (J_a - J_b) = \sum_{b=1}^{N} q_{ab} \tag{9.5.11}$$

In considering the above in terms of an *electrical analogy*, $(J_a - J_b)$ may be considered as the driving potential, or voltage, and $(A_a F_{ab})^{-1}$ as a *space* or *geometrical resistance*, Incropera and DeWitt (2002).

Combining Eqs (9.5.11) and (9.5.6) we get

$$q_a = \frac{E_{ba} - J_a}{(1 - \varepsilon_a)/\varepsilon_a A_a} = \sum_{b=1}^{N} \frac{(J_a - J_b)}{(A_a F_{ab})^{-1}} \tag{9.5.12}$$

Eq. (9.5.12) defines the radiation balance for the radiosity node associated with surface *a*. As shown in Fig. (9.5.2), the above relationship can be demonstrated through, once again, using an analogous electrical circuit. The rate of

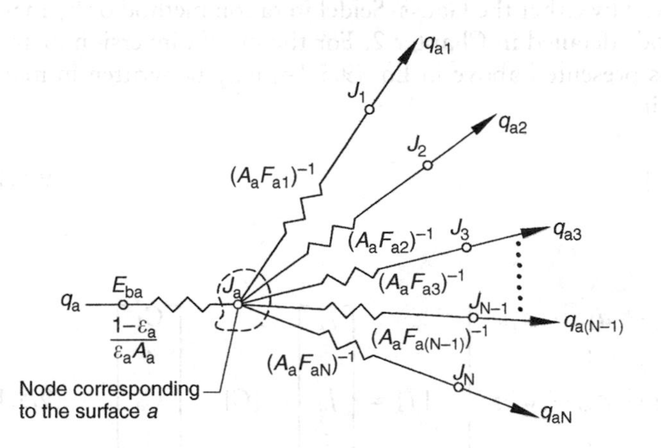

Figure 9.5.2 Network representation of the radiation heat exchange between surface '*A*' and all other surfaces within the enclosure.

radiation transfer, or current flow, through the materials surface resistance to its surface radiosity node, node a, is equal to the net rate of radiation transfer between node a and all other nodes of all other surfaces in the enclosure.

While it is more common for the surface temperatures to be known, in situations where, instead, the net radiation transfer rate is known for surface a, Eq. (9.5.11), rearranged to give the form below, is the more appropriate form of the radiation balance.

$$q_a = \sum_{b=1}^{N} \frac{(J_a - J_b)}{(A_a F_{ab})^{-1}} \tag{9.5.13}$$

Where the temperature of some surfaces is known, Eq. (9.5.12) is written for each of these, and, for the remainder, where q_a is known, Eq. (9.5.13) is applied. This produces a set of N linear algebraic equations which require to be solved for the N unknowns. Where J_a is known, Eq. (9.5.6) may be used to find q_a for each surface of known temperature, and T_a for each surface of known heat flux. The N resultant equations may be represented as follows,

$$\begin{bmatrix} a_{11}J_1 + \cdots + a_{1a}J_a + \cdots + a_{1N}J_N = C_1 \\ \vdots \\ a_{a1}J_1 + \cdots + a_{aa}J_a + \cdots + a_{aN}J_N = C_a \\ \vdots \\ a_{N1}J_1 + \cdots + a_{Na}J_a + \cdots + a_{NN}J_N = C_N \end{bmatrix} \tag{9.5.14}$$

and may be solved by either the Gauss–Seidel iteration method or by matrix inversion methods detailed in Chapter 2. For the matrix inversion method, the equations as presented above in Eq. (9.5.14) may be written in matrix form as shown in

$$[A][J] = [C] \tag{9.5.15}$$

where

$$[A] = \begin{bmatrix} a_{11} \cdots a_{1a} \cdots a_{1N} \\ \vdots \\ a_{a1} \cdots a_{aa} \cdots a_{aN} \\ \vdots \\ a_{N1} \cdots a_{Na} \cdots a_{NN} \end{bmatrix} \quad [J] = \begin{bmatrix} J_1 \\ \vdots \\ J_a \\ \vdots \\ J_N \end{bmatrix} \quad [C] = \begin{bmatrix} C_1 \\ \vdots \\ C_a \\ \vdots \\ C_N \end{bmatrix} \tag{9.5.16}$$

If we express the unknown radiosities as

$$\begin{bmatrix} J_1 = b_{11}C_1 + \cdots + b_{1a}C_a + \cdots + b_{1N}C_N \\ \vdots \\ J_a = b_{a1}C_1 + \cdots + b_{aa}C_a + \cdots + b_{aN}C_N \\ \vdots \\ J_N = b_{N1}C_1 + \cdots + b_{Na}C_a + \cdots + b_{NN}C_N \end{bmatrix} \tag{9.5.17}$$

They may be found from the inverse of matrix $[A]$, $[A]^{-1}$, with $[J] = [A]^{-1}[C]$, where

$$[A]^{-1} = \begin{bmatrix} b_{11} \cdots b_{1a} \cdots b_{1N} \\ \vdots \\ b_{a1} \cdots b_{aa} \cdots b_{aN} \\ \vdots \\ b_{N1} \cdots b_{Na} \cdots b_{NN} \end{bmatrix} \tag{9.5.18}$$

9.5.2 *Radiative heat transfer between gray, diffuse, and opaque surfaces in an enclosure*

The simplest form of enclosure is one in which there are only two surfaces, as shown in Fig. 9.5.3a, which exchange radiation with only each other. The analogous electric circuit is given in Fig. 9.5.3b. As there are only two surfaces, the net gain by one must equal the net loss by the other, thus

$$q_1 = -q_2 = q_{12} \tag{9.5.19}$$

For this simple case the solution is more easily obtained through working with the representative electrical network than with Eq. (9.5.12). By observation we see that the total resistance to radiative heat transfer between the two surfaces is the sum of their respective surface resistances and the

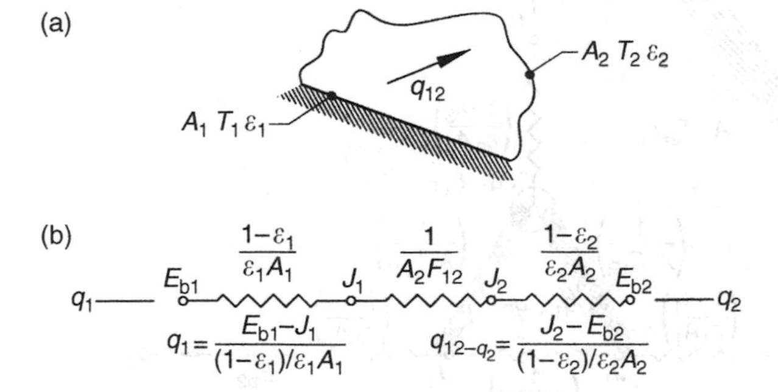

Figure 9.5.3 (a) The two surface enclosure; (b) electrical circuit analogue for the two surface enclosure.

geometrical resistance. Therefore,

$$q_{12} = \frac{E_b}{(1-\varepsilon_1)/\varepsilon_1 A_1 + 1/A_1 F_{12} + (1-\varepsilon_2)/\varepsilon_2 A_2} \tag{9.5.20}$$

The appropriate equations for some common two surface enclosures are given in Table 9.5.1.

For the case of a three surface enclosure, for the specific case of one surface re-radiating, the equivalent analogous electrical circuit is shown in Fig. 9.5.4. Here, the radiation transfer rate may be determined by applying Eq. (9.5.12) to each of the three surfaces. This will yield a set of three equations with three variables which may be solved in a similar manner to those in the preceding section.

Table 9.5.1 Radiant heat transfer relationships for special diffuse, gray, two surface enclosures

Geometry	
Large (infinite) parallel plates	$q_{12} = \dfrac{A\sigma(T_1^4 - T_2^4)}{1/\varepsilon_1 + 1/\varepsilon_2 - 1}$
Long (infinite) concentric cylinders (radiuses r_1 [inner] and r_2 [outer])	$q_{12} = \dfrac{A_1\sigma(T_1^4 - T_2^4)}{1/\varepsilon_1 + (1-\varepsilon_2)/\varepsilon_2\,(r_1/r_2)}$
Concentric spheres (radiuses r_1 [inner] and r_2 [outer])	$q_{12} = \dfrac{A_1\sigma(T_1^4 - T_2^4)}{1/\varepsilon_1 + (1-\varepsilon_2)/\varepsilon_2\,(r_1/r_2)^2}$
Small convex object in a large cavity (cavity area=A_2)	$q_{12} = A_1\sigma\varepsilon_1(T_1^4 - T_2^4)$

Note
For all the above cases, $F_{12} = 1$.

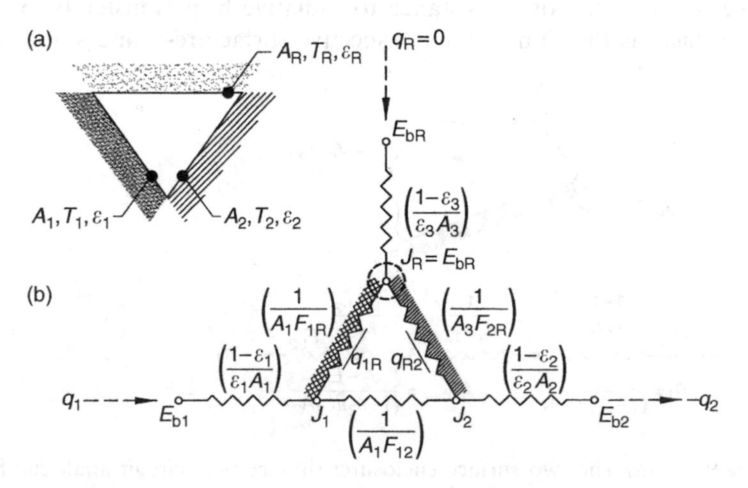

Figure 9.5.4 Three surface enclosure analogous electric circuit.

Where one of the surfaces of the three surface enclosure is well insulated on the outer face, and convection effects may be neglected on the inner, radiating, surface, the net radiation transfer from this surface will be equal to 0 ($G_R = J_R = E_{bR}$). Such a surface, common in many engineering applications is often termed a re-radiating surface. For this special case, again considering Eq. (9.5.12), $q_R = 0$, and the net radiation transfer from the other two surfaces must therefore be equal. Analysis of the network reveals that

$$q_1 = -q_2$$

$$= [E_{b1} - E_{b2}] \Big/ \left[\frac{1-\varepsilon_1}{\varepsilon_1 A_1} + \frac{1}{A_1F_{12} + [(1/A_1F_{1R}) + (1/A_2F_{2R})]^{-1}} + \frac{1-\varepsilon_2}{\varepsilon_2 A_2}\right] \tag{9.5.21}$$

Through solving the above, and applying Eq. (9.3.6) to both surfaces 1 and 2, the respective radiosities may be determined. The radiosity of surface R may then be found from

$$\frac{J_1 - J_R}{1/A_1F_{1R}} - \frac{J_R - J_2}{1/A_{21}F_{2R}} = 0 \tag{9.5.22}$$

As $J_R = E_{bR}$, the temperature of the re-radiating surface is easily determined from the Stefan–Boltzmann law, Eq. (9.1.3).

Example 9.5.1 Radiant heat exchange within an enclosure with two diffuse surfaces.

As part of a manufacturing process a flat plate, at a temperature of 900 K, is drilled to give a series of flat bottomed holes 1 cm in diameter, and 5 cm deep. If the plate is made from a diffuse gray material with surface emissivity of 0.7, determine the rate of radiant heat loss from each hole. If the effective emissivity, ε_e, of the hole is defined as the ratio of the radiant heat flux leaving hole to that would be emitted by a black body at the same temperature, having the same area as the hole aperture, determine the effective emissivity for the holes within the plate. Furthermore, if the depth of the hole were increased, what effect would this have on the effective emissivity. What is the limit of ε_e as depth increases.

Analysis

We must first assume that the hole aperture, area A_2, may be considered to be a black body at 0 K (sky). This implies that all radiation incident on this surface results from that emitted by the cavity, and escapes to the surroundings. Therefore, if $\varepsilon = 1, E_{b2} = J_2 = 0$.

It is possible to use the coaxial disc view factor relation to define the respective view factors. However, a simpler method is to use the reciprocity relation. If we define area A_1 as being the hole internal surface area, (base and cylindrical bore), by reciprocity $A_1F_{12} = A_2F_{21}$, and, as $F_{21} = 1$ and A_2 is simply the aperture area, F_{12} is easily determined.

As $E_{b2} = 0$, from the equivalent thermal circuit, the rate of radiation loss through the hole aperture is given by

$$q_1 = \frac{E_{b1}}{[(1-\varepsilon_1)/\varepsilon_1 A_1 + 1/A_1F_{12} + (1-\varepsilon_2)/\varepsilon_2 A_2]}$$

From the definition given above, the effective emissivity can be found from

$$\varepsilon_e = \frac{q_1}{A_2\sigma T_1^4}$$

Solution

Open the workbook Ex09-05-01.xls. The workbook has only one worksheet, namely, *Main*. Input the data given in **cells (B4:B9)**. Excel performs the calculations detailed in the above section. For the initial conditions, 2.9 W, **cell B18**, are lost by radiant heat transfer to the surroundings, and the effective emissivity is 0.98. The effect of increasing hole depth on ε_e can be studied by increasing the value in **cell B8**. As depth increases the effective emissivity is also seen to increase, that is, for $L/D = 20, \varepsilon_e = 0.995$.

Discussion

Where the integrity of design requires the use of a specific material, the effective surface emissivity may be increased through the drilling of small holes. As the depth of the hole increases, the term, $1 - \varepsilon_1/\varepsilon_1 A_1$ goes to zero, while $1/A_1F_{12}(= 1/A_2F_{21})$ predominates (as surface 2 is black with $\varepsilon = 1$, the term $1 - \varepsilon_2/\varepsilon_2 A_1 = 0$). Hence, as L increases, q_1 tends to $A_2F_{21}E_{b1}$, and ε_e tends to 1.

Example 9.5.2 Radiant heat exchange in a diffuse three surface enclosure with an adiabatic wall.

In the previous example we considered the exchange of radiant heat in an enclosure with one black and two diffuse surfaces. For the purpose of the analysis required, by reciprocity and a commonality of surface properties, the two diffuse surfaces could be considered as a single surface. This somewhat simplified the analysis. In the present example, we consider the case of a three surface enclosure, in which all surfaces are diffuse and gray, and have different emissivities, with one of the surfaces being well insulated. The problem is presented and the method of solution outlined to allow a fuller understanding of the effect and interaction of the various parameters involved.

A small kiln has the form of a truncated cone with a base diameter of 0.7 m (surface 1), vertical height of 1.0 m and upper diameter of 0.3 m, (surface 2).

All surfaces may be assumed to be diffuse-gray. The lateral wall (surface 3) is perfectly insulated with an emissivity of 0.3. If the surface of the kiln base has an emissivity of 0.7 and is maintained at 1000 K via an internal electrical heating element of rating 3 kW/m^2, determine the temperature of surfaces 2 and 3 if the emissivity of surface 2 is 0.5.

Analysis

As previously, we again require to perform a surface energy balance to afford a solution to the problem and determine the temperatures of surfaces 2 and 3. In Example 9.3.2 we were required to solve for three unknowns (J_1, J_2, and J_3) and used the matrix inversion method presented in Chapter 2. Presently, as one of the surfaces is adiabatic, (surface 3) and the heat flux from the kiln base is defined, the problem is simplified and we need only solve for two unknowns, J_1 and J_2.

For the three surface enclosure, the radiation surface energy balances are as follows

$$\frac{E_{b1} - J_1}{1 - \varepsilon_1/\varepsilon_1 A_1} = \frac{J_1 - J_2}{1/A_1F_{12}} + \frac{J_1 - J_3}{1/A_1F_{13}}$$

$$\frac{E_{b2} - J_2}{1 - \varepsilon_2/\varepsilon_2 A_2} = \frac{J_2 - J_1}{1/A_2F_{21}} + \frac{J_2 - J_3}{1/A_2F_{23}}$$

$$\frac{E_{b3} - J_3}{1 - \varepsilon_3/\varepsilon_3 A_3} = \frac{J_2 - J_1}{1/A_3F_{31}} + \frac{J_2 - J_3}{1/A_3F_{32}}$$

However, as

$$q_1 = \frac{E_{b1} - J_1}{1 - \varepsilon_1/\varepsilon_1 A_1}$$

and

$$q_3 = 0 \text{ (perfectly insulated surface)}$$

$$q_2 = -q_1$$

J_1 may be found directly via the fourth of the above equations, with J_2 and J_3 subsequently solved by rearranging the first and second (or second and third, etc.) and substituting the values obtained for q_1, q_2 (W), J_1(W/m^2), and the respective view factors and surface areas. The view factors can again be found from the coaxial parallel disc relation. (A simple formula for the area of the lateral surface can be derived.) From J_2 and J_3, with knowledge of the respective surface properties, the required temperatures can be defined.

Solution

Open the workbook Ex09-05-02.xls. The workbook comprises one worksheet, namely *Main*. Input the data given in **cells (B4:B11)**. Excel performs all necessary calculations and provides the required solution in cells (**B29** and **B30**). T_2 is found to be 776.4 K, while the temperature of the adiabatic surface is 983.2 K.

Discussion

From the data provided and problem posed, the view factors required for the solution of Example 9.5.1 were found by reciprocity. In the present example, while knowledge of the view factor F_{33} was not directly required for the solution, it has been included, cell **B25**. It is immediately apparent that it 'sees' more of itself than either of the kiln ends. In many thermal radiation problems it will be necessary to define all view factors, and, where, for a given surface, the emissivity varies across that surface, consider the effects of 'self' view on radiant heat transfer within the system.

9.6 Radiation shields

If we wish to reduce the radiation heat loss from a surface to its surroundings, such as, for example, from the outer walls of a furnace, (surface 1), to those of the building in which it is housed (surface 2) we can place a low emissivity (high reflectivity) material between them. Such a material is termed a radiation shield. If the radiation shield is placed adjacent to the furnace outer wall, and is of the same area, the view factor will effectively be equal to one. Without a shield, the net radiative heat transfer rate between the furnace and building walls is given by

$$q_{fs} = \frac{A_1(E_{b1} - E_{b2})}{1/\varepsilon_1 + 1/\varepsilon_2 - 1} \tag{9.6.1}$$

Including a shield (surface 3) adds additional resistance to heat transfer. Considering Fig. 9.6.1a, and the associated resistance network, Fig. 9.6.1b, the net radiative heat transfer rate between the furnace and building walls will now be given by

$$q_{12} = \frac{A_1\sigma(T_1^4 - T_2^4)}{1/\varepsilon_1 + 1/\varepsilon_2 + (1 - \varepsilon_3)/\varepsilon_3 + (1 - \varepsilon_4)/\varepsilon_4} \tag{9.6.2}$$

In this network representation, E_{b3} is shown as being equal to E_{b4}. This implied that the temperature of the inner and outer surfaces of the shield is the same. It can be seen from Eq. (9.6.2) that when the surfaces of the shield have a low emissivity, the associated resistance, being the third and fourth terms of the denominator is high. As with the shield present $q_{12} = q_{13} = q$, if T_1 and T_2 are known, it is possible to determine the temperature of the shield by expressing Eq. (9.6.1) for q_{13} or q_{32}. The presence of a radiation shield can also aid in suppressing convective heat loss.

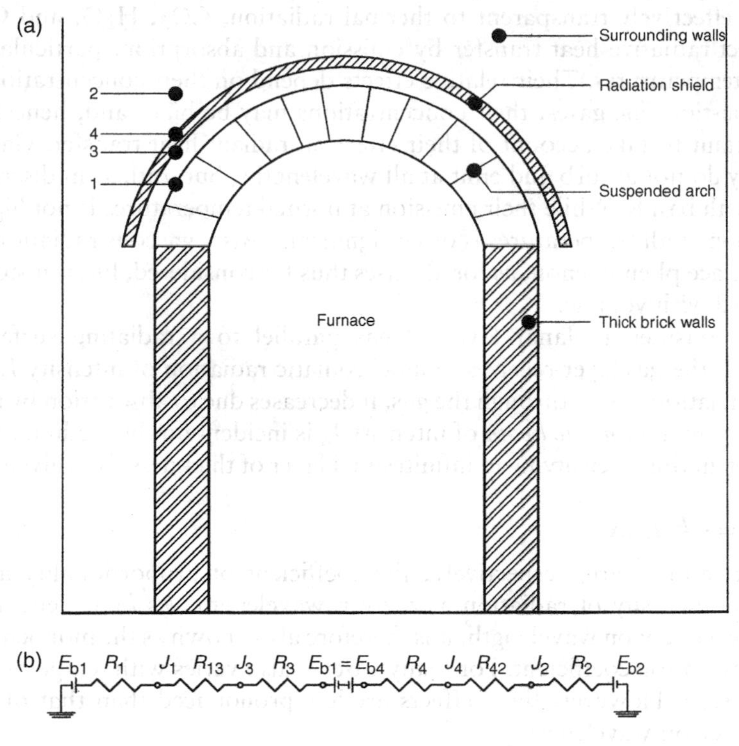

Figure 9.6.1 (a) An arched furnace with thermal radiation shield; (b) radiation shield equivalent electrical circuit.

9.7 Radiation in an absorbing, emitting medium

So far we have only considered radiative heat exchange between surfaces in the presence of a non-participating medium. We have assumed uniform radiosity and irradiation for isothermal gray surfaces that emit and reflect diffusely. In most cases these assumptions and the related equations presented may be applied to obtain accurate results for radiation transfer in an enclosure. However, the assumption of a non participating medium is only valid when the medium within the enclosure or between two surfaces comprises monatomic and/or diatomic symmetric molecules. We now briefly consider the radiation properties of participating media. For a more detailed and advanced discussion on the appropriate methods the reader is directed to the relevant literature (Hottel and Sarofim, 1967; Siegel and Howell, 1992).

9.7.1 Radiation properties of gases

This subject is gaining importance due to the global environmental problem of the green house effect and the role of CO_2 within that framework.

Considering air, while both N_2 and O_2 are diatomic symmetric molecules, and are effectively transparent to thermal radiation, CO_2, H_2O, and CO can affect radiative heat transfer by emission and absorption, particularly at high temperatures. Their relative effects depend on their concentrations. In combustion flue gases, their concentrations may be high, and, hence, it is important to take account of their effect on radiant heat transfer. Gases generally do not absorb and emit at all wavelengths, but, rather, in discrete wavelength bands. While their emission at normal temperatures is not high, absorption at all temperatures is often significant. Also, gaseous radiation is not a surface phenomenon, as for all cases thus far considered, but is instead associated with volume.

If we consider a planar layer of gas parallel to a radiating surface, Fig. 9.7.1, the gas layer receives monochromatic radiation of intensity $I_{\lambda,x}$. As the radiation passes through the gas, it decreases due to absorption by the gas. If a *monochromatic beam* of intensity I_λ is incident on the medium, the reduction in this intensity in an infinitesimal layer of thickness dx is given by

$$\mathrm{d}I_\lambda = -k_\lambda I_\lambda \,\mathrm{d}x \tag{9.7.1}$$

where k_λ and I_λ are, respectively, the coefficient of proportionality and the initial intensity of radiation at a given wavelength. As k_λ is generally highly dependent on wavelength, it is therefore also known as the monochromatic absorption coefficient. For many gases it also varies with temperature and pressure. However, these effects are less pronounced than that of its dependency on wavelength.

Rearranging Eq. (9.7.1), and integrating over the whole layer we get

$$\int_{I_{\lambda,0}}^{I_{\lambda,L}} \frac{\mathrm{d}I_\lambda(x)}{I_\lambda(x)} = -k_\lambda \int_0^L \mathrm{d}x \tag{9.7.2}$$

If k_λ is assumed to be independent of x then

$$\frac{I_{\lambda,L}}{I_{\lambda,0}} = \mathrm{e}^{-k_\lambda L} \tag{9.7.3}$$

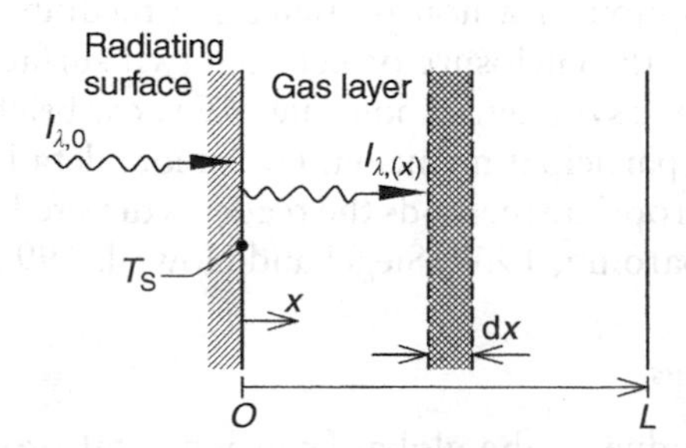

Figure 9.7.1 A planar layer of gas parallel to a radiating surface.

This exponential decay is called *Beer's Law*, and may be used to infer the total spectral absorptivity of the medium. Defining the transmissivity as

$$\tau_\lambda = \frac{I_{\lambda,L}}{I_{\lambda,0}} = e^{-k_\lambda L} \tag{9.7.4}$$

As the reflectivity is assumed to be zero, the absorptivity is therefore given by

$$\alpha_\lambda = 1 - \tau_\lambda = 1 - e^{-k_\lambda L} \tag{9.7.5}$$

If the surface is gray, with constant emissivity for all wavelengths, then

$$I_{\lambda,0} = \varepsilon I_{b\lambda}(T_S) \tag{9.7.6}$$

It therefore follows that

$$\int_0^\infty I_{\lambda,0}\, d\lambda = \varepsilon I_b(T_S) \tag{9.7.7}$$

In respect of the emissivity of the gas, it can be shown that the emission from the gas per unit area of the surface is expressed as

$$E_g = \varepsilon_g \sigma T_g^4 \tag{9.7.8}$$

where

$$\varepsilon_g(T_g) = \frac{\int_0^\infty \varepsilon_\lambda E_{b\lambda}(T_g)\, d\lambda}{E_b(T_g)} \tag{9.7.9}$$

However, as both the emissivity and absorptvity are highly wavelength dependent, and, as stated earlier, to a lesser extent dependent on temperature and pressure, the gas emissivity is generally obtained from a regression of available data.

9.7.2 *Radiation properties of water*

While the transmission of short wavelength radiation (visible spectrum) is high for some liquids, for most, their transmissivity is generally much lower for wavelengths greater than 2 μm. Many liquids are essentially opaque, with the incident radiation being absorbed in a thin surface layer. Emission from the interior of liquids is therefore generally ignored (Suryanarayana, 1995). As for emissivity of gases, the emissivity of water is also obtained from the regression of data for different pressures and temperatures.

Where the medium comprises both a gas, for example, CO_2, and water vapour, the total gas emissivity may be expressed as

$$\varepsilon_g = \varepsilon_w + \varepsilon_{CO_2} - \Delta\varepsilon \tag{9.7.10}$$

where $\Delta\varepsilon$, the correction factor, again a function of temperature and pressure, is obtained from the regression of experimental data based on the

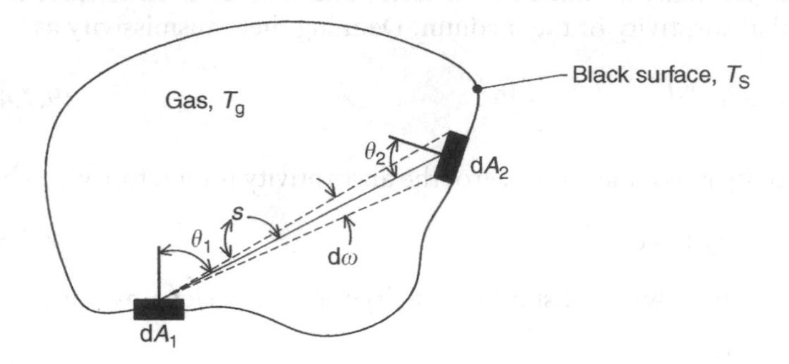

Figure 9.7.2 Surface incident energy for gaseous emission.

respective partial pressures of the participating gases. This correction factor accounts for the reduction in emission resulting from the mutual absorption of radiation by the two above mentioned species.

9.7.3 *Energy exchange between an enclosure and a gas*

Let us consider the case of radiant heat transfer from a gas at temperature T_g to a surrounding black surface at temperature T_S. All energy emitted by the gas that reaches the surface will be absorbed. The heat transfer rate will equal that emitted by the gas that reaches the surface less that emitted by the surface and absorbed by the gas. Considering Fig. 9.7.2, it can be shown that the total rate of energy incident on an infinitesimal area of the surface, dA_1, due to gaseous emission, may be expressed as

$$dQ_{ge} = \int_{A_S} \frac{\varepsilon_g(s, T_g) I_b(T_g) \cos\theta_1 \cos\theta_2 \, dA_2 \, dA_1}{s^2} \tag{9.7.11}$$

The total incident energy on the solid surface due to emission by the gas is given by

$$Q_{ge} = E_b(T_g) \int_{A_S} \int_{A_S} \frac{\varepsilon_g(s, T_g) \cos\theta_1 \cos\theta_2 \, dA_1 \, dA_2}{\pi s^2} \tag{9.7.12}$$

This may also be expressed as

$$Q_{ge} = E_b(T_{gs}) \varepsilon_g(L, T_g) A_S \tag{9.7.13}$$

Thickness L is the mean beam length of the gas. Rearranging Eqs (9.7.12) and (9.7.13) above we get

$$\varepsilon_g(L, T_g) = \frac{1}{A_S} \int_{A_S} \int_{A_S} \frac{\varepsilon_g(s, T_g) \cos\theta_1 \cos\theta_2 \, dA_1 \, dA_2}{\pi s^2} \tag{9.7.14}$$

Table 9.7.1 Mean beam length for selected common geometries

Geometry	Characteristic length	Beam length L_e
1. Infinite circular cylinder (radiation to curved surface)	Diameter '*D*'	0.95 *D*
2. Semi-infinite circular cylinder (radiation to base)	Diameter '*D*'	0.65 *D*
3. Circular cylinder of equal height and diameter (radiation to entire surface)	Diameter '*D*'	0.60 *D*
4. Sphere (radiation to surface)	Diameter '*D*'	0.65 *D*
5. Infinite parallel planes (radiation to planes)	Plane spacing '*L*'	1.80 *L*
6. Cube (radiation to any surface)	Side '*L*'	0.66 *L*
7. Arbitrary shape of volume '*V*' (radiation to surface of vol.)	Volume to Area ratio '*V*/*A*'	3.6 *V*/*A*

Eq. (9.7.14) may be considered to be the integrated mean emissivity of the gas. This may be equated to the emissivity of a gas of planar thickness *L*. Therefore, via Eq. (9.7.14), the mean beam length may be determined for different enclosures for different temperatures. Hence, the emissivity, average emissive power, and therefore the total emitted power may also be defined.

As we stated earlier, emissivity is a function of pressure. As it is also a function of the mean beam length it is often appropriate to consider it as function of the pressure/mean beam length product, PL. Table 9.7.1 gives details of the mean beam length for a few common geometries.

With knowledge of the PL product, we can calculate the emissivity for different temperatures. The results obtained, as demonstrated by Hottel and Egbert (1942), are easily presented in graphical form. If the total pressure is not equal to 1 bar, the values obtained for the emissivity from the above mentioned graphs should be multiplied by the appropriate pressure correction factor C, again obtained via graphical interpretation. For a mixture of water vapour and CO_2, for pressures unequal to 1 bar, Eq. (9.7.10) becomes

$$\varepsilon_g = C_W \varepsilon_W + C_{CO_2} \varepsilon_{CO_2} - \Delta\varepsilon \tag{9.7.15}$$

In considering the absorptivity of the mixture we must also take account of the temperature and wavelength dependence of α. We first compute the emissivity at the temperature of the solid surface, and then correct it through the following equations:

$$\alpha_g = \alpha_W + \alpha_{CO_2} - \Delta\alpha \tag{9.7.16}$$

where $\Delta\alpha = \Delta\varepsilon$

$$\alpha_W = C_W \left(\frac{T_g}{T_S}\right)^{0.45} \varepsilon_W \left(T_S, P_W L, \frac{T_S}{T_g}\right) \tag{9.7.17}$$

$$\alpha_{CO_2} = C_{CO_2} \left(\frac{T_g}{T_S}\right)^{0.65} \varepsilon_{CO_2} \left(T_S, P_{CO_2} L, \frac{T_S}{T_g}\right) \tag{9.7.18}$$

To evaluate the absorptivity, the emissivities of the component gases should be evaluated at the solid surface temperature T_S, corresponding to $PL(T_S/T_g)$, where P is the partial pressure of the gas.

Now, with knowledge of both the emissivity and absorptivity of the gas, for a black surface, the total rate of radiant energy emitted by the gas may be calculated from

$$q = \varepsilon_g A_S \sigma T_g^4 \tag{9.7.19}$$

and the net radiative heat transfer between the surface and the gas is obtained from

$$q_{net} = A_S \sigma (\varepsilon_g T_S^4 - \alpha_g T_S^4) \tag{9.7.20}$$

For a gray surface, computations become more complex. However, Hottel (1954) suggests that, for surfaces with $\varepsilon > 0.8$ the following relationship may be applied

$$q_{gray} = q_{black} \frac{\varepsilon + 1}{2} \tag{9.7.21}$$

Problems

Problem 9.1

Determine the energy flux emitted from a blackbody in the waveband $\lambda_{max} - 1$ to $\lambda_{max} + 1\,\mu m$ if the temperature of the blackbody surface is (a) 300 K, (b) 600 K, (c) 900 K, (d) 1500 K, and (e) 3000 K.

Problem 9.2

The incident solar radiation on a panel is 950 W/m^2. If the panel back is perfectly insulated, determine the panel surface temperature for (a) the panel surface being black, and

(b) $0 < \lambda < 1\,\mu m$, $\quad \varepsilon_\lambda = 0.8$

$1 < \lambda < 4.0\,\mu m$, $\quad \varepsilon_\lambda = 0.4$

Problem 9.3
The monochromatic reflectivity of a special black surface may be simplified as,

$$0 < \lambda < 1.5\,\mu\text{m}, \qquad \rho_\lambda = 0.8$$
$$1.5 < \lambda < 4.0\,\mu\text{m}, \qquad \rho_\lambda = 0.4$$
$$4.0\,\mu\text{m} < \lambda < \infty, \qquad \rho_\lambda = 0.75$$

Determine the total absorptivity of the surface for (a) solar radiation, and (b) irradiation from a black surface at 800 K.

Problem 9.4
A circular ice ring of diameter 32 m is housed within a hemispherical dome of radius 15 m. If the surface temperature of the ice is 0°C and that of the dome 17°C, calculate the net rate of radiant heat transfer from the dome to the rink.

Problem 9.5
Two coaxial discs are separated by a distance of $L = 0.25$ m. The lower disc has a diameter D_1 of 0.6 m and a surface temperature of 28°C. The upper disc also has a diameter of 0.6 m, but is however ring shaped with an inner diameter of 0.15 m having been removed. The temperature of the upper disc is 350°C. Assuming the discs to be black bodies, calculate the net radiative heat exchange between them.

References

Hottel, H. C. and Egbert, R. B. (1942) Radiant heat transmission from water vapour, *AIChE Trans.* 38, 531.

Hottel, H. C. and Sarofim, A. F. (1967) *Radiative Transfer*, McGraw-Hill, New York.

Hottel, H. C. (1954) Radiant heat transmission, in: McAdams, W. H. (ed.), *Heat Transmission*, 3rd edn, McGraw-Hill, New York.

Howell, J. R. (1982) *A Catalog of Radiation Configuration Factors*. McGraw-Hill, New York.

Incropera, F. P. and DeWitt, P. (2002) *Fundamentals of Heat and Mass Transfer*, 5th edn, John Wiley and Sons, New York.

Siegel, R. and Howell, J. R. (1992) *Thermal Radiation Heat Transfer*, 3rd edn, Hemisphere, Washington, D.C.

Sparrow, E. M. and Minkowycz, W. J. (1962) Heat transfer characteristics of several radiator finned-tube configurations, NASA, Washington, D.C. (NASA $TN_{D\text{-}1435}$).

Suryanarayana, N. V. (1995) *Engineering Heat Transfer*, West Publishing Company, New York.

10 Multi-mode heat transfer

The examples of combined heat transfer considered in the previous chapters were not restricted to those cases in which only one mode of transfer was operational. However, primary emphasis was placed on the application of conduction, convection or radiation and the relevant analytical development. The isolation of each mode helps us to identify the mode of heat transfer and to develop the analytical and computational skills required to estimate the heat transfer by each mode. This chapter will bring together the material developed in the previous chapters with view to present a comprehensive treatment of combined modes of heat transfer. This chapter thus represents the culmination of knowledge gained through study of Chapters 1–9. A firm grasp of the physics of conduction, radiation, and boundary layer theory is thus essential before proceeding with this chapter.

Thermal radiation heat transfer usually occurs in combination with convective heat transfer when the space adjacent to the surfaces is filled with any gaseous medium. One exception is however space applications where energy transfer can take place by conduction and/or radiation. If the convective heat transfer is by natural convection to gases or by forced convection to gases at high temperatures, radiation heat transfer cannot be neglected. By common experience the reader will acknowledge the fact that indeed in most situations, be they examples of heat transfer occurrence within a domestic household, or in the design of simple or complex engineering systems, thermal transmission takes place by a variety of combination of conduction, convection, and radiation.

Problems introduced in earlier chapters on conduction, convection, and thermal radiation emphasised the difference of approach required for their solution. As both conductive and convective heat transfer rates are related to the difference in temperatures, the resulting equations are linear and, hence, the solution is straightforward unless the thermal conductivity or the convective heat transfer coefficient is strongly temperature dependent. If thermal radiation heat transfer is also significant, the equations are then non-linear owing to the fact that radiation heat transfer is proportional to the difference in the fourth power of the temperatures.

A familiar example of transfer of heat by a combination of all of the three modes is the case of pipe carrying steam or hot water, suspended within a room. Such contraptions are common within a boiler plant room. The outside surface temperature of the pipe is not specified. Instead, the temperature of the water/steam substance flowing inside the pipe is the known parameter. The determination of heat loss from the pipe surface then requires a trial and error solution. The iterative solution of such problems converges less rapidly when the thermal radiation is included than when it is neglected, because the radiation coefficient is much more sensitive to changes in the surface temperature than is the convective coefficient. Needless to say that a calculator-based solution is quite laborious in such instances, although an attempt may be made by putting to good use the algorithms for solution of non-linear algebraic equations presented in Chapter 2.

The examples presented in this chapter represent an increasing complexity of analysis, that is, starting from a much simpler situation which can be analysed with an electronic calculator solutions are presented for more involved problems that warrant use of VBA subroutines and other sophisticated tools available within the Excel environment.

10.1 Simultaneous convection and radiation

Example 10.1.1 Heated plate placed in a horizontal or vertical position.

Consider a heated plate of 1 m length and 0.6 m width whose surface is maintained at a uniform temperature. The plate is placed inside a room whose walls, ceiling, and ambient air are maintained at a temperature of 20°C. Investigate the heat given off by this plate by means of convection and thermal radiation for the following cases:

(a) The plate is placed horizontally at a temperature of 90°C and transfers heat through natural convection and radiation.
(b) Repeat problem (a), but with a plate temperature of 500°C.
(c) The plate is placed vertically (along the length) at a temperature of 90°C and transfers heat via natural convection and radiation.
(d) Repeat problem (c), but with a plate temperature of 500°C.
(e) Ambient air blows over the plate (along the length) at a velocity of 30 m/s, the plate being heated to 90°C.
(f) Repeat problem (d), but with a plate temperature of 500°C.

Assume that the thermal radiation emissivity of the plate is 0.8 and that the walls have an emissivity of 1.

Analysis

Recall that the physical formulation for forced- and free-convection, and thermal radiation were respectively presented in Chapters 7–9. In the present problem we may base our calculations on material presented in Sections 7.2.1 and 8.2.1, respectively for external forced- and free-convection.

The solution strategy to be adopted is quite straight-forward, that is, the heat transfer via convection and radiation will be computed separately, then the sum of these obtained. For convection calculations, all transport properties are to be obtained at the film temperature (see Chapters 7 and 8).

Solution

Refer to workbook Ex10-01-01.xls which contains three worksheets – *'Compute'*, *'Properties'*, and *'Pasted information'*. The given data in worksheet *'Compute'* may be linked to library functions that are presented in that worksheet. It is important that after entering each given plate temperature in **cell B6** the user copies the output information in the worksheet *'Pasted information'* or else the information will be lost soon after any changes are made for subsequent cases.

Discussion

We note that:

- The horizontal plate, facing upwards dissipates energy at a slightly higher rate than the vertical plate.
- At lower temperatures, the convective component is almost equal to the radiative part of heat transfer.
- Forced convection at moderate velocities increases heat transfer over free convection, but the difference is only modest for high temperatures.

10.2 Simultaneous forced-convection and radiation

Example 10.2.1 Superheated steam production within a boiler tube.

Culp (1991) and Weston (1992) have provided classification of steam boilers. It is customary to classify these as 'fire-' or 'water-tube' steam boilers. A fire-tube boiler is essentially composed of a water-filled vessel containing a large number of tubes through which hot gases arising from combustion process flow. Heat is thus transferred from the hot gases to the water in the vessel. These types of boilers operate at relatively low pressures, the normal limiting pressure being around 1.7 MPa. Other than the low-operating pressures, the fire-tube design also suffers from potential problem of high-pressure water flowing into the combustion chamber should a tube failure occur. This type of plant is however, simple in design and less expensive. Figure 10.2.1(a) shows a schematic of a fire-tube boiler.

In the water-tube boiler, high-pressure water and steam flows from water/steam drum through tubes or in tube bundles mounted in the exhaust duct. A typical scheme is shown in Fig. 10.2.1(b). The tubes within this design range from 0.05 to 0.1 m in diameter and can handle pressures of up to 35 MPa.

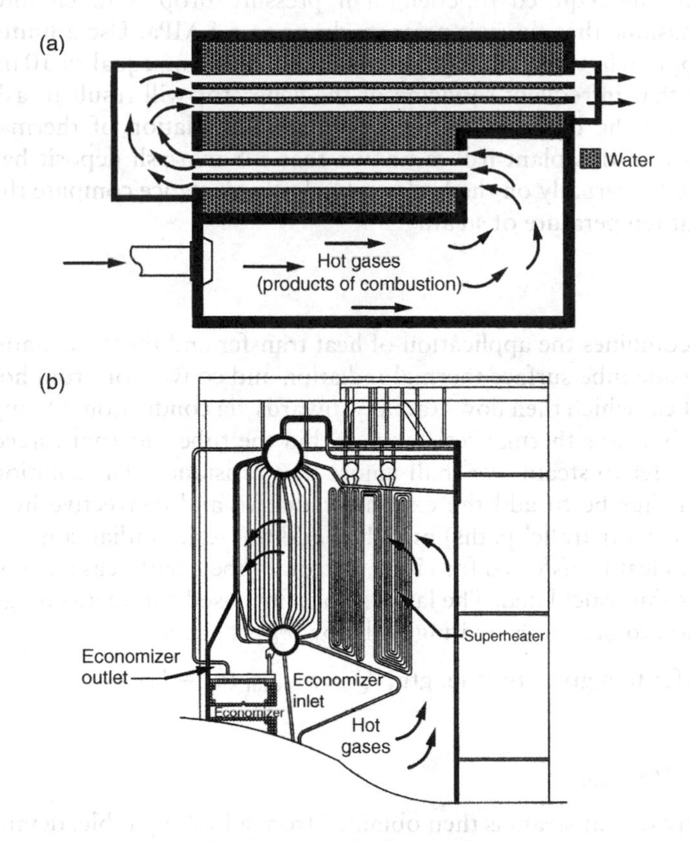

Figure 10.2.1 (a) Schematic of a fire-tube and (b) water-tube boiler.

The water-tube steam generators may further be classified as either free- or forced-convection systems. Free-convection systems are not suitable for operational pressures greater than 16.5 MPa, as the density differences between saturated water and steam decreases to a point where the buoyancy force required for circulation becomes too small. The forced-convection systems are easily operable at elevated pressures and thus offer higher plant thermal efficiencies.

Consider a simplified representation of a single boiler tube within the superheater section of a fire-tube steam generator. Products of combustion flow upwards at a temperature of 1600 K while the generator walls are maintained at 1400 K. The convection heat transfer coefficient is estimated to be 100 W/m^2 K. Steam enters the tube at a condition of 5.9 MPa and 306°C.

1 Treating the entire tube-length as an aggregate system, obtain the tube length required to raise the temperature of steam to 577°C. Assume that there is no drop in pressure between the inlet and outlet sections.

2 Re-estimate the required tube-length if pressure drop is taken into account. Assume that the exit pressure drops to 5.8 MPa. Use a finite-element approach, with the length of each element being equal to 10 m.
3 It is likely that infrequent servicing of the generator will result in ash deposition on the boiler tubes. Evaluate the degradation of thermal performance of this plant if it is known that a 5-mm ash deposit has accumulated externally on the boiler tube. In this instance compute the drop in exit temperature of steam.

Analysis

This problem combines the application of heat transfer and thermodynamics. On the outside tube surface thermal radiation and convection from hot gases impinge heat which then flows radially inwards via conduction. Owing to a significantly lower thermal resistance within the tube (internal forced convection transfer to steam) we shall ignore that resistance. The solution strategy would thus be to add the external radiative and convective heat transfer coefficients (parallel paths) and then add the series radial conduction resistance. Heat transferred for either the total tube length (case (a)) or each segment is thus calculated. The latter step is then used within an energy balance equation to obtain the exiting enthalpy (h_e) of steam,

Heat transfer to a given tube length, $Q = m_{steam}\ (h_e - h_i)$.

Thus,

$$h_e = h_i + Q/m_{steam}$$

The temperature of exit steam is then obtained from a look-up table, details of which are given below.

Solution

Excel workbook Ex10-02-01.xls presents the solution to this multi-mode heat transfer problem. There are six worksheets within this workbook. The worksheet *'statement'* presents a colour schematic of the given problem along with the given data, while worksheets *'cases (a) + (b)'* and *'case (c)'* present the solutions to the respective parts of the problem. The remainder of the three worksheets provide thermodynamic and transport property data, the former having been obtained from the 'Allprops' software that is included in the companion CD.

Computations for case (a) are relatively straightforward and are provided in **row 12**. The radiative transfer is computed and added to the convective heat transfer in **cell E12**. The total length of the boiler tube required is found to be 51.6 m. Case (b) requires the use of finite elements, with the pressure drop, gain in enthalpy, and rise of temperature computed for each 10 m sections of the tube. The answer in this case is found to be 50 m for the total length of the tube.

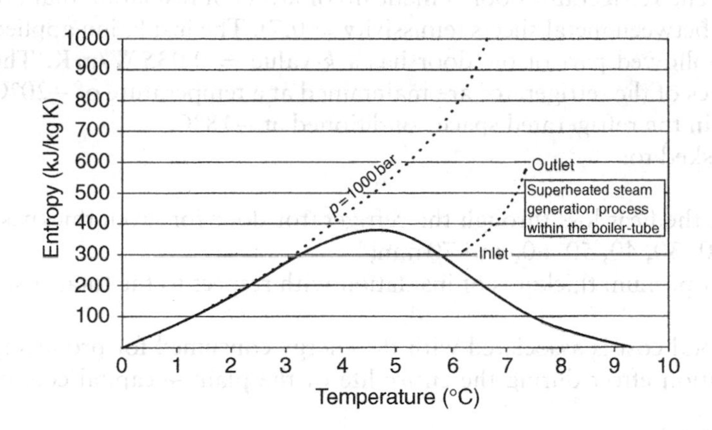

Figure 10.2.2 Temperature-entropy chart for steam including the boiler-tube superheating process.

The effect of ash deposit on the boiler tube performance is investigated in worksheet *'case (c)'*. The tube diameter with and without the ash deposit is shown in **cells B2** and **B10** respectively. The exit temperature from the tube is now downgraded to 468°C.

It is also possible to show the above thermodynamic process for steam generation on a *T–s* chart. This is shown in Fig. 10.2.2. Using the 'Allprops' software provided in the CD it is possible to create coincident saturation '*T*' and '*s*' values for any one of the forty substances database included within Allprops. One such workbook is Steamchart1.xls. **Cells (A768:B773)** contain data for the boiler tube steam generation process. The latter cell range can be expanded or reduced as the problem dictates. All other data, required for producing the *T–s* curve and the present $p = 1000$ bar boundary need not be interfered with.

Discussion

Allprops enables the user to generate other thermodynamic charts such as *h–s* or *p–h* diagrams and heat exchanger processes displayed to scale.

The user may wish to perform further numerical experimentation by means of reworking above cases, but with the use of LMTD (Eq. (7.2.73)).

10.3 Simultaneous free-convection and radiation

Example 10.3.1 Heat exchanges occurring from a refrigerator door.

A warehouse refrigerator door has a height of 2 m and width of 2 m. The ambient air temperature of the warehouse is 30°C while the weighted-average temperature of the warehouse walls and windows may be assumed to be 35°C. The refrigerator door is made up of layers of insulation that are sandwiched between metal sheets (emissivity = 0.7). The insulation applied within the hollowed part of the door has a k-value = 0.035 W/m K. The inside surfaces of the refrigerator are maintained at a temperature of −20°C with the air in the refrigerated space conditioned at −18°C.

You are asked to:

- estimate the heat loss through the refrigerator door for given thickness of 10, 20, 30, 40, 50, 60, and 70 mm;
- find the optimum thickness of insulation with respect to the total cost.

Note: The total cost is associated with the energy consumed for producing the refrigeration effect during the entire life of the plant + capital cost of insulation.

Additional data:

- first cost of refrigeration equipment = 1,000 Euro/kW refrigerating capacity,
- overall COP of refrigeration plant = 2,
- insulation cost = 24 Euro/m^2 for each 10 mm thickness of insulation,
- plant life = 10 years.

Analysis

With reference to the thermal circuit presented within Fig. 10.3.1 and worksheet *'Diagram'*, the governing equations for this problem are:

$$h_o(T_{\infty,o} - T_{do}) + \varepsilon\sigma(T_{so}^4 - T_{do}^4) = (T_{do} - T_{di})/R_{c,d}$$

and

$$h_i(T_{di} - T_{\infty,i}) + \varepsilon\sigma(T_{di}^4 - T_{si}^4) = (T_{do} - T_{di})/R_{c,d}$$

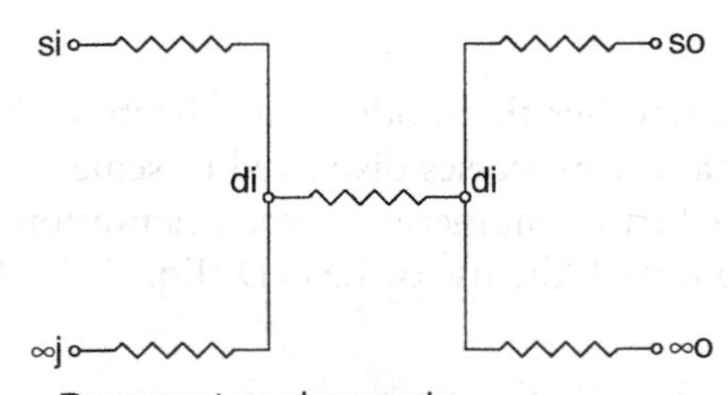

Figure 10.3.1 Thermal circuit for warehouse refrigerator door (Example 10.3.1).

Note
Notation for nodes provided within example text.

Thus there are two equations and two unknowns, T_{di} and T_{do}. The solution for the above non-linear equations is further complicated by the fact that evaluation of h_i and h_o is dependent on thermophysical properties for air that are in turn functions of T_{di} and T_{do}.

Let us assign the following notation for ease of computation within Excel.

$$\text{Term1} = h_o(T_{\infty,o} - T_{do}) + \varepsilon\sigma(T_{so}^4 - T_{do}^4)$$

$$\text{Term2} = (T_{do} - T_{di})/R_{c,d}$$

$$\text{Term3} = h_i(T_{di} - T_{\infty,i}) + \varepsilon\sigma(T_{di}^4 - T_{si}^4)$$

$R_{c,d}$ is the thermal conductive resistance = insulation thickness/thermal conductivity.

It is now possible to set up an optimisation problem as follows:

$$\text{Sum of square of errors, SSE} = (\text{Term1} - \text{Term2})^2 + (\text{Term3} - \text{Term2})^2$$

SSE represents the objective function that needs to be minimised against the independent variables T_{do} and T_{di}.

Solution

Open the workbook Ex10-03-01.xls. Note that there are three worksheets in this workbook – '*Diagram*', '*Compute*', and '*Properties*' and these respectively include the thermal circuit and governing equation, all computations, and transport property look-up table for air substance.

The **Solver** facility within Excel may be invoked to obtain solution to the above problem and this is demonstrated in the workbook Ex10-03-01.xls. For each given thickness, w and for each assumed value of T_{do} and T_{di} sequential computations of the respective outdoor and indoor (with respect to the refrigerated space) film temperature and free convection heat transfer coefficient h_o and h_i are carried out. These are respectively shown in **cells (C15:G21)** and **(I15:M21)**. This leads to evaluation of the above three 'Terms' and SSE. The solution presented here is the final converged value for each insulation thickness. The user ought to 'disturb' this worksheet by changing values in **cells (B15:B21)** and **(H15:H21)** and trying the **Solver** facility (using commands: **Tools, Solver**). The dialog box for SSE of **row 15** is shown in Fig. 10.3.2.

Discussion

The user is referred to Chapter 2 for further information on the **Solver** function. A word of caution is necessary at this stage. The **Solver** function offers a choice of Newton and Conjugate gradient methods for optimisation. Both methods require that the initial (assumed) values of the independent parameters, T_{di} and T_{do} in the present context, are close to the final root.

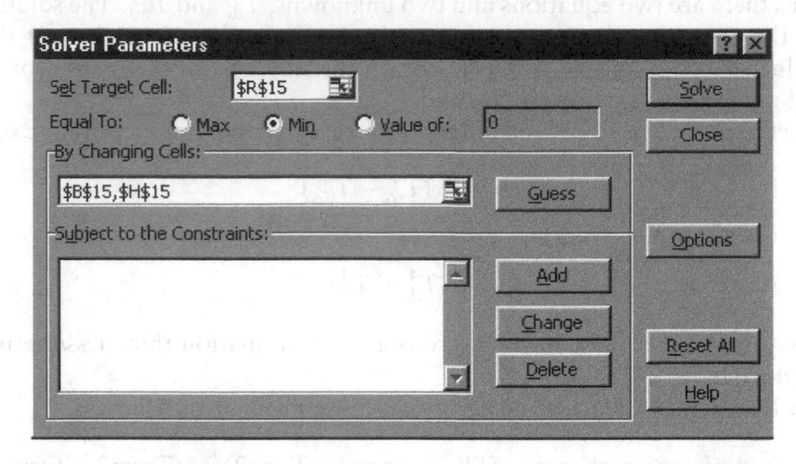

Figure 10.3.2 Solver dialog box for function minimisation (Example 10.3.1).

Hence some manipulation is required if the user experiences lack of convergence. However, this is easily done by trying out various combinations of T_{di} and T_{do} that result in a low value of SSE (of the order of double-digits in the present case).

For further discussion on the above optimisation techniques the reader is referred to Muneer (1988) and other relevant references given in Chapter 2.

Once convergence has been obtained for each of the given insulation thickness the rest of the analysis may easily be completed on the total cost using additional data provided above. A dynamically linked graph enables ease of observing the cost trends and locating the optimum value of insulation that is presently found to be 50 mm.

The user may wish to explore the influence of parameters such as COP on the optimum insulation thickness.

10.4 Simultaneous conduction, forced-convection, and radiation

Example 10.4.1 Air-to-water heat exchange.

Figure 10.4.1 shows the schematic of a water carrying tube in cross-flow. Hot water circulating within the tube is exchanging heat with cross-flowing air at 25°C. At a given station the temperature of water within the tube is at 80°C. Compute the overall heat transfer coefficient that represents the series path constituting of the tube-side and outside convective boundary layers as well as the tube conductive resistance.

Given data: Tube diameters: inside (D_i) = 0.02 m, outside (D_o) = 0.025 m. Velocity of air in cross flow = 20 m/s. Mass flow rate of water (m_w) = 0.111 kg/s.

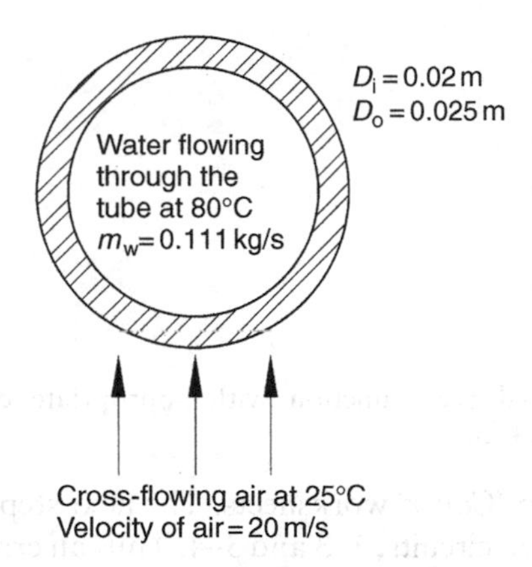

Figure 10.4.1 Schematic of a water carrying tube in cross-flow.

Analysis

This problem is similar to Example 10.2.1 with the additional complexity of computation of convection within the tube. We need to obtain the three resistances that are in a series path – external forced-convection across the tube, followed by conduction through the tube and forced, internal convection inside the tube.

We shall use Sections 7.2.3 and 7.3.11, respectively for the external- and internal-convection together with Section 3.7.3 for conduction transfer.

As mentioned above in this particular exercise access to two sets of transport property tables is required, that is, air and water. While transport properties for water may easily be read-off for the given bulk temperature of 80°C, properties for air have to be obtained for film temperature which is unknown. Hence an iterative solution procedure has to be adopted, the starting point of which is to assume the tube's outside temperature, t_{wo}. The Excel **Goal Seek** function is useful in this respect and its use is demonstrated here.

Solution

Open workbook Ex10-04-01.xls which reveals five worksheets (from right to left): *'Properties–Water'*, *'Properties–Air'*, *'Cover–Water'*, *'Cover–Air'*, and *'Combined'*. The *'Cover'* named worksheets are front-ends for computing the heat transfer coefficients for the respective fluids by conversing with the appropriate *'Property'* worksheets. Note that the only unknown in this problem is the outside tube wall temperature, assumed to be 50°C. The **Goal Seek** facility within Excel may be invoked to obtain solution to this problem. The procedure is as follows.

Refer to worksheet *'Combined'* that includes a sketch for the present problem. The resistances 1–3 (series sum of 1–2 and 2–3) and 3–4 are obtained

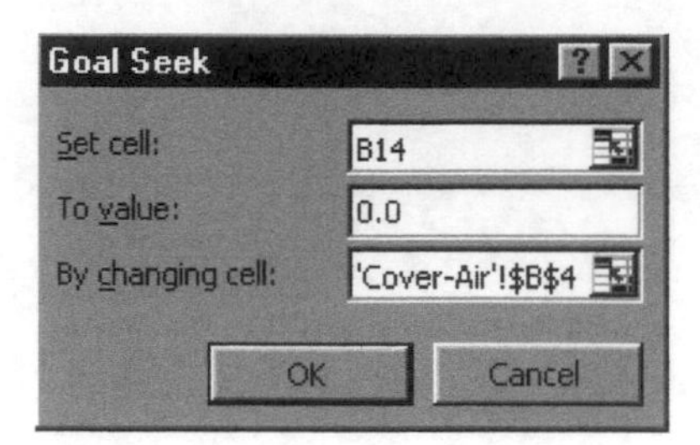

Figure 10.4.2 Dialogue box for **Goal Seek** function with appropriate cell selection (Example 10.4.2).

from information presented in the two *'Cover'* worksheets. The next step is to estimate the heat flows within the two circuits, 1–3 and 3–4. Thus an error function is established which represents the absolute value of the difference between the above heat flows. When convergence has been reached for the assumed value of t_{wo} the two heat flows q1–3 and q3–4, given in **cells B11 and B12**, equalise. Hence the objective function in **cell B14**, computed as the absolute value of the difference between q1–3 and q3–4, has to be minimised with an ideal value being equal to nil.

The procedure may be activated by selecting **Goal Seek** function from **Tools** menu. The dialog box that consequentially appears has to be filled as shown in Fig. 10.4.2.

Once the 'OK' button is clicked the **cell B14** in worksheet *'Combined'* returns a nil value. We note that the value of the overall heat transfer coefficient is 7.9 W/m^2 K. It may also be noted that the converged value of t_{wo} is 77.2°C.

Discussion

The user may wish to explore the influence of the velocity of air and water flow rate on the value of the overall heat transfer coefficient and t_{wo}.

10.5 Simultaneous conduction, free-convection, and radiation

Example 10.5.1 Heat transfer across pipe insulation.

Saturated steam at 200°C flows in a thin-walled, 5-cm horizontal tube. The tube is insulated with 5-cm thick rock wool and covered with a thin painted cladding of emissivity = 0.8. The surrounding air and surfaces are at 25°C. The heat loss from this system is to be estimated.

Analysis

This problem is similar to Example 10.4.1 with the outside forced convection replaced by free convection and the inside tube surface temperature being a given value. We shall use the analysis of Section 3.7.3 for conduction, followed by the material presented in Section 8.2.5 for free-convection. The thermal circuit is shown within the workbook that is introduced in the following paragraph. The governing equation is:

Heat transferred by conduction through insulation = combined convection–radiation transfer from insulation surface to surrounding air and surfaces.

Solution

Open the workbook Ex10-05-01.xls that contains two worksheet, '*Cover*' and '*Properties*'. Activate the '*Cover*' worksheet by clicking on its name tab. Starting with an assumed value of 40°C (**cell B12**) for the exposed surface temperature of insulation (T_i), we note that the error function (**cell D14**) has a magnitude of 30.8. Once again the Excel **Goal Seek** function may be used for solving this problem and this is demonstrated in the workbook Ex10-05-01.xls.

The user may try to vary the thickness of insulation and see its effect on T_i and the heat loss.

Discussion

With respect to Examples 10.4.1 and 10.5.1 it may be possible that if the assumed value of the unknown temperature has a considerable departure from the root, convergence may not take place. In such instances it is advisable either to provide a higher value in the 'Set cell' box or provide other, more suitable value for T_i. Alternatively, the user may abandon the **Goal Seek** function and use **Solver** facility instead. The latter is inherently a more stable means of solving non-linear equations.

10.6 Simultaneous conduction, free-convection, and radiation within enclosures

Example 10.6.1 Heat loss from double-glazed windows of varying designs.

Increased concern for the depletion of fossil fuels and sustainability demands has brought energy efficiency in sharp focus. Within buildings a considerable amount of energy is consumed for conditioning the space to appropriate levels of thermal comfort. Traditionally, windows have been identified as thermally weak elements of older building structures. The advent of double-glazing and more recently triple-glazing has attempted to address this problem. Recall that in Chapter 8 the heat transport mechanism within double-glazing was presented.

Muneer *et al.* (2000) have presented a detailed account of the state of research on window heat transfer. Basically, the thermal energy transfer from the complete window is divided into three regions – the centre-glazing, edge effects of the glazing, and frame heat loss. A complete treatment of this subject is beyond the scope of the present text and hence attention is focussed herein on the centre-glazing heat transmission.

Consider a double-glazing with a cavity spacing of 16 mm. One of the two panes of glass may or may not be coated with a low-emissivity material such as silver. The cavity space may have air or indeed it may be filled with one of the inert gases that are readily available – argon, krypton, or xenon. Obtain the heat loss from the following double-glazing designs:

(a) Cavity gas: air. Uncoated glass
(b) Cavity gas: air. Silver coating on glass
(c) Cavity gas: argon. Uncoated glass
(d) Cavity gas: argon. Silver coating on glass
(e) Cavity gas: krypton. Uncoated glass
(f) Cavity gas: krypton. Silver coating on glass
(g) Cavity gas: xenon. Uncoated glass
(h) Cavity gas: xenon. Silver coating on glass.

Critically evaluate the economics and environmental impact (carbon dioxide emissions) for the above cases and hence identify the cost optimum. The following additional data are provided:

- Respective cost of inert gases argon, krypton, and xenon: 0.10, 10, and 30 Euro/l;
- Cost of window frame: 200 Euro;
- Cost of uncoated glass: 10 Euro/m^2;
- Cost of silver coating: 10 Euro/m^2;
- Heating degree-days: 2500 (Box 10.6.1 presents information on degree-days);
- Cost of electricity for space heating: 0.10 Euro/kWh;
- Environmental cost: 4 kWh/Kg fossil fuel burnt at the power station and 3 kg CO_2/kg fossil fuel;
- Infrared radiation emissivity data: Float glass = 0.88, silver = 0.05;
- Assume respective indoor and outdoor ambient temperature to be 18 and 0°C.

Analysis

Recall that in Chapter 8 models for free-convection within a cavity space were presented. We shall employ the Muneer and Han model given by Eq. (8.3.15). Thermal radiation exchange within infinite, parallel planes may be obtained along the lines of material presented in Chapter 9. Since we are currently dealing with a parallel heat transfer path, the two quantities – free-convection and thermal radiation shall be added to obtain the total window heat loss.

Box 10.6.1 The concept of degree-days

Degree-days are a measure of the variation of outside temperatures, which enables building designers and users to determine how the energy consumption of a building is related to the weather. When temperature maintained inside a building is higher than the temperature outside, a spontaneous heat loss through the building fabric will take place. On any day the external temperature is not constant. It is usually coolest during the night and warmest during the day. External temperature is thus cyclic with a daily maximum and minimum. The figure below shows the variation of outside temperature in Edinburgh, Scotland on 1–2 January 1992. The hatched area shown in the figure would thus represent the heating degree-days for this period. Note that 18.3°C, corresponding to the horizontal line shown in the figure, is widely accepted as the recommended indoor comfort temperature. Total annual heating degree-days are the mean number of degrees by which the outside temperature on a given day is less than the base temperature, added up for all the days in the period.

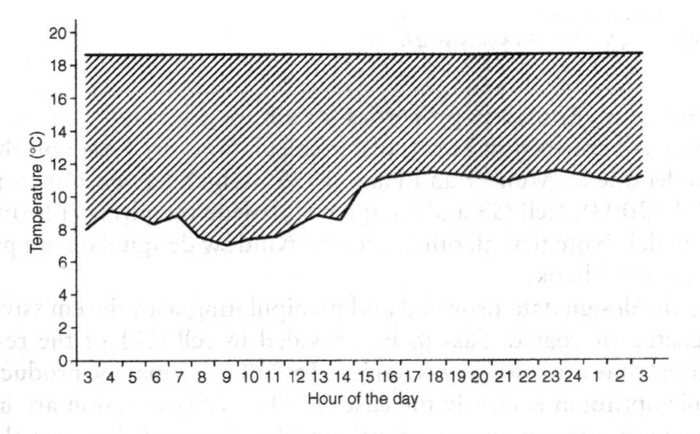

Figure 10.6.1 Outdoor temperature in Edinburgh, 1–2 January 1992.

Solution

Reference is made to workbook Ex10-06-01.xls, an extract from the above-cited reference. Open the latter mentioned workbook that contains five worksheets. The four worksheets bearing the gas names are the *'front ends'* for the respective glazing using a given in-fill gas, while the worksheet *'Calc'* is the background computational engine for all of those four worksheets. Attention is now drawn towards the *'Calc'* worksheet. A blue divider separates the transport property database and computations for the

Table 10.6.1 Window heat loss and economic analysis

Case	*U-value* (W/m^2K)	*Capital cost*[a] (Euro)	*Running cost*[b] (Euro)	*Total cost* (Euro)	CO_2 *emissions*[b] (Tonnes)
(a) Cavity gas: air. Uncoated glass	2.74	220.0	328.8	548.8	2.47
(b) Cavity gas: air. Silver coating on glass	1.42	230.0	170.4	400.4	1.28
(c) Cavity gas: argon. Uncoated glass	2.66	221.6	319.2	540.8	2.39
(d) Cavity gas: argon. Silver coating on glass	1.14	231.6	136.8	368.4	1.03
(e) Cavity gas: krypton. Uncoated glass	2.54	380.0	304.8	684.8	2.29
(f) Cavity gas: krypton. Silver coating on glass	0.94	390.0	112.8	502.8	0.85
(g) Cavity gas: xenon. Uncoated glass	2.48	700.0	297.6	997.6	2.23
(h) Cavity gas: xenon. Silver coating on glass	0.78	710.0	93.6	803.6	0.70

Notes
a Based on unit glazing area.
b Calculated over a period of twenty years.

respective gases. Transport properties for air are generated using regression equations. All convection heat transfer calculations are based on the reference model due to Muneer and Han (1996), which has been incorporated by CIBSE (2003). **Cell G3** and Chapter 8 provide the physical formulation of this model. Note that all other relevant window design data are provided within the workbook.

Using the design data provided and manipulating with the emissivity data for uncoated or coated glass to be provided in **cell C11** of the respective worksheets that provide the gas names, Table 10.6.1 may be produced. The economic optimum is clearly the case (d). The CO_2 emission are however increasingly becoming an important consideration and these are shown in the last column of the referred table.

Discussion

At a first glance it may appear that case (h) is the best case scenario in this respect. However, a complete life cycle assessment consideration of the window that includes the environmental cost of producing trace gases shifts the balance in favour of case (f). Box 10.6.2 highlights the basic processes for undertaking an environmental audit (Fig. 10.6.2).

For further information on this subject the reader may wish to consult Fernie and Muneer (1996) and Muneer *et al.* (2000).

Box 10.6.2 Life cycle assessment

Life cycle analysis is defined as: 'A process to evaluate the environmental burdens associated with a product or activity by quantifying the energy and materials used, wastes released to the environment, and assessing the energetic impacts. The assessment includes the entire life cycle encompassing extracting and processing of the raw materials, manufacturing; distribution, use, reuse, maintenance, recycling and/or final disposal.

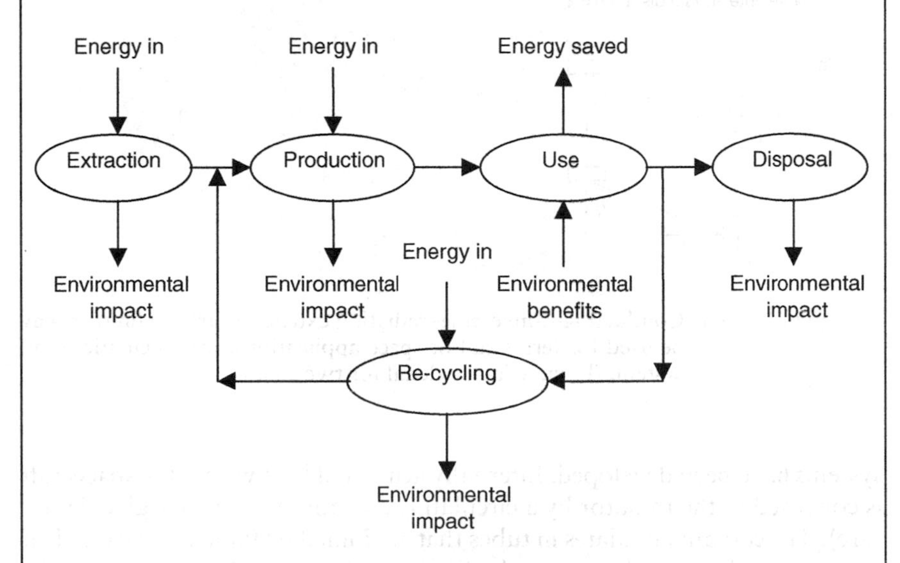

Figure 10.6.2 Audit process undertaken within life cycle assessment.

10.7 Simultaneous conduction and radiation, with or without free- or forced-convection

Example 10.7.1 Radiating fins – terrestrial and space applications.

Recall that in Chapter 3 an analytical treatment for one-dimensional fins was presented. Fins are typically considered to be conducting–convecting systems and this assumption is valid so long as the convection heat transfer coefficient and the fin operational temperatures are high. However, in this example the above assumption shall be closely scrutinised.

In outer space, the only means of rejecting heat is that of thermal radiation. For this purpose heat rejection space radiators have been developed. The relevant material was presented in Chapter 9. Many varieties of such

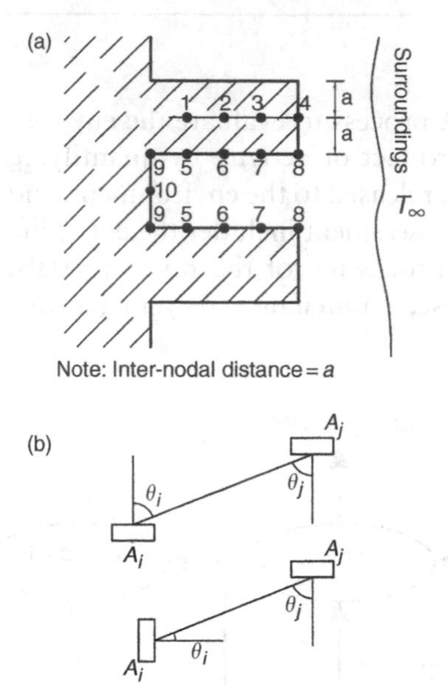

Figure 10.7.1 Conducting–convecting–radiating extended surface: fin that may be used for terrestrial or space applications: (a) schematic of fin system; (b) view factor detail for two cases.

systems have been developed. Internally generated heat within the spacecraft is conveyed to the radiator by a circulating coolant (e.g. a water-glycol mixture). The coolant circulates in tubes that are joined by finned surfaces. Heat from the coolant conducts into the fin and is dissipated by radiation to the surroundings. The finned surface may be subjected to an external irradiation from a source such as the Sun. The basic question to be answered is: given a radiating fin of known geometry, maintained at a known base temperature, what is the rate at which heat is dissipated by the fin to the surroundings. One would also be interested to know the temperature distribution of such radiating systems.

Consider the fin system shown in Fig. 10.7.1. You are asked to obtain the two-dimensional temperature distribution by means of estimating the eight nodal temperatures for the following four cases:

(a) Assume that the fin is a conducting–convecting system and hence use a matrix solution procedure as explained in Chapter 2. Use $h = 600\,\text{W/m}^2\,\text{K}$, $k = 10\,\text{W/mK}$. The temperature of the fin-base (T_b) and surroundings (T_S) are respectively given to be 45°C and 25°C.
(b) The same as case (a), but obtain the solution via the **Solver** facility available within Excel.

(c) Now consider the fin to be a conducting–convecting–radiating system. Hence obtain the solution via **Solver**.
(d) Same as case (c), but for $h = 0$ and $T_S = 1\,\text{K}$ (space application).

Analysis

Using the formulae for generating the two-dimensional finite-difference equations presented in Chapters 2 and 4 for conduction–convection systems, the following algebraic relationships may be produced:

- **Node 1**

  $$T_b + T_2 + T_5 + T_5 - 4T_1 = 0$$

  Hence, $T_b + T_2 + 2T_5 - 4T_1 = 0$

- **Node 2**

  $$T_1 + T_3 + 2T_6 - 4T_2 = 0$$

- **Node 3**

  $$T_2 + T_4 + 2T_7 - 4T_3 = 0$$

- **Node 4**

  $$(2T_3 + T_8 + T_8) + \frac{2h\Delta x}{K}T_\infty - 2\left(\frac{h\Delta x}{K} + 2\right)T_4 = 0$$

- **Node 5**

  $$2T_1 + T_6 + T_b + \frac{2h\Delta x}{K}T_\infty - 2\left(\frac{h\Delta x}{K} + 2\right)T_5 = 0$$

- **Node 6**

  $$2T_2 + T_5 + T_7 + \frac{2h\Delta x}{K}T_\infty - 2\left(\frac{h\Delta x}{K} + 2\right)T_6 = 0$$

- **Node 7**

  $$2T_3 + T_6 + T_8 + \frac{2h\Delta x}{K}T_\infty - 2\left(\frac{h\Delta x}{K} + 2\right)T_7 = 0$$

- **Node 8**

  $$(T_7 + T_4) + 2\frac{h\Delta x}{K}T_\infty - 2\left(\frac{h\Delta x}{K} + 1\right)T_8 = 0 \qquad (10.7.1)$$

Table 10.7.1 Shape (or configuration) factor matrix for Example 10.7.1

	5	6	7	8	9	10	∞
5	$\frac{1}{4}$	$\frac{4}{25}$	$\frac{1}{16}$	$\frac{2}{169}$	$\frac{4}{25}$	$\frac{1}{4}$	$1-\sum_{j=5}^{10} F_{5j}$
6	$\frac{4}{25}$	$\frac{1}{4}$	$\frac{4}{25}$	$\frac{1}{16.2}$	$\frac{1}{16}$	$\frac{2}{25}$	$1-\sum_{j=5}^{10} F_{6j}$
7	$\frac{1}{16}$	$\frac{4}{25}$	$\frac{1}{4}$	$\frac{2}{25}$	$\frac{4}{169}$	$\frac{3}{100}$	$1-\sum_{j=5}^{10} F_{7j}$
8	$\frac{4.2}{169}$	$\frac{1.2}{16}$	$\frac{4.2}{25}$	$\frac{1}{4.2}$	$\frac{2}{100}$	$\frac{4.2}{289}$	$1-\sum_{j=5}^{10} F_{8j}$

Notes

a All matrix elements to be divided by 'πa'.

b The values in the extreme right-hand column are, as shown by the formula in the upper most of that column, equal to the difference of unity less the sum of the values in the corresponding row.

Solution

Open workbook Ex10-07-01.xls and note that it contains three worksheets, '*Data*', '*Cond&conv*', and '*Calc*'. **Cells (A1:H8)** within the '*Cond&conv*' worksheet contain coefficients for T_1–T_8 as presented in equations suite 10.7.1. Note that these coefficients are generated directly using the basic data presented in worksheet '*Data*'. Hence there is no need for any reader input with the exception of the basic data discussed above. Any change in any of the design parameters will directly influence the temperature profile solution. The above coefficient matrix is inverted in **cells (A10:H17)** (shown in red), and then multiplied with the constant matrix (blue-font numbers) to provide the desired temperatures that are shown in pink font. This solution procedure is straightforward, as it does not involve radiation. Hence we have only a linear system of equations to deal with. Note that to solve for a finer grid size, the user will have to set-up new equations and this will inevitably change the size of the matrices in worksheet '*Cond&conv*'.

Now we turn our attention to rather more involved analysis for including thermal radiation exchange for the surface nodes, 4–8. First of all we shall establish a procedure to obtain radiation shape factors for the surface finite-elements for the latter surface nodes. In this respect Table 10.7.1 for the shape factors is produced using the analysis presented below. This information is copied within '*Data*' worksheet of the workbook.

Recall Eq. (9.3.14), presented in Chapter 9, for obtaining the view factor for thermal radiation exchange

$$F_{ij} = \frac{1}{A_i}\int_{A_j}\int_{A_i}\frac{\cos\theta_j}{\pi R^2}\,dA_i\,dA_j$$

The blackbody radiation exchange then is

$$Q_i = A_i F_{ij}\,\sigma(T_i^4 - T_j^4)$$

If A_i is sufficiently small, then θ_i, θ_j, and R are independent of their position on A_i.

If both A_i and A_j are sufficiently small, then

$$F_{ij} = \Delta A_j \frac{\cos\theta_i \cos\theta_j}{\pi R^2}$$

Note: If A_i and A_j are parallel (see Fig. 10.7.1), then $\theta_i = \theta_j$.

If A_i and A_j are in $\perp$ planes, then $\cos\theta_j = \sin\theta_i$

Using the above information and the reciprocity relationship presented in Chapter 9, Table 10.7.1 may now be constructed.

We are now in a position to write the energy balance for each of the Nodes 4–8. Note that nodes 5–8 are of a similar nature and hence it would suffice to present equations for 4 and 5. In any case the workbook Ex10-07-01.xls presents every single equation required for this example.

Now attention is drawn towards worksheet *'Calc'*. This worksheet is divided into four sections – Nodal temperatures, Energy flows for each node (further divided in conduction + convection and thermal radiation), Modulus of error (modulus of the sum of energy entering a given node via conduction + convection and radiation) and finally, the sum of the latter moduli. This latter function needs to be minimised using an optimised set of nodal temperatures. In this worksheet the user needs to provide initial values of nodal temperatures in **cells (B4:E5)**. The final converged nodal (celsius) temperatures are then viewable in **cells (B10:E11)**.

Once again, the **Solver** facility within Excel may be invoked to obtain solution to the above problem and this is demonstrated in the workbook Ex10-07-01.xls. To compare the **Solver** generated solution with that obtained in case (a), the value of the Stefan–Boltzmann constant, σ is set to nil in **cell B4** of worksheet *'Data'*. This action amounts to 'switching-off' thermal radiation exchange to and from the fin to the surroundings as well as other fin nodes. The workbook has been saved in this thermal radiation 'switched-off' mode.

The solution for this case (b) is included in Table 10.7.2. Note that with the exception of a slight difference in temperature of Node 7 due to the approximate nature of **Solver** solution procedure, good concordance is to be seen when compared with the matrix solution (case (a)).

The solution for case (c) is now easily obtained by switching on thermal radiation by keying-in the value of σ as 5.67×10^{-8} in **cell B4** of worksheet

Table 10.7.2 Results for Example 10.7.1: comparison of temperatures for cases (a) to (d)

Case		
	Nodal scheme	b 5 6 7 8 b 1 2 3 4 b 5 6 7 8
a	Conduction + convection (Matrix solution)	45.0 39.3 35.7 33.5 32.2 45.0 40.0 36.4 34.0 32.6 45.0 39.3 35.7 33.5 32.2
b	Conduction + convection (**Solver** solution, $\sigma = 0$)*	45.0 39.3 35.7 33.4 32.2 45.0 40.0 36.4 34.0 32.6 45.0 39.3 35.7 33.4 32.2
c	Conduction + convection + radiation (**Solver** solution)**	45.0 38.7 34.3 31.5 30.7 45.0 39.4 35.1 32.3 31.1 45.0 38.7 34.3 31.5 30.7
d	Conduction + radiation (**Solver** solution, $h = 0$ and $T_S = 1\,\text{K}$)	45.0 22.9 3.0 −10.1 −11.3 45.0 24.3 6.4 −4.7 −5.2 45.0 22.9 3.0 −10.1 −11.3

Notes
* Slight difference in temperature of Node 7 due to approximate nature of Solver solution procedure.
** Lowered temperatures of all nodes due to fin-tip radiation.

'Data'. Note that in this case lowered temperatures for all nodes result due to fin-tip radiation to the surroundings. Likewise, solution to case (d) is obtained by invoking the **Solver** macro once again. Note that when the convection heat transfer coefficient, h is set to nil the worksheet *'Cond&conv'* indicates that the entire fin is at a uniform temperatures that is equal to the fin-base temperature, as one would expect in the absence of convection. This is a confirmation of the validity of the latter worksheet as it incorporates energy exchange due to conduction and convection alone.

Discussion

It is quite likely that the **Solver** may not converge in one attempt. Hence, several trials using different seed values of nodal temperatures, provided by the user in **cells** (**A4:E5**) within worksheet *'Calc'*, may be required.

Problems

Problem 10.1

An alcohol-in-glass thermometer with a bulb having an emissivity, ε of 0.88 is hung inside a forced-air heated building. The thermometer bulb indicates a temperature (T_{bulb}) reading of 18°C. The convection heat transfer coefficient, h_c from the ellipsoid-shaped thermometer bulb may be assumed to be 5 W/m^2 K.

Consider the multi-mode exchange of heat between the thermometer bulb, ambient air, and the building walls. Obtain the true temperature of the ambient air, T_a for the following two cases:

(a) The building walls are well insulated and are at a temperature, T_w of 16°C.
(b) The building walls are poorly insulated and are at a temperature, T_w of 10°C.

Hint: The thermal balance for the thermometer bulb may be written as,

$$h_c(T_a - T_{\text{bulb}}) = \sigma\varepsilon(T_{\text{bulb}}^4 - T_w^4)$$

Problem 10.2

A coal tar covered flat roof of a building has a short-wave solar absorptivity, of 0.96 and an absorptivity for long-wave radiation = 0.7. The roof is exposed to wind generated convection, h_c the temperature at which wind blows being equal to 10°C. The sky temperature = 5°C while the wind speed, V = 3 m/s. Assuming that the roof is heavily insulated, obtain the surface temperature of the roof.

Note: The wind-generated convection, $h_c = 4 + 4V$

Problem 10.3

A long, steam-carrying pipe has an external surface temperature of 100°C. With the view to reduce the heat loss from the pipe surface a thermal radiation shield is placed around the pipe. The air-gap between the pipe and the shield may vary, but operational experience dictates that the minimum value of this gap must be 5 mm. The pipe and inner/outer surfaces of the shield have respective emissivity of 0.9 and 0.08. All room surfaces and its ambient air are at a temperature of 18°C.

Consider combined free-convection and thermal radiation exchange between the pipe and the shield and between shield and the room surfaces as well as ambient air. Obtain a parametric solution for the heat loss from the pipe as a function of increasing air-gap. Would there be an optimum if the combined cost of energy supply and material that is used for the shield is considered? You may wish to obtain that optimum for suitably assumed values of the above two costs.

Note: Thermal radiation exchange between the pipe (1) and the shield (2) $= \sigma A_1(T_1^4 - T_2^4)/[(1/\varepsilon_1) + (A_1/A_2)(1/\varepsilon_2 - 1)]$.

Free-convection regressions for concentric cylinders are provided in Chapter 8.

Problem 10.4

The total rate of heat transfer through a double-glazing window system can be calculated knowing the separate heat transfer contributions of the centre-glass, edge-seal, and frame. The most widely used spacers in multiple glazed units are made of aluminium that obviously, significantly increase conductive heat transfer between the lower and top part of inner and outer glazing. This phenomenon is known as cold-bridging.

A number of double-glazed window designs were tested by Muneer *et al.* (1997) to investigate the temperature stratification along the height of the inner pane. A detailed cross-section of one such window edge-seal is shown in Fig. 10.P.1. Obtain the heat loss by means of conduction through the spacer materials, given the thickness of glass and aluminium spacer are respectively, 4 and 0.2 mm.

Note: All thermal property data are given within the figure. You may assume that the temperature of the inner (left) and outer pane are respectively

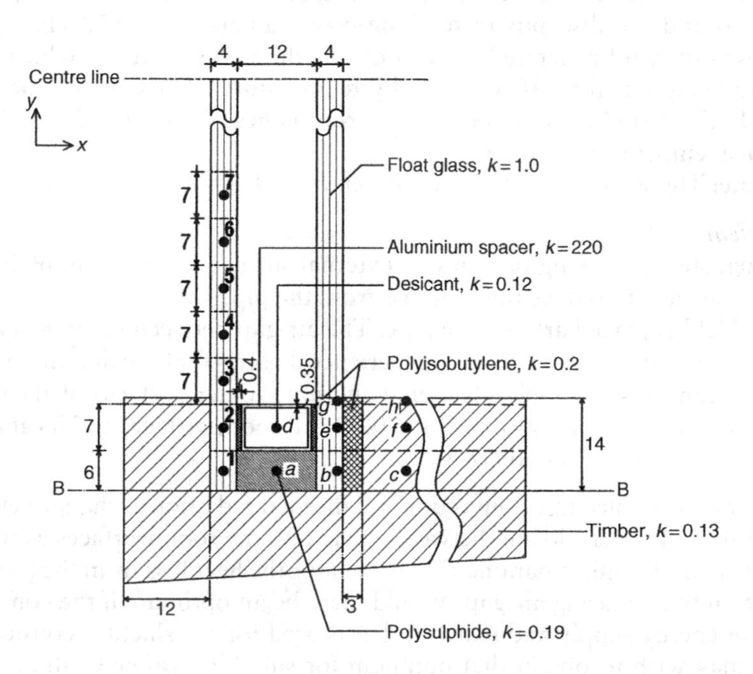

Figure 10.P.1 Design schematic and nodal arrangement for window design.

Note

All dimensions are in mm and k values in W/m K.

20°C and 2°C. Consider only the bottom 35 mm of the glazing as this has been shown to be the region of influence of the spacer (Muneer *et al.*, 1997).

Problem 10.5
Recall that in Example 10.7.1: Radiating fins for terrestrial and space applications a fundamental analysis for the view factors for finite elements of the fin was undertaken. Redo that example using a double pitch for the fins. The distance between any two adjacent fins will therefore be 2a. Compare your results with those obtained for the data given for Example 10.7.1.

References

CIBSE (2003) *Guide on Heat Transfer*, Chartered Institute of Building Services Engineers, London.

Culp, A. W. (1991) *Principles of Energy Conversion*, McGraw-Hill, New York.

Fernie, D. and Muneer, T. (1996) Monetary, energy and environmental cost of infill gases for double glazings, *Building Services Eng. Res. Tech.* 17(1), 43–45.

Muneer, T. and Han, B. (1996) Multiple glazed windows: design charts, *Building Services Eng. Res. Tech.* 17(4), 223–229.

Muneer, T., Abodahab, N., and Gilchrist, A. (1997) Combined conduction, convection, and radiation heat transfer model for double glazed windows, *Building Services Eng. Res. Tech.* 18, 183.

Muneer, T., Abodahab, N., Weir, G., and Kubie, J. (2000) *Windows in Buildings*, Reed-Elsevier, Oxford, UK.

Muneer, T. (1988) Comparison of optimisation methods for non-linear, least squares minimisation., *Int. J. Math. Educ. Sci. Tech.* 19(1), 192–197.

Weston, K. C. (1992) *Energy Conversion*, West Publishing Company, St Paul, USA.

Appendices

Table A.1 Thermal conductivity of metals and alloys

Data for metals (W/m K)	*Temperature* (°C)					
	0	100	200	300	400	500
Aluminium	202	206	215	230	249	268
Brass	97	104	109	114	116	
Cast iron	55	52	48	45	43	
Copper (pure)	388	377	372	367	363	358
Graphite (longitudinal)	168	151	132	114	100	92
Lead	35	33	31	31		
Nickel	62	59	57	55		
Silver	419	412				
Steel (mild)		45	45	43	40	38
Tin	62	59	57			
Wrought iron		55	52	48	45	40
Zinc	112	111	107	102	93	

Data for alloys (W/m K)	*Temperature* (°C)	*Thermal conductivity*
Antimony	0	18.3
Bismuth	100	6.7
Cadmium	100	90.3
Gold	100	294.2
Iron wrought	18	58.8
Steel (1% C)	18	45.3
Mercury	0	8.3
Platinum	18	69.6
Nickle silver	0	29.2

Table A.2 Thermal conductivity of building and insulating materials (W/m K)

Material	*Density* (kg/m^3)	*Temperature* (°C)	*Thermal conductivity*
Aerogel, silica, opacified	136	120	0.0
Aluminium foil	3	38	0.0
Chrome brick (32% Cr_2O_3 by mass)	3204	200	1.2
Diatomaceous earth, natural, across strata	444	204	0.9
Calcium carbonate, natural	2595	30	2.2
Charcoal flakes	191	80	0.1
Cotton wool	80	30	0.0
Dolomite	2675	50	1.7
Fiber insulating board	237	21	0.0
Gypsum, molded and dry	1249	20	0.4
Ice	921	0	2.2
Limestone	1650	24	0.9
Magnesia, powdered	796	47	0.6
Mineral wood	151	30	0.0
Porcelain	0	200	1.5
Rubber, hard	1198	0	0.1
Sand, dry	1515	20	0.3
Snow	556	0	0.5
Wallboard, insulating type	237	21	0.0
Pine, white	545	15	0.2
Wool, animal	111	30	0.0

Table A.3 Thermal conductivity of soils (W/m K)

Type of soil	*Dry density* (kg/m^3)						
	1602	1762	1922	1442	1762	1442	1602
	Thermal conductivity						
Fine crushed quartz	1.73	2.30					
Graded Ottawa sand	1.44	2.02					
Chena river gravel	0.00	1.30	1.87				
Crushed granite	0.80	1.07	1.44				
Crushed trap rock	0.73	0.90	1.00				
Northway fine sand	0.64	0.80			1.23		
Hely clay	0.57			0.80	1.30	1.07	1.44
Fairblank silt loam				0.73	1.30	1.07	1.44

Table A.4 Thermal conductivity of liquids (W/m K)

Acetone	0.177
Benzene	0.159
Decane	0.147
Ether	0.138
Heptane	0.140
Octane	0.144
Paraldehydem	0.145
Propyl alcohol	0.171
Sodium chloride	0.571
Sulfuric acid	0.363
Sulfur dioxide	0.222
Water	0.594

Table A.5 Thermal conductivity of gases and vapours at 0°C (W/m K)

Acetone	0.010
Air	0.024
Ammonia	0.022
Argon	0.016
Carbon dioxide	0.015
Carbon monoxide	0.023
Chlorine	0.007
Dichlorodifluoromethane	0.008
Helium	0.142
Hydrogen	0.167
Methane	0.030
Alcohol	0.014
Neon	0.004
Nitrogen	0.024
Oxygen	0.025
Propane	0.015
Sulfur dioxide	0.009

Table A.6 Specific heat

(a) Solids at 100°C (kJ/kg K)	
Material	
Lead	0.132
Zinc	0.401
Aluminium	0.932
Silver	0.239
Gold	0.131
Copper	0.394
Nickle	0.474
Iron	0.488
Cobalt	0.452
Quartz	0.863
(b) Organic liquids (kJ/kg K)	
Compound	
Acetal	1.955
Castor oil	2.135
Dodecylene	1.905
Gasoline	2.093
Hexylene	2.110
Kerosene	2.093
Machine oil	1.675
Paraffin oil	2.177
Propane	2.412
Trichlorethane	1.114
(c) Liquified gases (kJ/kg K)	
Liquid	
Ammonia (−60)	4.40
Carbondioxide (63 bar and −50°C)	1.95
Chlorine (−205°C)	0.96
Hydrogen (−258°C)	7.33
Nitric oxide (−158)	2.43
Nitrogen (−209)	1.99
Oxygen (−216)	1.67
Sulfur dioxide (20)	1.37

Table A.7 Prandtl number data for gases and vapours at 1 bar and 100°C

Air, hydrogen	0.690
Ammonia	0.860
Argon	0.660
Carbon dioxide, methane	0.750
Carbon monoxide	0.720
Helium	0.710
Nitric oxide, Nitrous oxide	0.720
Nitrogen, oxygen	0.700
Steam	1.060

Index

9780367395155